Principles of
Geotechnical Engineering

Principles of Geotechnical Engineering

Contributors

Silvia Garcia et al.

AURIS
Reference

www.aurisreference.com

Principles of Geotechnical Engineering

Contributors: Silvia Garcia et al.

Published by Auris Reference Limited

www.aurisreference.com

United Kingdom

Principles of Geotechnical Engineering

ISBN: 978-1-78154-912-4

British Library Cataloguing in Publication Data
A CIP record for this book is available from the British Library

Printed in the United Kingdom

Exclusively distributed by CBS Publishers & Distributors Pvt. Ltd.

Sales & Distribution Rights only for India, Pakistan, Bangladesh, Sri Lanka, Nepal and Bhutan. This book is not to be sold outside these territories.

Contents

List of Abbreviations

AIST	Advanced Industrial Science and Technology
ANNs	Artificial neural networks
CBR	California bearing ratio
CBR	Californian Bearing ration
CH4	Compound Methane
CPT	Cone penetration test
CLMS	controlled-low strength materials
cCBR	cyclic California bearing ratio
CRR	cyclic resistance ratio
ERT	Electrical resistivity tomography
EDS	Energy dispersive spectra microprobe
EEMD	Ensemble empirical mode decomposition
FEPA	Federal Environment Protection Agency
GP	Genetic programming
GSI	Geospatial Information Authority of Japan
GGBFS	Ground granulated blast furnace slag
HHT	Hilbert-Huang transform
IET	Innovations in Engineering and Technology
NIED	National Research Institute for Earth Science and Disaster Prevention
OEPA	Ohio Environmental Protection Agency
PEs	Processing elements
RSL	Recompacted Soil Liners
SEM	Scanning electron microscope
SSRS	Secondary steel making operation produces stainless steel reducing slag
SIPEEDIT	Seismic refraction tomography data in the current study,
SSRT	Shallow seismic-refraction tomography
SIRT	Simultaneous iterative reconstruction technique
SPT	Standard Penetration Test
TFA	Theory of time-frequency-amplitude
UCT	Unconfined Compression Test characteristics
XRD	X-ray diffraction
XRF	X-ray fluorescence
CT	Classification tree
EC	Evolutionary computing
FL	Fuzzy logic
GA	Genetic algorithms
NN	Neural networks
PR	Probabilistic reasoning
SC	Soft computing

List of Contributors

Silvia Garcia
Geotechnical Department, Institute of Engineering, National University of Mexico, Mexico

Elisabeth Olivia Logmo
Laboratory of Subsurface, Department of Earth Sciences, Faculty of Science, University of Douala, Douala, Cameroon

Gilbert François Ngon Ngon
Laboratory of Subsurface, Department of Earth Sciences, Faculty of Science, University of Douala, Douala, Cameroon

Williams Samba
Laboratory of Subsurface, Department of Earth Sciences, Faculty of Science, University of Douala, Douala, Cameroon

Michel Bertrand Mbog
Laboratory of Environmental Geology, Department of Earth Sciences, Faculty of Dschang, Dschang, Cameroon

Jacques Etame
Laboratory of Subsurface, Department of Earth Sciences, Faculty of Science, University of Douala, Douala, Cameroon

Meissa Fall
Laboratoire de Mécanique et Modélisation, UFR Sciences de l'Ingénieur, University of Thies, Thies, Senegal

Déthiè Sarr
Laboratoire de Mécanique et Modélisation, UFR Sciences de l'Ingénieur, University of Thies, Thies, Senegal

Makhaly Ba
Laboratoire de Mécanique et Modélisation, UFR Sciences de l'Ingénieur, University of Thies, Thies, Senegal

Etienne Berbinau
RAZEL sa, Christ de Sarclay, 3 rue René razel, Orsay cedex, Orsay, France

Jean-Louis Borel
RAZEL sa, Christ de Sarclay, 3 rue René razel, Orsay cedex, Orsay, France

Mapathé Ndiaye
Laboratoire de Mécanique et Modélisation, UFR Sciences de l'Ingénieur, University of Thies, Thies, Senegal

Cheikh H. Kane
Laboratoire de Mécanique et Modélisation, UFR Sciences de l'Ingénieur, University of Thies, Thies, Senegal
Laboratoire de Mécanique et Modélisation, UFR Sciences de l'Ingénieur, University of Thies, Thies, Senegal

Wojciech Sas
Water Centre Laboratory, Faculty of Civil and Environmental Engineering, Warsaw University of Life Sciences, 02-787 Warsaw, Poland

Andrzej Głuchowski
Department of Geotechnical Engineering, Faculty of Civil and Environmental Engineering, Warsaw University of Life Sciences, 02-787 Warsaw, Poland

Maja Radziemska
Department of Environmental Improvement, Faculty of Civil and Environmental Engineering, Warsaw University of Life Sciences, 02-787 Warsaw, Poland

Justyna Dzięcioł
Water Centre Laboratory, Faculty of Civil and Environmental Engineering, Warsaw University of Life Sciences, 02-787 Warsaw, Poland

Alojzy Szymański
Department of Geotechnical Engineering, Faculty of Civil and Environmental Engineering, Warsaw University of Life Sciences, 02-787 Warsaw, Poland

C. C. Iwuji
Department of Geology, Federal University of Technology, Owerri, Nigeria

O. C. Okeke
Department of Geology, Federal University of Technology, Owerri, Nigeria

B. C. Ezenwoke
Department of Geology, Federal University of Technology, Owerri, Nigeria

C. C. Amadi
Department of Geology, Federal University of Technology, Owerri, Nigeria

H. Nwachukwu
Department of Geology, Federal University of Technology, Owerri, Nigeria

Alhussein A. Basheer
Geomagnetic and Geoelectric Department, National Research Institute of Astronomy and Geophysics, Helwan, Cairo, Egypt

Abdelnasser M. Abdelmotaal
Geomagnetic and Geoelectric Department, National Research Institute of Astronomy and Geophysics, Helwan, Cairo, Egypt

Hany S. Mesbah
Geomagnetic and Geoelectric Department, National Research Institute of Astronomy and Geophysics, Helwan, Cairo, Egypt

Khamis K. Mansour
Geomagnetic and Geoelectric Department, National Research Institute of Astronomy and Geophysics, Helwan, Cairo, Egypt

Hirofumi Toyota
Department of Civil and Environmental Engineering, Nagaoka University of Technology, Nagaoka, Niigata 940-2118, Japan

Susumu Takada
Department of Civil and Environmental Engineering, Nagaoka University of Technology, Nagaoka, Niigata 940-2118, Japan

Anthony R. Moran
CH2M Hill, 7927 Nemco Way, Suite 120, Brighton, MI 48116, USA

Hiroshan Hettiarachchi
Department of Civil Engineering, Lawrence Technological University, 21000 West Ten Mile Road, Southfield, MI 48075, USA

Yang Changwei
School of Civil Engineering, Key Laboratory of Transportation Tunnel En-

gineering, Ministry of Education, Southwest Jiaotong University, Chengdu 610031, China

Su Tianbao
Henan University of Urban Construction, Pingdingshan, Henan 467036, China

Zhang Jianjing
School of Civil Engineering, Key Laboratory of Transportation Tunnel Engineering, Ministry of Education, Southwest Jiaotong University, Chengdu 610031, China

Du Lin
School of Civil Engineering, Key Laboratory of Transportation Tunnel Engineering, Ministry of Education, Southwest Jiaotong University, Chengdu 610031, China

Silvia García
Geotechnical Department, Institute of Engineering, National University of Mexico, Mexico

Ogbonnaya Igwe
Department of Geology, University of Nigeria, Nsukka, Nigeria

Mohamed A. Shahin
Department of Civil Engineering, Curtin University of Technology, Perth, WA 6845, Australia

Mark B. Jaksa
School of Civil, Environmental and Mining Engineering, University of Adelaide, Adelaide, SA 5005, Australia

Holger R. Maier
School of Civil, Environmental and Mining Engineering, University of Adelaide, Adelaide, SA 5005, Australia

Preface

Geotechnical engineering is the branch of civil engineering concerned with the engineering behavior of earth materials. The text *Principles of Geotechnical Engineering* presents the fundamentals of geotechnical engineering and offers an overview of soil properties and mechanics together with coverage of field practices and basic engineering procedure. A cognitive look at geotechnical earthquake engineering has been presented in first chapter. The objective of second chapter is to associate the geotechnical characteristic to the mineralogical and chemical compositions of the clay occurrences of the Missole II deposit in order to evaluate its suitability for manufacturing of construction materials and ceramics. Third chapter focuses on evolution of lateritic soils geotechnical parameters during a multi-cyclic optimum modified proctor (OPM) compaction and correlation with road traffic. The aim of fourth chapter is to investigate the chemical and selected significant geotechnical parameters of steel slag as the alternative materials used in road construction. Geological and engineering perspectives on earth resources exploitation and sustainable development have been proposed in fifth chapter. The objective of sixth chapter is to determine the dynamic characteristics and geotechnical parameters at the proposed site using seismic refraction and electrical resistively techniques. Seventh chapter discusses on geotechnical distinction of landslides induced by near-field earthquakes in Niigata, Japan. Eighth chapter describes a laboratory investigation conducted on mined clay from Appalachian Ohio to determine how and why the standard sampling and/or processing methods can affect the grain-size distributions. New developments in geotechnical earthquake engineering have been outlined in ninth chapter. In tenth chapter, geotechnical aspects of earthquake engineering under a soft examination are covered. Recent advances and future challenges for artificial neural systems in geotechnical engineering applications have been introduced in last chapter.

Chapter 1

A COGNITIVE LOOK AT GEOTECHNICAL EARTHQUAKE ENGINEERING: UNDERSTANDING THE MULTIDIMENSIONALITY OF THE PHENOMENA

Silvia Garcia[1]

[1]Geotechnical Department, Institute of Engineering, National University of Mexico, Mexico

INTRODUCTION

Not even windstorm, earth-tremor, or rush of water is a catastrophe.A catastrophe is known by its works; that is, to say, by the occurrence of disaster. So long as the ship rides out the storm, so long as the city resists the earth-shocks, so long as the levees hold, there is no disaster.It is the collapse of the cultural protections that constitutes the proper disaster.

Essentially, disasters are human-made. For a catastrophic event, whether precipitated by natural phenomena or human activities, assumes the state of a disaster when the community or society affected fails to cope. Earthquake hazards themselves do not necessarily lead to disasters, however intense, inevitable or unpredictable, translate to disasters only to the extent that the population is unprepared to respond, unable to deal with, and, consequently, severely affected. Seismic disasters could, in fact, be reduced if not prevented. With today's advancements in science and technology, including early warning and forecasting of the natural phenomena, together with innovative approaches and strategies for enhancing local capacities, the impact of earthquake hazards somehow could be predicted and mitigated, its detrimental effects on populations reduced, and the communities adequately protected.

After each major earthquake, it has been concluded that the experienced ground motions were not expected and soil behavior and soil-structure interaction were not properly predicted. Failures, associated to inadequate design/construction and to lack of phenomena comprehension, obligate further code reinforcement and research. This scenario will be repeated after each earthquake. To overcome this issue, *Earthquake Engineering* should change its views on the present methodologies and techniques toward more scientific, doable, affordable, robust and adaptable solutions.

A competent modeling of engineering systems, when they are affected by seismic activity, poses many difficult challenges. Any representation designed for reasoning about models of such systems has to be flexible enough to handle various degrees of complexity and uncertainty, and at the same time be sufficiently powerful to deal with situations in which the input signal may or may not be controllable.Mathematically-based models are developed using scientific theories and concepts that just apply to particular conditions. Thus, the core of the model comes from assumptions that for complex systems usually lead to simplifications (perhaps oversimplifications) of the problem phenomena. It is fair to argue that the representativeness of a particular theoretical model largely depends on the degree of comprehension the developer has on the behavior of the actual engineering problem. Predicting natural-phenomena characteristics like those of earthquakes, and thereupon their potential effects at particular sites, certainly belong to a class of problems we do not fully understand. Accordingly, analytical modeling often becomes the bottleneck in the development of more accurate procedures. As a consequence, a strong demand for advanced modeling an identification schemes arises.

Cognitive Computing CC technologies have provided us with a unique opportunity to establish coherent seismic analysis environments in which uncertainty and partial data-knowledge are systematically handled. By seamlessly combining learning, adaptation, evolution, and fuzziness, CC complements current engineering approaches allowing us develop a more comprehensive and unified framework to the effective management of earthquake phenomena. Each CC algorithm has well-defined labels and could usually be identified with specific scientific communities. Lately, as we improved our understanding of these algorithms' strengths and weaknesses, we began to leverage their best features and developed hybrid algorithms that indicate a new trend of co-existence and integration between many scientific communities to solve a specific task.

In this chapter geotechnical aspects of earthquake engineering under a cognitive examination are covered. Geotechnical earthquake engineering, an area that deals with the design and construction of projects in order to resist the effect of earthquakes, requires an understanding of geology, seismology and earthquake engineering. Furthermore, practice of geotechnical earthquake engineering also requires consideration of social, economic and political factors. Via the development of cognitive interpretations of selected topics:

i) spatial variation of soil dynamic properties, ii) attenuation laws for rock sites (seismic input), iii) generation of artificial-motion time histories, iv) effects of local site conditions (site effects), and iv) evaluation of liquefaction susceptibility, CC techniques (Neural Networks NNs, Fuzzy Logic FL and Genetic Algorithms GAs) are presented as appealing alternatives for integrated data-driven and theoretical procedures to generate reliable seismic models.

GEOTECHNICAL EARTHQUAKE HAZARDS

The author is well aware that standards for geotechnical seismic design are under development worldwide. While there is no need to "reinvent the wheel" there is a requirement to adapt such initiatives to fit the emerging safety philosophy and demands. This investigation also strongly endorses the view that "guidelines" are far more desirable than "codes" or "standards" disseminated all over seismic regions. Flexibility in approach is a key ingredient of geotechnical engineering and the cognitive technology in this area is rapidly advancing. The science and practice of geotechnical earthquake engineering is far from mature and need to be expanded and revised periodically in coming years. It is important that readers and users of the computational models presented here familiarize themselves with the latest advances and amend the recommendations herein appropriately.

This document is not intended to be a detailed treatise of latest research in geotechnical earthquake engineering, but to provide sound guidelines to support rational cognitive approaches. While every effort has been made to make the material useful in a wider range of applications, applicability of the material is a matter for the user to judge. The main aim of this guidance document is to promote consistency of cognitive approach to everyday situations and, thus, improve geotechnical-earthquake aspects of the performance of the built safe-environment.

A "SOFT" INTERPRETATION OF GROUND MOTIONS

After a sudden rupture of the earth's crust (caused by accumulating stresses, elastic strain-energy) a certain amount of energy radiates from the rupture as seismic waves. These waves are attenuated, refracted, and reflected as they travel through the earth, eventually reaching the surface where they cause ground shaking. The principal geotechnical hazards associated with this event are fault rupture, ground shaking, liquefaction and lateral spreading, and landsliding. Ground shaking is one of the principal seismic hazards that causes extensive damage to the built environment and failure of engineering systems over large areas. Earthquake loads and their effects on structures are directly related to the intensity and duration of ground shaking. Similarly, the level of ground deformation, damage to earth structures and ground failures are closely related to the severity of ground shaking.

In engineering evaluations, three characteristics of ground shaking are typically considered: i) the amplitude, ii) frequency content and iii) significant duration of shaking (time over which the ground motion has relatively significant amplitudes).These characteristics of the ground motion at a given site are affected by numerous complex factors such as the source mechanism, earthquake magnitude, rupture directivity, propagation path of seismic waves, source distance and effects of local soil conditions. There are many unknowns and uncertainties associated with these issues which in turn result in significant uncertainties regarding the characteristics of the ground motion and earthquake loads.

If the random nature of response to earthquakes (aleatory uncertainty) cannot be avoided [1,2], it is our limited knowledge about the patterns between seismic events and their manifestations -ground motions- at a site (epistemic uncertainty) that must be improved thorough more scientific seismic analyses. A strategic factor in seismic hazard analysis is the ground motion model or attenuation relation. These attenuation relationships has been developed based on magnitude, distance and site category, however, there is a tendency to incorporate other parameters, which are now known to be significant, as the tectonic environment, style of faulting and the effects of topography, deep basin edges and rupture directivity. These distinctions are recognized in North America, Japan and New Zealand [3-6], but ignored in most other regions of the world [7]. Despite recorded data suggest that ground motions depend, in a significant way, on these aspects, these inclusions did not have had a remarkable effect on the predictions confidence and the geotechnical earthquake engineer prefers the basic and clear-cut approximations on those that demand a *blind* use of coefficients or an intricate determination of soil/ fault conditions.

A key practice in current aseismic design is to develop design spectrum compatible time histories. This development entails the modification of a time history so that its response spectrum matches within a prescribed tolerance level, the target design spectrum. In such matching it is important to retain the phase characteristics of the selected ground motion time history. Many of the techniques used to develop compatible motions do not retain the phase [8]. The response spectrum alone does not adequately characterize specific-fault ground motion. Near-fault ground motions must be characterized by a long period pulse of strong motion of a fairly brief duration rather than the stochastic process of long duration that characterizes more distant ground motions. Spectrum compatible with these specific motions will not have these characteristics unless the basic motion being modified to ensure compatibility has these effects included. Spectral compatible motions could match the entire spectrum but the problem arises on finding a "real" earthquake time series that match the specific nature of ground motion. For nonlinear analysis of structures, spectrum compatible motions should also correspond to the particular energy input [9], for this reason, designers should be cautious about using spectrum compatible motions when estimating the displacements of embankment dams and earth structures under strong shaking, if the acceptable performance of these structures is specified by criteria based on tolerable displacements.

Another important seismic phenomenon is the liquefaction. Liquefaction is associated with significant loss of stiffness and strength in the shaken soil and consequent large ground deformation. Particularly damaging for engineering structures are cyclic ground movements during the period of shaking and excessive residual deformations such as settlements of the ground and lateral spreads. Ground surface disruption including surface cracking, dislocation, ground distortion, slumping and permanent deformations, large settlements and lateral spreads are commonly observed at liquefied sites. In sloping ground and backfills behind retaining structures in waterfront areas, liquefaction often results in large permanent ground displacements in the down-slope direction or towards waterways (lateral spreads). Dams, embankments and sloping ground near riverbanks where certain shear strength is required for stability under gravity loads are particularly prone to such failures. Clay soils may also suffer some loss of strength during shaking but are not subject to boils and other "classic" liquefaction phenomena. For intermediate soils, the transition from "sand like" to "clay-like" behavior depends primarily on whether the soil is a matrix of coarse grains with fines contained within the pores or a matrix of plastic fines with coarse grained "filler". Recent papers by Boulanger and Idriss [10, 11] are helpful in clarifying issues surrounding the liquefaction and strain softening of different soil types during strong ground shaking. Engineering judgment based on good quality investigations and data interpretation should be used for classifying such soils as liquefiable or non-liquefiable.

Procedures for evaluating liquefaction, potential and induced lateral spread, have been studied by many engineering committees around the world. The objective has been to review research and field experience on liquefaction and recommended standards for practice. Youd and Idriss [12] findings and the liquefaction-resistance chart proposed by Seed et al. [13] in 1985, stay as standards for practice. They have been slightly modified to adjust new registered input-output conditions and there is a strong tendency to recommend i) the adoption of the cone penetration test CPT, standard penetration test SPT or the shear wave velocities for describing the *in situ* soil conditions [14] and

ii) the modification of magnitude factors used to convert the critical stress ratios from the liquefaction assessment charts (usually developed for M7:5) to those appropriate for earthquakes of diverse magnitudes [12, 15].

COGNITIVE COMPUTING

Cognitive Computing CC as a discipline in a narrow sense, is an application of computers to solve a given computational problem by imperative instructions; while in a broad sense, it is a process to implement the instructive intelligence by a system that transfers a set of given information or instructions into expected behaviors. According to theories of cognitive informatics [16-18], computing technologies and systems may be classified into the categories of imperative, autonomic, and cognitive from the bottom up. Imperative computing is a traditional and passive technology based on stored-program controlled behaviors for data processing [19-24]. An autonomic computing is goal-driven and self-decision-driven technologies that do not rely on instructive and procedural information [25-28]. Cognitive computing is more intelligent technologies beyond imperative and autonomic computing, which embodies major natural intelligence behaviors of the brain such as thinking, inference, learning, and perceptions.

Cognitive computing is an emerging paradigm of intelligent computing methodologies and systems, which implements computational intelligence by autonomous inferences and perceptions mimicking the mechanisms of the brain. This section presents a brief description on the theoretical framework and architectural techniques of cognitive computing beyond conventional imperative and autonomic computing technologies. Cognitive models are explored on the basis of the latest advances in applying computational intelligence. These applications of cognitive computing are described from the aspects of cognitive search engines, which demonstrate how machine and computational intelligence technologies can drive us toward autonomous knowledge processing.

Computational Intelligence: Soft Computing Technologies

The *computational intelligence* is a synergistic integration of essentially three computing paradigms, viz. neural networks, fuzzy logic and evolutionary computation entailing probabilistic reasoning (belief networks, genetic algorithms and chaotic systems) [29]. This synergism provides a framework for flexible information processing applications designed to operate in the real world and is commonly called *Soft Computing SC* [30]. Soft computing technologies are robust by design, and operate by trading off precision for tractability. Since they can handle uncertainty with ease, they conform better to real world situations and provide lower cost solutions.

The three components of soft computing differ from one another in more than one way. Neural networks operate in a numeric framework, and are well known for their learning and generalization capabilities. Fuzzy systems [31] operate in a linguistic framework, and their strength lies in their capability to handle linguistic information and perform approximate reasoning. The evolutionary computation techniques provide powerful search and optimization methodologies. All the three facets of soft computing differ from one another in their time scales of operation and in the extent to which they embed *a priori* knowledge.

Figure 1 shows a general structure of Soft Computing technology. The following main components of SC are known by now: fuzzy logic FL, neural networks NN, probabilistic reasoning PR, genetic algorithms GA, and chaos theory ChT (Figure 1). In SC FL is mainly concerned with imprecision and approximate reasoning, NN with learning, PR with uncertainty and propagation of belief, GA with global optimization and search and ChT with nonlinear dynamics. Each of these computational paradigms (emerging reasoning technologies) provides us with complementary reasoning and searching methods to solve complex, real-world problems. In large scope, FL, NN, PR, and GA are complementary rather that competitive [32-34]. The interrelations between the components of SC, shown in Figure 1, make the theoretical foundation of Hybrid Intelligent Systems. As noted by L. Zadeh: "… the term hybrid intelligent systems is gaining currency as a descriptor of systems in which FL, NC, and PR are used in combination. In my view, hybrid intelligent systems are the wave of the future" [35]. The use of Hybrid Intelligent Systems are leading to the development of numerous manufacturing system, multimedia system, intelligent robots, trading systems, which exhibits a high level of MIQ (machine intelligence quotient).

Comparative Characteristics Of Sc Tools

The constituents of SC can be used independently (fuzzy computing, neural computing, evolutionary computing etc.), and more often in combination [36, 37, 38- 40, 41]. Based on independent use of the constituents of Soft Computing, fuzzy technology, neural technology, chaos technology and others have been recently applied as emerging technologies to both industrial and non-industrial areas.

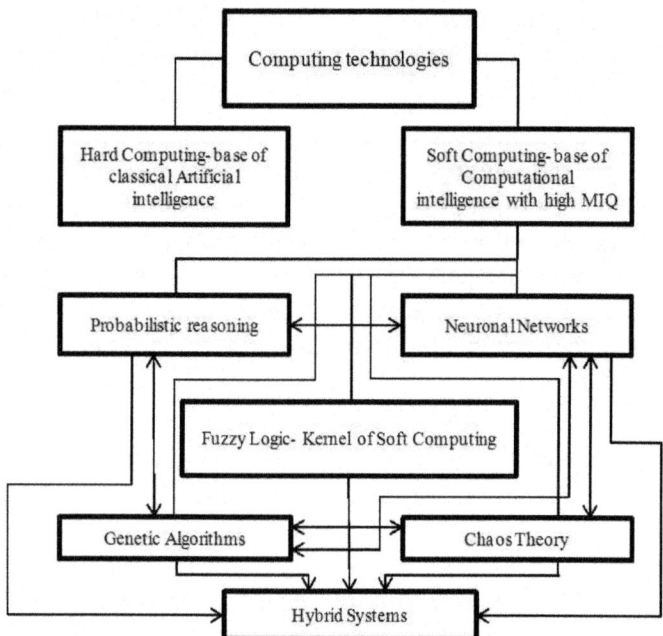

Figure 1: Soft Computing Components.

Fuzzy logic is the leading constituent of Soft Computing. In Soft Computing, fuzzy logic plays a unique role. FL serves to provide a methodology for computing [36]. It has been successfully applied to many industrial spheres, robotics, complex decision making and diagnosis, data compression, and many other areas. To design a system processor for handling knowledge represented in a linguistic or uncertain numerical form we need a fuzzy model of the system. Fuzzy sets can be used as a universal approximator, which is very important for modeling unknown objects. If an operator cannot tell linguistically what kind of action he or she takes in a specific situation, then it is quite useful to model his/her control actions using numerical data. However, fuzzy logic in its so called *pure form* is not always useful for easily constructing intelligent systems. For example, when a designer does not have sufficient prior information (knowledge) about the system, development of acceptable fuzzy rule base becomes impossible. As the complexity of the system increases, it becomes difficult to specify a correct set of rules and membership functions for describing adequately the behavior of the system. Fuzzy systems also have the disadvantage of not being able to extract additional knowledge from the experience and correcting the fuzzy rules for improving the performance of the system.

Another important component of Soft Computing is neural networks. Neural networks NN viewed as parallel computational models, are parallel fine-grained implementation of non-linear static or dynamic systems. A very important feature of these networks is their adaptive nature, where «learning by example» replaces traditional «programming» in problems solving. Another key feature is the intrinsic parallelism that allows fast computations. Neural networks are viable computational models for a wide variety of problems including pattern classification, speech synthesis and recognition, curve fitting, approximation capability, image data compression, associative memory, and modeling and control of non-linear unknown systems [42, 43]. NN are favorably distinguished for efficiency of their computations and hardware implementations. Another advantage of NN is generalization ability, which is the ability to classify correctly new patterns. A significant disadvantage of NN is their poor interpretability. One of the main criticisms addressed to neural networks concerns their black box nature [35].

Evolutionary Computing EC is a revolutionary approach to optimization. One part of EC—genetic algorithms—are algorithms for global optimization. Genetic algorithms GAs are based on the mechanisms of natural selection and genetics [44]. One advantage of genetic algorithms is that they effectively implement parallel multi-criteria search. The mechanism of genetic algorithms is simple. Simplicity of operations and powerful computational effect are the two main advantages of genetic algorithms. The disadvantages are the problem of convergence and the absence of strong theoretical foundation. The requirement of coding the domain of the real variables' into bit strings also seems to be a drawback of genetic algorithms. It should be also noted that the computational speed of genetic algorithms is low.

Because in this investigation PR and ChT are not exploited, they are not going to be explained. For the interested reader [41] is recommended. Table 1 presents the comparative characteristics of the components of Soft Computing. For each component of Soft Computing there is a specific class of problems, where the use of other components is inadequate.

Intelligent Combinations Of Sc

As it was shown above, the components of SC complement each other, rather than compete. It becomes clear that FL, NC and GA are more effective when used in combinations. Lack of interpretability of neural networks and poor learning capability of fuzzy systems are similar problems that limit the application of these tools. Neurofuzzy systems are hybrid systems which try to solve this problem by combining the learning capability of connectionist models with the interpretability property of fuzzy systems. As it was noted above, in case of dynamic work environment, the automatic knowledge base correction in fuzzy systems becomes necessary. On the other hand, artificial neural networks are successfully used in problems connected to knowledge acquisition using learning by examples with the required degree of precision.

Incorporating neural networks in fuzzy systems for fuzzification, construction of fuzzy rules, optimization and adaptation of fuzzy knowledge base and implementation of fuzzy reasoning is the essence of the Neurofuzzy approach.

Table 1: Central characteristics of Soft Computing technologies

	Fuzzy Sets	Artificial Neural Networks	Evolutionary Computing, GA	Probabilistic Reasoning	Chaotic computing
Weaknesses	•Knowledge acquisition •Learning	•Black Box interpretability	•Coding •Computational speed	•Limitation of the axioms of Probability Theory •Lack of complete knowledge •Copmputational complexity	•Computational complexity •Chaos identification complexity
Strengths	•Interpretability •Transparency •Plausibility •Graduality •Modeling •Reasoning •Tolerance to imprecision	•Learning •Adaptation •Fault tolerance •Curve fiting •Generalization ability •Approximation ability	Computational efficiency •Global optimization	•Rigorous framework •Well understanding	•Nonlinear dynamics simulation •Discovering chaos in observed data (with noise) •Determinig the predictability •Prediction startegies formulation

The combination of genetic algorithms with neural networks yields promising results as well. It is known that one of main problems in development of artificial neural systems is selection of a suitable learning method for tuning the parameters of a neural network (weights, thresholds, and structure). The most known algorithm is the "error back propagation" algorithm. Unfortunately, there are some difficulties with "back propagation". First, the effectiveness of the learning considerably depends on initial set of weights, which are generated randomly. Second, the "back propagation", like any other gradient-based method, does not avoid local minima. Third, if the learning rate is too slow, it requires too much time to find the solution. If, on the other hand, the learning rate is too high it can generate oscillations around the desired point in the weight space. Fourth, "back propagation" requires the activation functions to be differentiable. This condition does not hold for many types of neural networks. Genetic algorithms used for solving many optimization problems when the "strong" methods fail to find appropriate solution, can be successfully applied for learning neural networks, because they are free of the above drawbacks.

The models of artificial neurons, which use linear, threshold, sigmoidal and other transfer functions, are effective for neural computing. However, it should be noted that such models are very simplified. For example, reaction of a biological axon is chaotic even if the input is periodical. In this aspect the more adequate model of neurons seems to be chaotic. Model of a chaotic neuron can be used as an element of chaotic neural networks. The more adequate results can be obtained if using fuzzy chaotic neural networks, which are closer to biological computation. Fuzzy systems with If-Then rules can model non-linear dynamic systems and capture chaotic attractors easily and accurately. Combination of Fuzzy Logic and Chaos Theory gives us useful tool for building system's chaotic behavior into rule structure. Identification of chaos allows us to determine predicting strategies. If we use a Neural Network Predictor for predicting the system's behavior, the parameters of the strange attractor (in particular fractal dimension) tell us how much data are necessary to train the neural network. The combination of Neurocomputing and Chaotic computing technologies can be very helpful for prediction and control.

The cooperation between these formalisms gives a useful tool for modeling and reasoning under uncertainty in complicated real-world problems. Such cooperation is of particular importance for constructing perception-based intelligent information systems. We hope that the mentioned intelligent combinations will develop further, and the new ones will be proposed. These SC paradigms will form the basis for creation and development of Computational Intelligence.

COGNITIVE MODELS OF GROUND MOTIONS

The existence of numerous databases in the field of civil engineering, and in particular in the field of geotechnical earthquake, has opened new research lines through the introduction of analysis based on soft computing. Three methods are mainly applied in this emerging field: the ones based on the Neural Networks NN, the ones created using Fuzzy Sets FS theory and the ones developed from the Evolutionary Computation [45].

The SC hybrids used in this investigation are directed to tasks of prediction (classification and/or regression). The central objective is obtaining numerical and/or categorical values that mimic input-output conditions from experimentation and in situ measurements and then, through the recorded data and accumulated experience, predict future behaviors. The examples presented herein have been developed by an engineering committee that works for generating useful guidance to geotechnical practitioners with geotechnical seismic design. This effort could help to minimize the perceived significant and undesirable variability within geotechnical earthquake practice. Some urgency in producing the alternative guidelines was seen, after the most recent earthquakes disasters, as being necessary with a desire to avoid a long and protracted process. To this end, a two stage approach was suggested with the first stage being a cognitive interpretation of well-known procedures with appropriate factors for geotechnical design, and a posterior step identifying the relevant philosophy for a new geotechnical seismic design.

Spatial Variation Of Soil Dynamic Properties

The spatial variability of subsoil properties constitutes a major challenge in both the design and construction phases of most geo-engineering projects. Subsoil investigation is an imperative step in any civil engineering project. The purpose of an exploratory investigation is to infer accurate information about actual soil and rock conditions at the site. Soil exploration, testing, evaluation, and field observation are well-established and routine procedures that, if carried out conscientiously, will invariably lead to good engineering design and construction. It is impossible to determine the optimum spacing of borings before an investigation begins because the spacing depends not only on type of structure but also on uniformity or regularity of encountered soil deposits. Even the most detail soil maps are not efficient enough for predicting a specific soil property because it changes from place to place, even for the same soil type. Consequently interpolation techniques have been extensively exploited. The most commonly used methods are kriging and co-kriging but for better estimations they require a great number of measurements available for each soil type, what is generally impossible.

Based on the high cost of collecting soil attribute data at many locations across landscape, new interpolation methods must be tested in order to improve the estimation of soil properties. The integration of GIS and Soft Computing SC offers a potential mechanism to lower the cost of analysis of geotechnical information by reducing the amount of time spent understanding data. Applying GIS to large sites, where historical data can be organized to develop multiple databases for analytical and stratigraphic interpretation, originates the establishment of spatial/chronological efficient methodologies for interpreting properties (soil exploration) and behaviors (in situ measured). GIS-SC modeling/simulation of natural systems represents a new methodology for building predictive models, in this investigation NN and GAs, nonparametric cognitive methods, are used to analyze physical, mechanical and geometrical parameters in a geographical context. This kind of spatial analysis can handle uncertain, vague and incomplete/redundant data when modeling intricate relationships between multiple variables. This means that a NN has not constraints about the spacing (minimum distance) between the drill holes used for building (training) the SC model. The NNs-GAs acts as computerized architectures that can approximate nonlinear functions of several variables, this scheme represent the relations between the spatial patterns of the stratigraphy without restrictive assumptions or excessive geometrical and physical simplifications.

The geotechnical data requirements (geo-referenced properties) for an easy integration of the SC technologies are explained through an application example: a geo-referenced three-dimensional model of the soils underlying Mexico City. The classification/prediction criterion for this very complex urban area is established according to two variables: the cone penetration resistance qcqc (mechanical property) and the shear wave velocity VsVs (dynamic property). The expected result is a 3D-model of the soils underlying the city area that would eventually be improved for a more complex and comprehensive model adding others mechanical, physical or geometrical geo-referenced parameters.

Cone-tip penetration resistances and shear wave velocities have been measured along 16 bore holes spreaded throughout the clay deposits of Mexico City (Figure 2). This information was used as the set of examples inputs (latitude, longitude and depth) → output (qcqc /VsVs). The analysis was carried out in an approximate area of 125 km2km2 of Mexico City downtown. It is important to point out that 20% of these patterns (sample points and complete variables information) are not used in the training stage; they will be presented for testing the generalization capabilities of the closed system components (once the training is stopped).

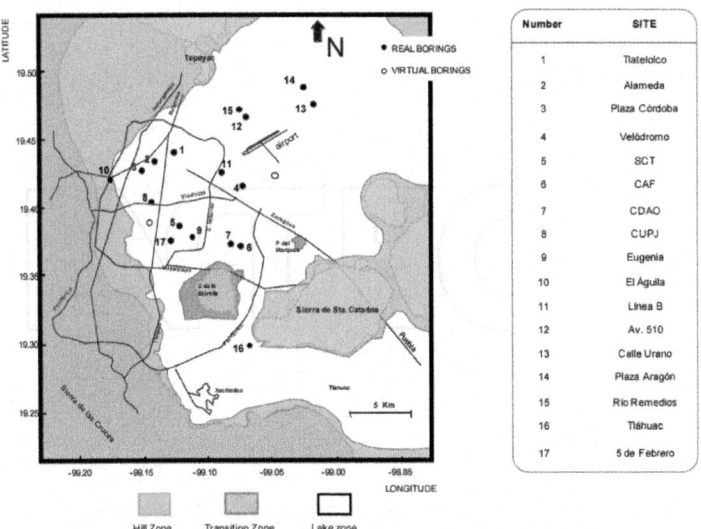

Figure 2: Mexico City Zonation.

In the 3D-neurogenetic analysis, the functions

$qc=\{qc(X,Y,Z)\}/Vs=\{Vs(X,Y,Z)\}qc=\{qc(X,Y,Z)\}/Vs=\{Vs(X,Y,Z)\}$ are to be approximated using the procedure outlined below:

Generate the database including identification of the site [borings or stations] (X,Y –geographical coordinates, Z –depth, and a CODE –ID number), elevation reference (meters above de sea level, m.a.s.l.), thickness of predetermined structures (layers), and additional information related to geotechnical zoning that could be useful for results interpretation.

Use the database to train an initial neural topology whose weights and layers are tuned by an evolutive algorithm (see [46] for details), until the minimum error between calculated and measured values $qc=fNN(X,Y,Z)\}/Vs=\{fNN(X,Y,Z)\}qc=fNN(X,Y,Z)\}/Vs=\{fNN(X,Y,Z)\}$ is achieved (Figure 3a). The generalization capabilities of the optimal 3D neural model are tested presenting real work cases (information from borings not included in the training set) to the net. Figure 3b presents the comparison between the measured $qcqc$, $VsVs$ values and the NN calculations for testing cases. Through the neurogenetic results for unseen situations we can conclude that the procedure works extremely well in identifying the general trend in materials resistance (stiffness). The correlation between NN calculations and "real" values is over 0.9.

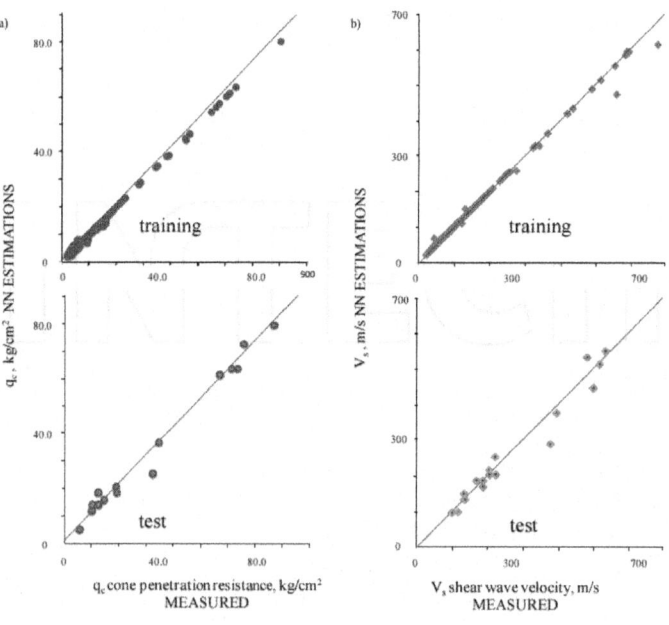

Figure 3: Neural estimations of mechanical and dynamic parameters.

For visual environment requirements a grid is constructed using raw information and neurogenetic estimations for defining the spatial variation of properties (Figure 4). The 3D view of the studied zone represents an easier and more understandable engineering system. The 3D neurogenetic-database also permits to display property-contour lines for specific depths. Using the neurogenetic contour maps, the spatial distribution of the mechanical/dynamic variables can be visually appreciated. The 3D model is able to reflect the stratigraphical patterns (Figure 5), indicating that the proposed networks are effective in site characterization with remarkable advantages if comparing with geostatistical approximations: it is easier to use, to understand and to develop graphical user interfaces. The confidence and practical advantages of the defined neurogenetic layers is evident. Precision of predictions depends on neighborhood structure, grid size, and variance response, but based on the results we can conclude that despite of the grid cell (size) is not too small the spatial correlation extends beyond the training neighborhood, but the higher confidence is obviously only within.

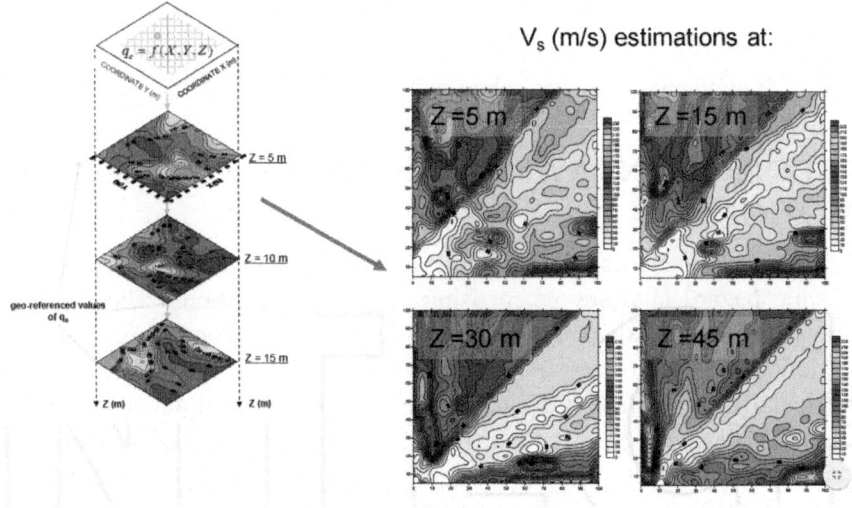

Figure 4: Neural response.

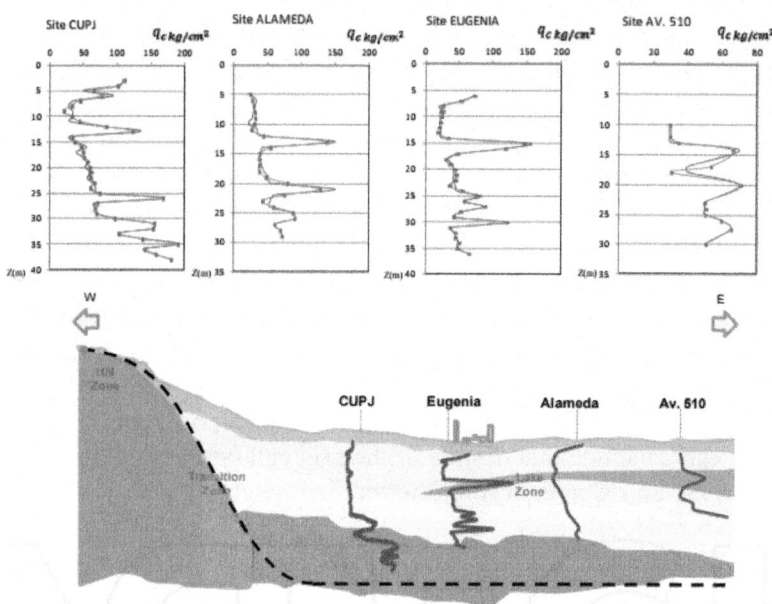

Figure 5: Stratigraphy sequence obtained using the 3D Neural estimations.

ATTENUATION LAWS FOR ROCK SITE (OUTCROPPING MOTIONS)

Source, path, and local site response are factors that should be considered in seismic hazard analyses when using attenuation relations. These relations, obtained from statistical regression, are derived from strong motion recordings to define the occurrence of an earthquake with a specific magnitude at a particular distance from the site. Because of the uncertainties inherent in the variables describing the source (e.g. magnitude, epicentral distance, focal depth and fault rupture dimension), the difficulty to define broad categories to classify the site (e.g. rock or soil) and our lack of understanding regarding wave propagation processes and the ray path characteristics from source to site, commonly the predictions from attenuation regression analyses are inaccurate. As an effort to recognize these aspects, multiparametric attenuation relations have been proposed by several researchers [47-53]. However, most of these authors have concluded that the governing parameters are still source, ray path, and site conditions. In this section an empirical NN formulation that uses the minimal information about magnitude, epicentral distance, and focal depth for subduction-zone earthquakes is developed to predict the peak ground acceleration PGA and spectral accelerations SaSa at a rock-like site in Mexico

City.

The NN model was training from existing information compiled in the Mexican strong motion database. The NN uses earthquake moment magnitudeMwMw, epicentral distanceEDED, and focal depth FDFD from hundreds of events recorded during Mexican subduction earthquakes (Figure 6) from 1964 to 2007. To test the predicting capabilities of the neuronal model, 186 records were excluded from the data set used in the learning phase. Epicentral distance EDED is considered to be the length from the point where fault-rupture starts to the recording site, and the focal depth FDFD is not declared as mechanism classes, the NN should identify the event type through the FDFD crisp value coupled with the others input parameters [54, 47, 55], The interval of MwMw goes from 3 to 8.1 approximately and the events were recorded at near (a few km) and far field stations (about 690 km). The depth of the zone of energy release ranged from very shallow to about 360 km.

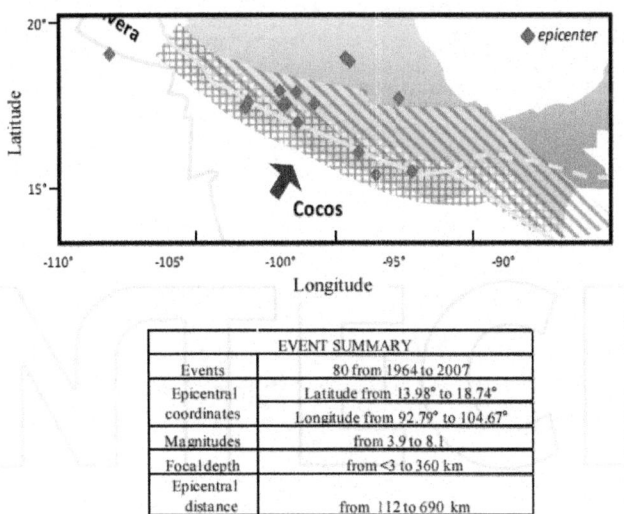

EVENT SUMMARY	
Events	80 from 1964 to 2007
Epicentral coordinates	Latitude from 13.98° to 18.74°
	Longitude from 92.79° to 104.67°
Magnitudes	from 3.9 to 8.1
Focal depth	from <3 to 360 km
Epicentral distance	from 112 to 690 km

Figure 6: Earthquakes characteristics

Modeling of the data base has been performed using backpropagation learning algorithm. Horizontal (mutually orthogonalPGAh1PGAh1, N-S component, andPGAh2PGAh2, E-W component) and vertical components (PGAvPGAv) are included as outputs for neural mapping. After trying many topologies, the best horizontal and vertical modules with quite acceptable approximations were the simpler alternatives (BP backpropagation, 2 hidden layers/15 units or nodes each). The neuronal attenuation model for {Mw,ED,FD}→{PGAh1,PGAh2,PGAv}

{Mw,ED,FD}→{PGAh1,PGAh2,PGAv} was evaluated by performing testing analyses. The predictive capabilities of the NNs were verified by comparing the estimated PGA's to those induced by the 186 events excluded from the original database (data for training stage).

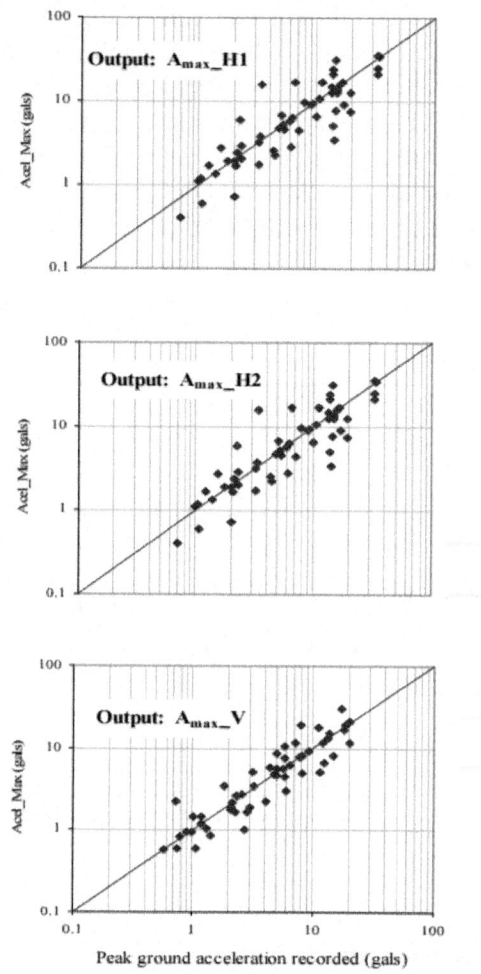

Figure 7: Some examples of measured and NN-estimated PGA values.

In Figure 7are compared the computed PGA's during training and testing stages to the measured values. The relative correlation factors ($R2 \not\cong 0.97 R2 \not\cong 0.97$), obtained in the training phase, indicate that those topologies selected as optimal behave consistently within the full range of intensity, distances and focal depths depicted by the patterns. Once the networks converge to the selected stop criterion, learning is finished and each of these black-boxes become a nonlinear multidimensional functional. Following this procedure 20 NN are trained to evaluate de SaSa at different response spectra periods (from T= 0.1 s to T= 5.0 s with DT=0.25 s). Forecasting of the spectral components is reliable enough for practical applications.

In Figure 8 two case histories correspond to large and medium size events are shown, the estimated values obtained for these events using the relationships proposed by Gómez, Ordaz &Tena [56], Youngs et al. [47], Atkinson and Boore [55] –proposed for rock sites– and Crouse et al. [51] –proposed for stiff soil sites– and the predictions obtained with the PGAh1−h2PGAh1-h2modules are shown. It can be seen that the estimation obtained with Gómez, Ordaz y Tena [56] seems to underestimate the response for the large magnitude event. However, for the lower magnitude event follows closely both the measured responses and NN predictions. Youngs et al. [47] attenuation relationship follows closely the overall trend but tends to fall sharply for long epicentral distances.

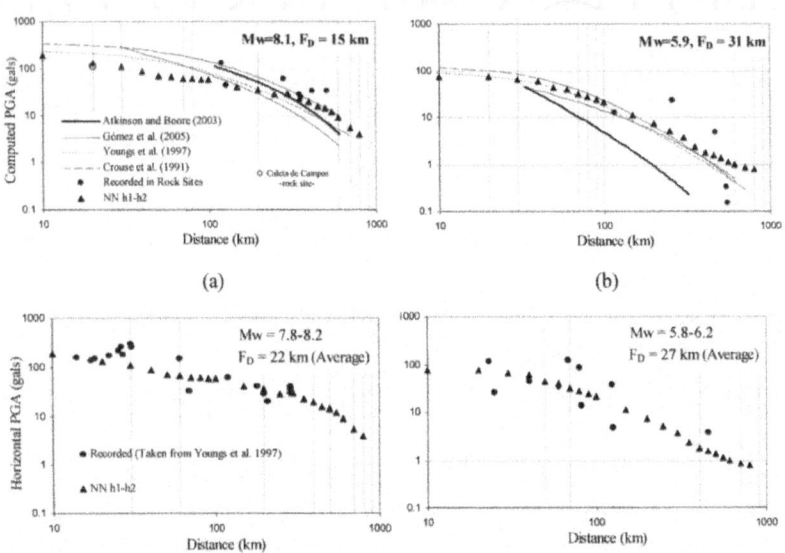

Figure 8: Attenuation laws comparisons.

Furthermore, it should be stressed the fact that, as can be seen in Figure 9 the neural attenuation model is capable to follow the general behavior of the measure data expressed as spectra while the traditional functional approaches are not able to reproduce. A neural sensitivity study for the input variables was conducted for the neuronal modules. The results are strictly valid only for the data base utilized, nevertheless, after several sensitivity analyses conducted changing the database composition, it was found that the following trend prevails; theMwMw would be the most relevant parameter then would follow EDED coupled with FDFD. However, for near site events the epicentral distance could become as relevant as the magnitude, particularly, for the vertical component and for minor earthquakes (M low) the FDFD becomes very transcendental.

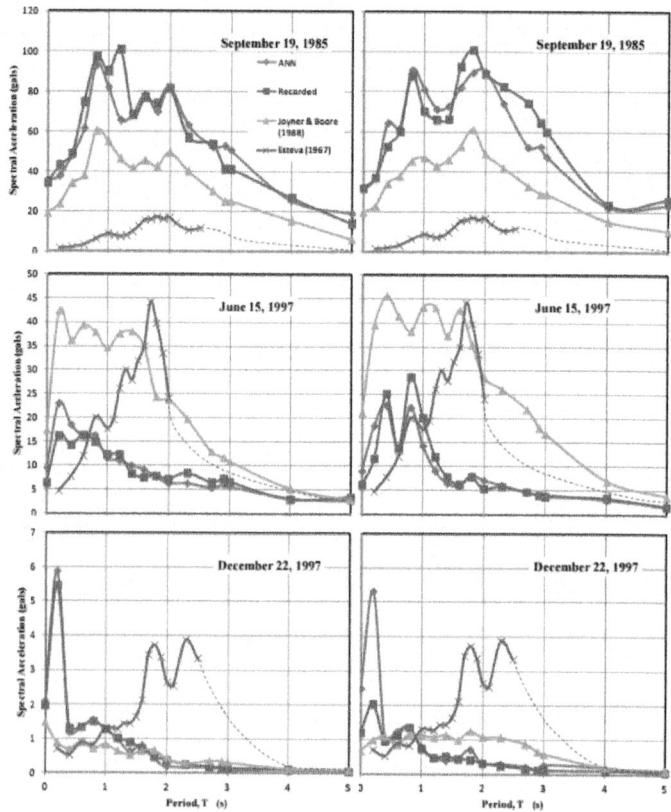

Figure 9: Response Spectra: NN-calculated vs traditional functions.

Through$\{Mw,ED,FD\} \rightarrow \{PGAhi,Sa\}\{Mw,ED,FD\} \rightarrow \{PGAhi,Sa\}$ mapping, this neuronal approach offers the flexibility to fit arbitrarily complex trends in magnitude and distance dependence and to recognize and select among the tradeoffs that are present in fitting the observed parameters within the range of magnitudes and distances present in data. This approach seems to be a promising alternative to describe earthquake phenomena despite of the limited observations and qualitative knowledge of the recording stations geotechnical site conditions, which leads to a reasoning of a partially defined behavior.

Generation Of Artificial Time Series: Accelerograms Application

For nonlinear seismic response analysis, where the superposition techniques do not apply, earthquake acceleration time histories are required as inputs. Virtually all seismic design codes and guidelines require scaling of selected ground motion time histories so that they match or exceed the controlling design spectrum within a period range of interest. Considerable variability in the characteristics of the recorded strong-motions under similar conditions may still require a characterization of future shaking in terms of an ensemble of accelerograms rather than in terms of just one or two "typical" records. This situation has thus created a need for the generation of synthetic (artificial) strong-motion time histories that simulate realistic ground motions from different points of views and/or with different degrees of sophistication. To provide the ground motions for analysis and design, various methods have been developed: i) frequency-domain methods where the frequency content of recorded signals is manipulated [57-60] and ii) time-domain methods where the recorded ground motions amplitude is controlled [61, 62]. Regardless of the method, first, one or more time histories are selected subjectively, and then scaling mechanisms for spectrum matching are applied. This is a trial and error procedure that leads artificial signals very far from real-earthquake time series.

In this investigation a Genetic Generator of Signals is presented. This genetic generator is a tool for finding the coefficients of a pre-specified functional form, which fit a given sampling of values of the dependent variable associated with particular given values of the independent variable(s). When the genetic generator is applied to synthetic accelerograms construction, the proposed tool is capable of i) searching, under specific soil and seismic conditions (within thousands of earthquake records) and recommending a desired subset that better match a target design spectrum, and ii) through processes that mimic mating, natural selection, and mutation, producing new generations of accelerograms until an optimum individual is obtained. The procedure is fast and reliable and results in time series that match any type of target spectrum with minimal tampering and deviation from recorded earthquakes characteristics.

The objective of the genetic generator, when applied to synthetic earthquakes construction, is to produce compatible artificial signals with specific design spectra. In this model specific seismic (fault rupture, magnitude, distance, focal depth) and site characteristics (soil/ rock) are the first set of inputs. They are included to take into consideration that a typical strong motion record consists of a variety of waves whose contribution depends on the earthquake source mechanism (wave path) and its particular characteristics are influenced by the distance between the source and the site, some measure of the size of the earthquake, and the surrounding geology and site conditions; and that the design spectra can be an envelope or integration of many expected ground motions that are possible to occur in certain period of time, or the result of a formulation that involves earthquake magnitude, distance and soil conditions. The second set of inputs consist of the target spectrum, the period range for the matching, lower- and upper-bound acceptable values for scaling signal shape, and a collection of GAs parameters (a population size, number of generations, crossover ratio, and mutation ratio). The output is the more success individual with a chromosome array generated from "real" accelerograms parents (a set of).

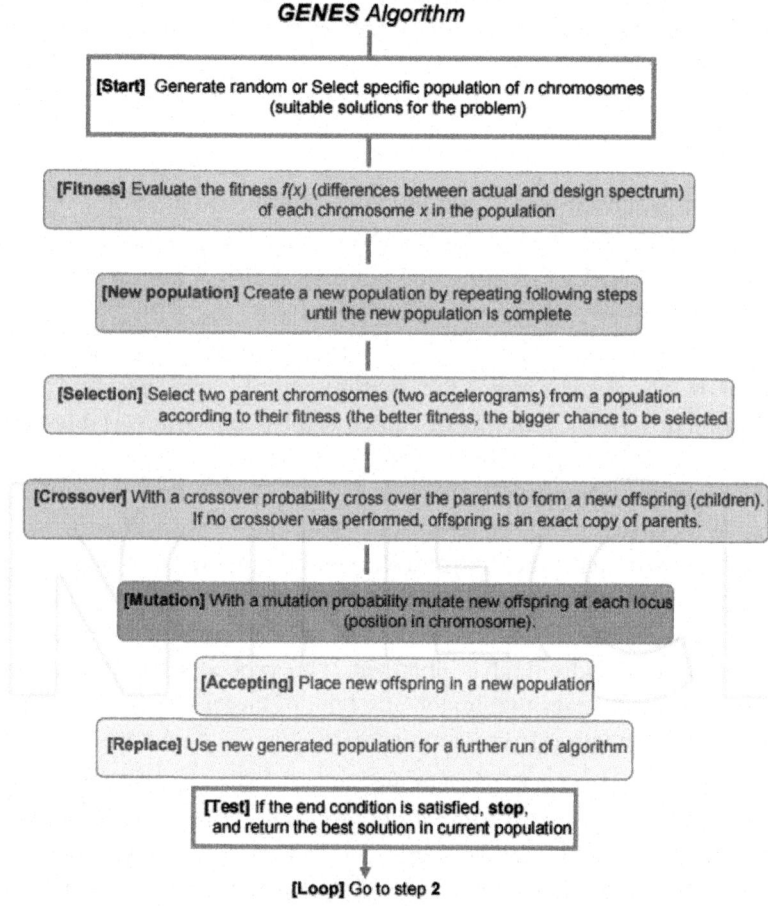

Figure 10: Genetic Generator: flow diagram.

The algorithm (see Figure 10) is started with a set of solutions (each solution is called a chromosome). A solution is composed of thousands of components or genes (accelerations recorded at the time), each one encoding a particular trait. The initial solutions (original population) are selected based on the seismic parameters at a site (defined previously by the user): fault mechanism, moment magnitude, epicentral distance, focal depth, geotechnical and geological site classification, depth of sediments.

If the user does not have a priori seismic/site knowledge, the genetic generator could select the initial population randomly (Figure 11). Once the model has found the seed-accelerogram(s) or chromosome(s), the space of all feasible solutions can be called accelerograms space (state space). Each point in this search space represents one feasible solution and can be "marked" by its value or fitness for the problem. The looking for a solution is then equal to a looking for some extreme (minimum or maximum) in the space.

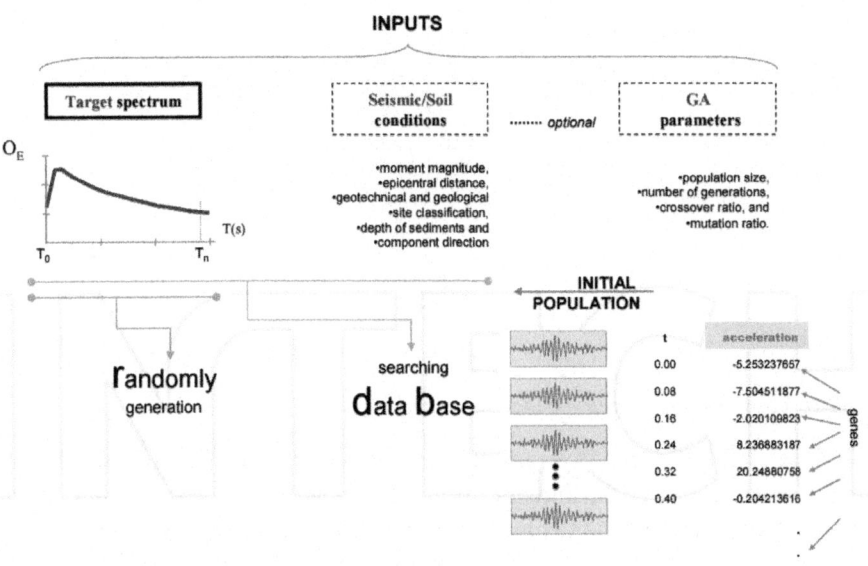

Figure 11: Genetic Generator: working phase diagram.

According to the individuals' fitness, expressed by difference between the target design spectrum and the chromosome response spectrum, the problem is formulated as the minimization of the error function, Z, between the actual and the target spectrum in a certain period range. Solutions with highest fitness are selected to form new solutions (offspring). During reproduction, the recombination (or crossover) and mutation permits to change the genes (accelerations) from parents (earthquake signals) in some way that the whole new chromosome (synthetic signal) contains the older organisms attributes that assure success. This is repeated until some user's condition (for example number of populations or improvement of the best solution) is satisfied (Figure 12).

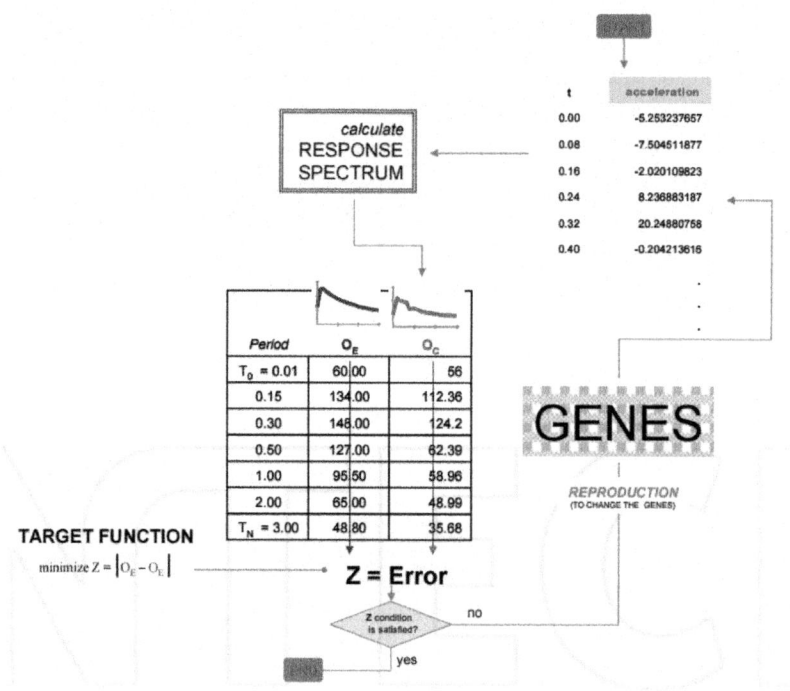

Figure 12: Iteration process of the Genetic Generator.

One of the genetic advantages is the possibility of modifying on line the image of the expected earthquake. While the genetic model is running the user interface shows the chromosome per epoch and its response spectra in the same window, if the duration time, the highest intensities interval or the ØtØt are not convenient for the user's interests, these values can be modified without retraining or a change on model' structure.

In Figure 13 are shown three examples of signals recovered following this methodology. The examples illustrate the application of the genetic methodology to select any number of records to match a given target spectrum (only the more successful individuals for each target are shown in the figure). It can be noticed the stability of the genetic algorithm in adapting itself to smooth, code or scarped spectrum shapes. The procedure is fast and reliable as results in records match the target spectrum with minimal deviation. The genetic procedure has been applied successfully to generate synthetic ground motions having different amplitudes, duration and combinations of moment magnitude and epicentral distance. Although the variations in the target spectra, the genetic signals maintain the nonlinear and nonstationary characteristics of real earthquakes. It is still under development an additional toolbox that will permit to use advanced signal analysis instruments because, as it has been demonstrated [63] [64], studying nonstationary signals through Fourier or response spectra is not convenient for all applications.

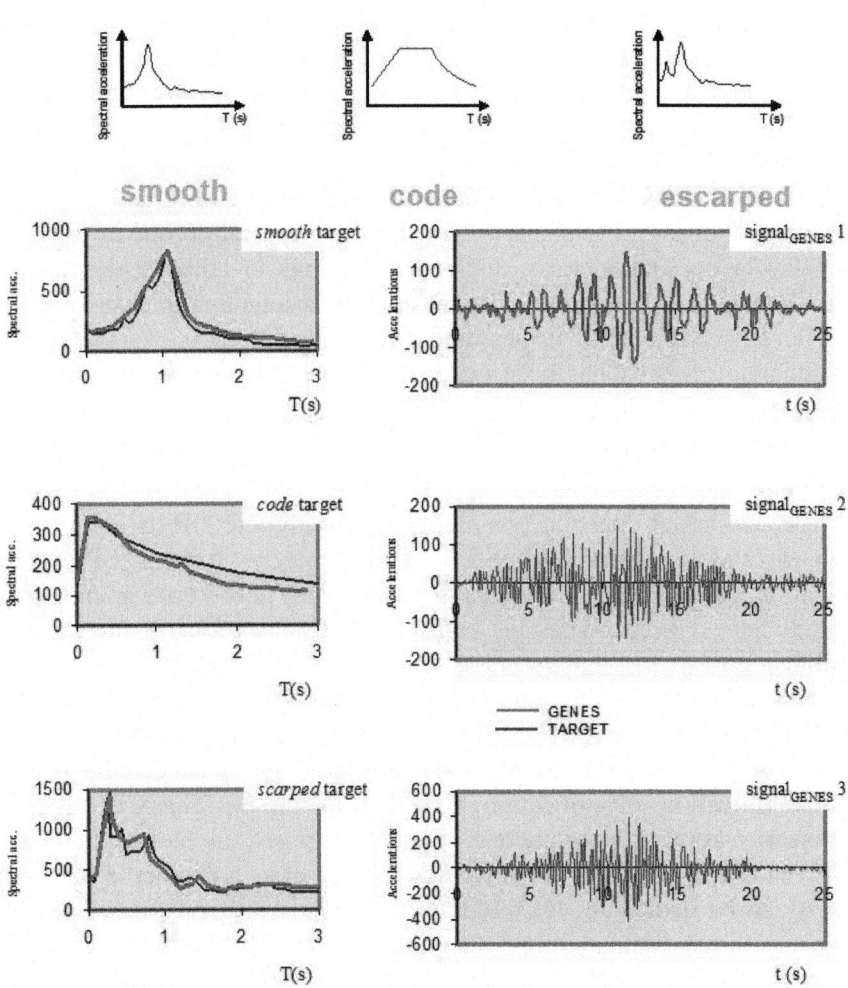

Figure 13: Some Generator results: accelerograms application.

Effects Of Local Site Conditions On Ground Motions

Geotechnical and structural engineers must take into account two fundamental characteristics of earthquake shaking: 1) how ground shaking propagates through the Earth, especially near the surface (site effects), and 2) how buildings respond to this ground motion. Because neither characteristic is completely understood, the seismic phenomenon is still a challenging research area.

Site effects play a very important role in forecasting seismic ground responses because they may strongly amplify (or deamplify) seismic motions at the last moment just before reaching the surface of the ground or the basement of man-made structures. For much of the history of seismological research, site effects have received much less attention than they should, with the exception of Japan, where they have been well recognized through pioneering work by Sezawa and Ishimoto as early as the 1930's [65]. The situation was drastically changed by the catastrophic disaster in Mexico City during the Michoacan, Mexico earthquake of 1985, in which strong amplification due to extremely soft clay layers caused many high-rise buildings to collapse despite their long distance from the source. The cause of the astounding intensity and long duration of shaking during this earthquake is not well resolved yet even though considerable research has been conducted since then, however, there is no room for doubt that the primary cause of the large amplitude of strong motions in the soft soil (lakebed) zone relative to those in the hill zone is a site effect of these soft layers.

The traditional data-analysis methods to study site effects are all based on linear and stationary assumptions. Unfortunately, in most soil systems, natural or manmade ones, the data are most likely to be both nonlinear and nonstationary. Discrepancies between calculated responses (using code site amplification factors) and recent strong motion evidence point out serious inaccuracies may be committed when analyzing amplification phenomena. The problem might be due partly because of the lack of understanding regarding the fundamental causes in soil response but also a consequence of the distorted soil amplification quantification and the incomplete characterization of nonlinearity-induced nonstationary features exposed in motion recordings [66]. The objective of this investigation is to illustrate a manner in which site effects can be dealt with for the case of Mexico City soils, making use of response spectra calculated from the motions recorded at different sites during extreme and minor events (see Figure 6). The variations in the spectral shapes, related to local site conditions, are used to feed a multilayer neural network that represent a very advantageous nonlinear-amplification relation. The database is composed by registered information earthquakes affecting Mexico City originated by different source mechanisms.

The most damaging shocks, however, are associated to the subduction of the Cocos Plate into the Continental Plate, off the Mexican Pacific Coast. Even though epicentral distances are rather large, these earthquakes have recurrently damaged structures and produced severe losses in Mexico City. The singular geotechnical environment that prevails in Mexico City is the one most important factor to be accounted for in explaining the huge amplification of seismic movements [67-70].

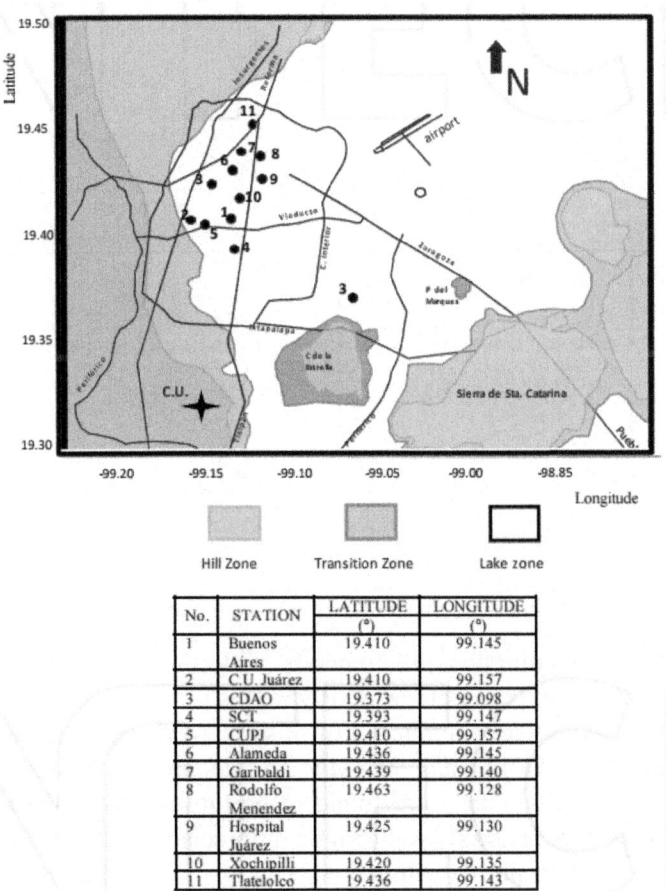

No.	STATION	LATITUDE (°)	LONGITUDE (°)
1	Buenos Aires	19.410	99.145
2	C.U. Juárez	19.410	99.157
3	CDAO	19.373	99.098
4	SCT	19.393	99.147
5	CUPJ	19.410	99.157
6	Alameda	19.436	99.145
7	Garibaldi	19.439	99.140
8	Rodolfo Menendez	19.463	99.128
9	Hospital Juárez	19.425	99.130
10	Xochipilli	19.420	99.135
11	Tlatelolco	19.436	99.143

Figure 14: Accelerographic stations used in this study.

The soils in Mexico City were formed by the deposition into a lacustrine basin of air and water transported materials. From the view point of geotechnical engineering, the relevant strata extend down to depths of 50 m to 80 m, approximately. The superficial layers formed the bed of a lake system that has been subjected to dessication for the last 350 years. Three types of soils may be broadly distinguished: in Zone I, firm soils and rock-like materials prevail; in Zone III, very soft clay formations with large amounts of microorganisms interbedded by thin seams of silty sands, fly ash and volcanic glass are found; and in Zone II, which is a transition between zones I and III, sequences of clay layers an coarse material strata are present (Figure 14).

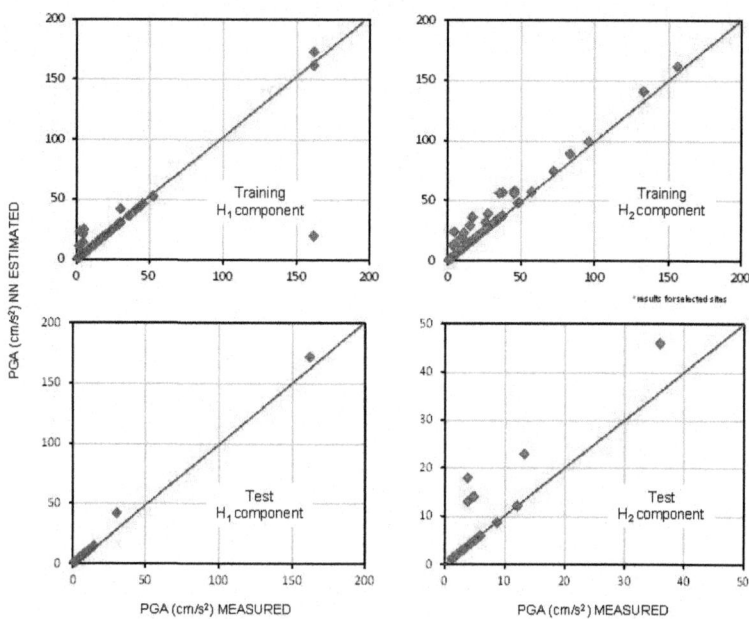

Figure 15: Neural estimations for PGA in Lake Zone sites.

Due to space limitations, reference is made only to two seismic events: the June 15, 1999 and the October 9, 1995. This module was developed based in a previous study (see section 4.2 of this Chapter) where the effect of the parametersEDED, FDFD and MwMw on the ground motion attenuation from epicentre to the site, were found to be the most significant [71]. The recent disaster experience showed that the imprecision that is inherent to most variables measurements or estimations makes crucial the consideration of subjectivity to evaluate and to derive numerical conclusions according to the phenomena behavior. The neuronal training process starts with the training of four input variables booked:EDED, FDFD and MwMw. The output linguistic variables are PGAh1PGAh1 (peak ground acceleration horizontal, component 1) and ,PGAh2,PGAh2 (peak ground acceleration, horizontal component 2) registered in a rock-like site in Zone I .The second training process is linked *feed-forward* with the previous module (PGA for rock-like site) and the new seismic inputs are Seismogenic Zone and PGA_{rock} and the Latitude and Longitude coordinates are the geo-referenced position needed to draw the deposition variation into the basin. This neuro-training runs one step after the first training phase and until the minimum difference between theSaSa and the neuronal calculations is attained. In Figure 15 some results from training and testing modes are shown.

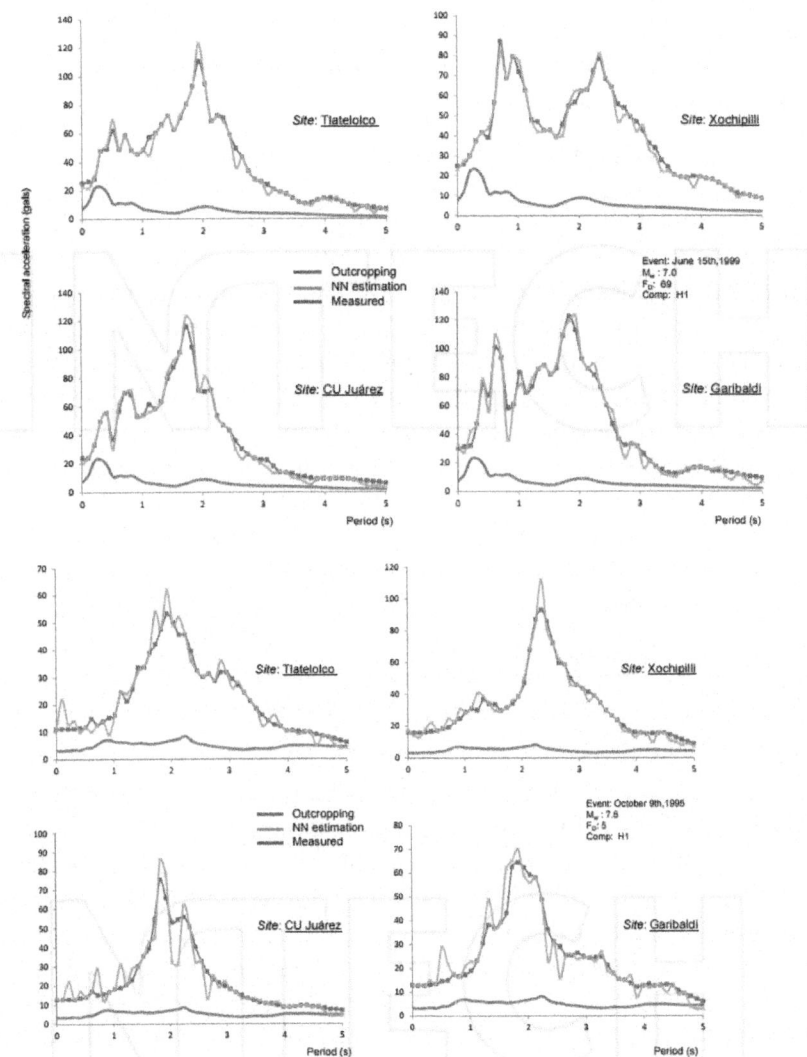

Figure 16: Spectral accelerations in some Lake-Zone sites: measured vs NN.

This second NN represents the geo-referenced amplification ratio that take into consideration the topographical, geotechnical and geographical conditions, implicit in the recorded accelerograms.The results of these two NNs are summarized in Figure 16. These graphs show the predicting capabilities of the neural system comparing the measured values with those obtained in neural-working phase. It can be observed a good correspondence throughout the full distance and magnitude range for the seismogenic zones considered in this study for the whole studied area (Lake Zone).

Liquefaction Phenomena: Potential Assessment and Lateral Displacements Estimation

Over the past forty years, scientists have conducted extensive research and have proposed many methods to predict the occurrence of liquefaction. In the beginning, undrained cyclic loading laboratory tests had been used to evaluate the liquefaction potential of a soil [72] but due to difficulties in obtaining undisturbed samples of loose sandy soils, many researchers have preferred to use in situtests [73]. In a semi-empirical approach the theoretical considerations and experimental findings provides the ability to make sense out of the field observations, tying them together, and thereby having more confidence in the validity of the approach as it is used to interpolate or extrapolate to areas with insufficient field data to constrain a purely empirical solution. Empirical field-based procedures for determining liquefaction potential have two critical constituents: i) the analytical framework to organize past experiences, and ii) an appropriate in situ index to represent soil liquefaction characteristics. The original simplified procedure [74] for estimating earthquake-induced cyclic shear stresses continues to be an essential component of the analysis framework. The refinements to the various elements of this context include improvements in the in-situ index tests (e.g., SPT, CPT, BPT,VsVs), and the compilation of liquefaction/no-liquefaction cases.

The objective of the present study is to produce an empirical machine learning ML method for evaluating liquefaction potential. ML is a scientific discipline concerned with the design and development of algorithms that allow computers to evolve behaviours based on empirical data, such as from sensor data or databases. Data can be seen as examples that illustrate relations between observed variables. A major focus of ML research is to automatically learn to recognize complex patterns and make intelligent decisions based on data; the difficulty lies in the fact that the set of all possible behaviours given all possible inputs is too large to be covered by the set of observed examples (training data). Hence the learner must generalize from the given examples, so as to be able to produce a useful output in new cases. In the following two ML tools, Neural Networks NN and Classification Trees CTs, are used to evaluate liquefaction potential and to find out the liquefaction control parameters, including earthquake and soil conditions. For each of these parameters, the emphasis has been on developing relations that capture the essential physics while being as simplified as possible. The proposed cognitive environment permits an improved definition of i) seismic loading or cyclic stress ratio CSR, and ii) the *resistance* of the soil to triggering of liquefaction or cyclic resistance ratio CRR.

The factor of safety FS against the initiation of liquefaction of a soil under a given seismic loading is commonly described as the ratio of cyclic resistance ratio (CRR), which is a measure of liquefaction resistance, over cyclic stress ratio (CSR), which is a representation of seismic loading that causes liquefaction, symbolically, FS=CRR/CSRFS=CRR/CSR. The reader is referred to Seed and Idriss[74], Youd et al. [75], and Idriss and Boulanger[76] for a historical perspective of this approach. The term CSR CSR=f(0.65,ꝶvo,amax,ꝶ›vo,rd,MSF)CS R=f0.65,ꝶvo,amax,ꝶvo›,rd,MSFis function of the vertical total stress of the soil ꝶvoꝶvo at the depth considered, the vertical effective stress ꝶ›voꝶvo›, the peak horizontal ground surface acceleration amaxamax, a depth-dependent shear stress reduction factor rdrd (dimensionless), a magnitude scaling factor MSFMSF (dimensionless). For CRR, different in situ-resistance measurements and overburden correction factors are included in its determination; both terms operate depending of the geotechnical conditions. Details about the theory behind this topic in Idriss and Boulanger, [76] and Youd et al. [75].

Many correction/adjustment factors have been included in the conventional analytical frameworks to organize and to interpret the historical data. The correction factors improve the consistency between the geotechnical/seismological parameters and the observed liquefaction behavior, but they are a consequence of a constrained analysis space: a 2D plot [CSR *vs.* CRR] where regression formulas (simple equations) intend to relate complicated nonlinear/multidimensional information. In this investigation the ML methods are applied to discover unknown, valid patterns and relationships between geotechnical, seismological and engineering descriptions using the relevant available information of liquefaction phenomena (expressed as empirical prior knowledge and/or input-output data). These ML techniques "work" and "produce" accurate predictions based on few logical conditions and they are not restricted for the mathematical/analytical environment. The ML techniques establish a *natural* connection between experimental and theoretical findings.

Following the format of the simplified method pioneered by Seed and Idriss [74], in this investigation a nonlinear and adaptative *limit state* (a fuzzy-boundary that separates liquefied cases from nonliquefied cases) is proposed (Figure 17). The database used in the present study was constructed using the information included in Table 2 and it was compiled by Agdha et al., [77], Juang et al., [78], Juang [79], Baziar, [80] and Chern and Lee [81]. The cases are derived from cone penetration tests CPT, and shear wave velocities VsVs measurements and different world seismic conditions (U.S., China, Taiwan, Romania, Canada and Japan). The soils types ranges from clean sand and silty sand to silt mixtures (sandy and clayey silt). Diverse geological and geomorphological characteristics are included. The reader is referred to the citations in Table 2 for details.

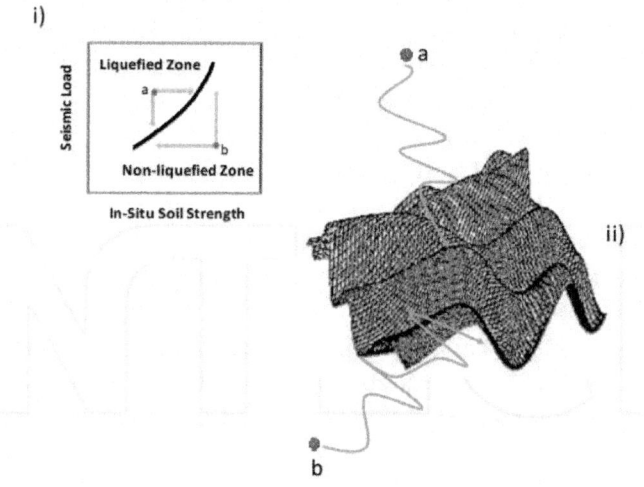

Figure 17: An schematic view of the nonlinear-liquefaction boundary.

The ML formulation uses Geotechnical (qcqc, VsVs, Unit weight, Soil Type, Total vertical stresses, Effective vertical stresses, Geometrical (Layer thickness, Water Level Depth, Top Layer Depth) andSeismological (Magnitude, PGA) input parameters and the output variable is "Liquefaction?" and it can take the values "YES/NO" (Figure 17). Once the NN is trained the number of cases that was correctly evaluated was 100% and applied to "unseen" cases (separated for testing) less than 10% of these examples were not fitted. The CT has a minor efficiency during the training showing 85% of cases correctly predicted, but when the CT runs on the unseen patterns its capability is not diminished and it asserts the same proportion. From these findings it is concluded that the neuro system is capable of predicting the in situ measurements with a high degree of accuracy but if improvement of knowledge is necessary or there are missed, vague even contradictory values in the analyzed case, the CT is a better option.

Table 2: Database for liquefaction analysis

Set	Input Parameters	Number .of Patterns	Ref.
A	Z, Z_{NAF}, H, Soil Class, Geomorphological units, Geological units, Site amplification, a_{max}	56	Fatemi-Agdha et al., 1988
B	Z, q_c, F_s, σ_0, σ_0', a_{max}, M	21	Juang et al., 1999
C	Z, q_c, F_s, σ_0, σ_0', a_{max}, M	242	Juang, 2003
D	D_{50}, a_{max}, σ_0', σ_0, M, F_s, q_c, SPT, Z	170	Baziar, 2003
E	M, σ_0, σ_0', q_c, a_{max}	466	Chern and Lee, 2009
F	Z_{NAF}, Z, H, σ_0, σ_0', Soil Class, V_s	80	Andrus and Stokoe, 1997; 2000
	Total:	**1035**	

Figure 18 shows the pruned liquefaction trees (two, one runs using qcqc values and the other through theVsVs measurements) with YES/ NO as terminal nodes. In the Figure 19, some examples of tree reading are presented. The trees incorporate soil type dependence through the resistance values (qcqc, andVsVs) and fine content, and it is not necessary to label the material as "sand" or "silt". The most general geometrical branches that split the behaviors are the Water table depth and the Layer thickness but only when the soil description is based onVsVs, whenqcqc, serves as rigidity parameter this geometrical inputs are not explicit exploited.

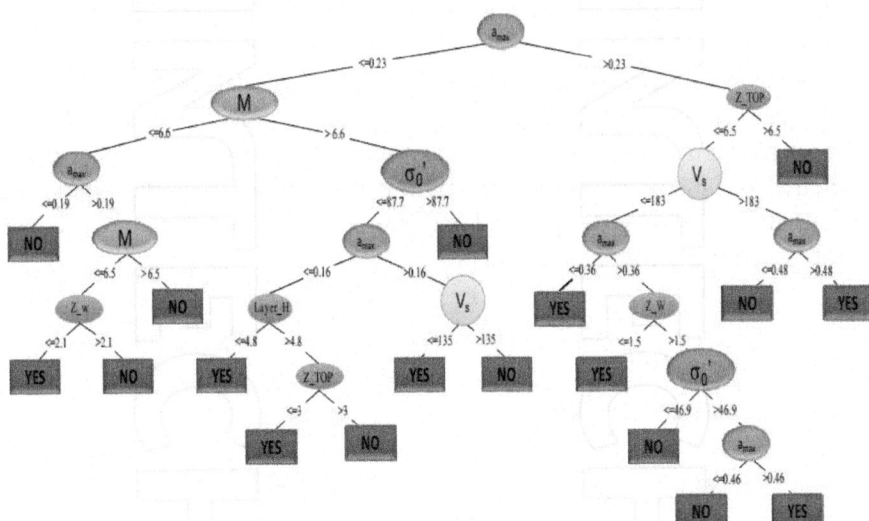

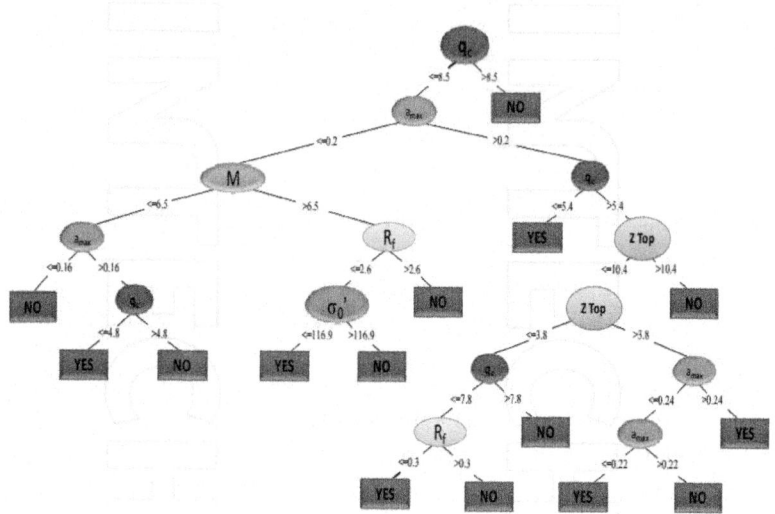

Figure 18: Classification tree for liquefaction potential assessment.

This finding can be related to the nature of the measurement: the cone penetration value contains the effect of the saturated material while the shear wave velocities need the inclusion of this situation explicitly. Without potentially confusing regression strategies, the liquefaction trees results can be seen as an indication of how effectively the ML model maps the assigned predictor variables to the response parameter. Using data from all regions and wide parameters ranges, the prediction capabilities of the neural network and classification trees are superior to many other approximations used in common practice, but the most important remark is the generation of meaningful clues about the reliability of physical parameters, measurement and calculation process and practice recommendations.

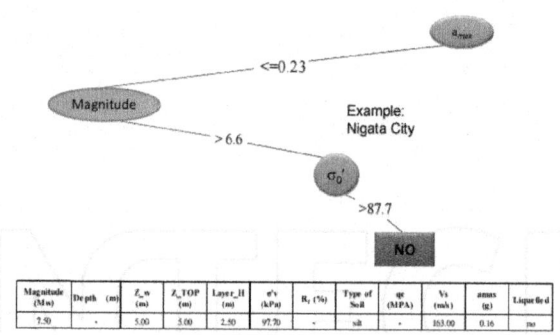

Magnitude (Mw)	Depth (m)	Z_{cn} (m)	Z_{TOP} (m)	Layer_H (m)	σ'_v (kPa)	R_f (%)	Type of Soil	qc (MPA)	Vs (m/s)	amax (g)	Liquefied
7.50	-	5.00	5.00	2.50	97.70	-	silt	-	103.00	0.16	no

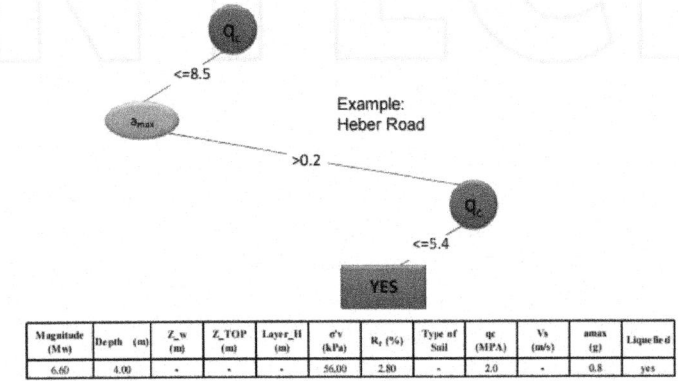

Magnitude (Mw)	Depth (m)	Z_w (m)	Z_TOP (m)	Layer_H (m)	$\sigma'v$ (kPa)	R_f (%)	Type of Soil	qc (MPA)	Vs (m/s)	amax (g)	Liquefied
6.60	4.00	.	.	.	56.00	2.80	.	2.0	.	0.8	yes

Figure 19: CT classification examples.

The intricacy and nonlinearity of the phenomena, an inconsistent and contradictory database, and many subjective interpretations about the observed behavior, make SC an attractive alternative for estimation of liquefaction induced lateral spread. NEFLAS [82], NEuroFuzzy estimation of liquefaction induced LAteral Spread, profits from fuzzy and neural paradigms through an architecture that uses a fuzzy system to represent knowledge in an interpretable manner and proceeds from the learning ability of a neural network to optimize its parameters. This blending can constitute an interpretable model that is capable of learning the problem-specific prior knowledge.

NEFLAS is based on the Takagi-Sugeno model structure and it was constructed according the information compiled by Bartlett and Youd [83] and extended later by Youd et al. [84]. The output considered in NEFLAS is the horizontal displacements due to liquefaction, dependent of moment magnitude, the PGA, the nearest distance from the source in kilometers; the free face ratio, the gradient of the surface topography or the slope of the liquefied layer base, the cumulative thickness of saturated cohesionless sediments with number of blows (modified by overburden and energy delivered to the standard penetration probe, in this case 60%), the average of fines content, and the mean grain size.

One of the most important NEFLAS advantages is its capability of dealing with the imprecision, inherent in geoseismic engineering, to evaluate concepts and derive conclusions. It is well known that engineers use words to classify qualities ("strong earthquake", "poor graduated soil" or "soft clay" for example), to predict and to validate "first principle" theories, to enumerate phenomena, to suggest new hypothesis and to point the limits of knowledge. NEFLAS mimics this method. See the technical quantity "magnitude" (earthquake input) depicted in Figure 20. The degree to which a crisp magnitude belongs to LOW, MEDIUM or HIGH linguistic label is called the degree of membership.

Based on the figure, the expression, "the magnitude is LOW" would be true to the degree of 0.5 for aMwMwof 5.7. Here, the degree of membership in a set becomes the degree of truth of a statement.

On the other hand, the human logic in engineering solutions generates sets of behavior rules defined for particular cases (parametric conditions) and supported on numerical analysis. In the neurofuzzy methods the human concepts are re-defined through a flexible computational process (training) putting (empirical or analytical) knowledge into simple "if-then" relations (Figure 20). The fuzzy system uses 1) variables composing the antecedents (premises) of implications; 2) membership functions of the fuzzy sets in the premises, and 3) parameters in consequences for finding simpler solutions with less design time.

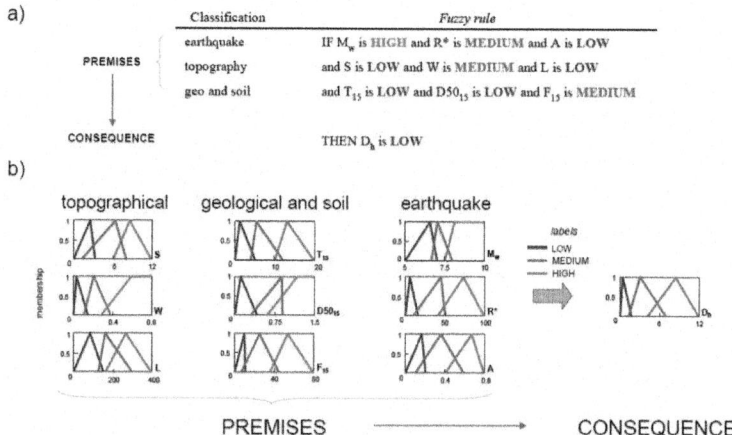

Figure 20: Neurofuzzy estimation of lateral spread.

NEFLAS considers the character of the earthquake, topographical, regional and geological components that influence lateral spreading and works through three modules: Reg-NEFLAS, appropriate for predicting horizontal displacements in geographic regions where seismic hazard surveys have been identified; Site- NEFLAS, proper for predictions of horizontal displacements for site-specific studies with minimal data on geotechnical conditions and Geotech-NEFLAS allows more refined predictions of horizontal displacements when additional data is available from geotechnical soil borings. The NEFLAS execution on cases not included in the database (Figure 21.b and Figure 21.c) and its higher values of correlation when they are compared with evaluations obtained from empirical procedures permit to assert that NEFLAS is a powerful tool, capable of predicting lateral spreads with high degree of confidence.

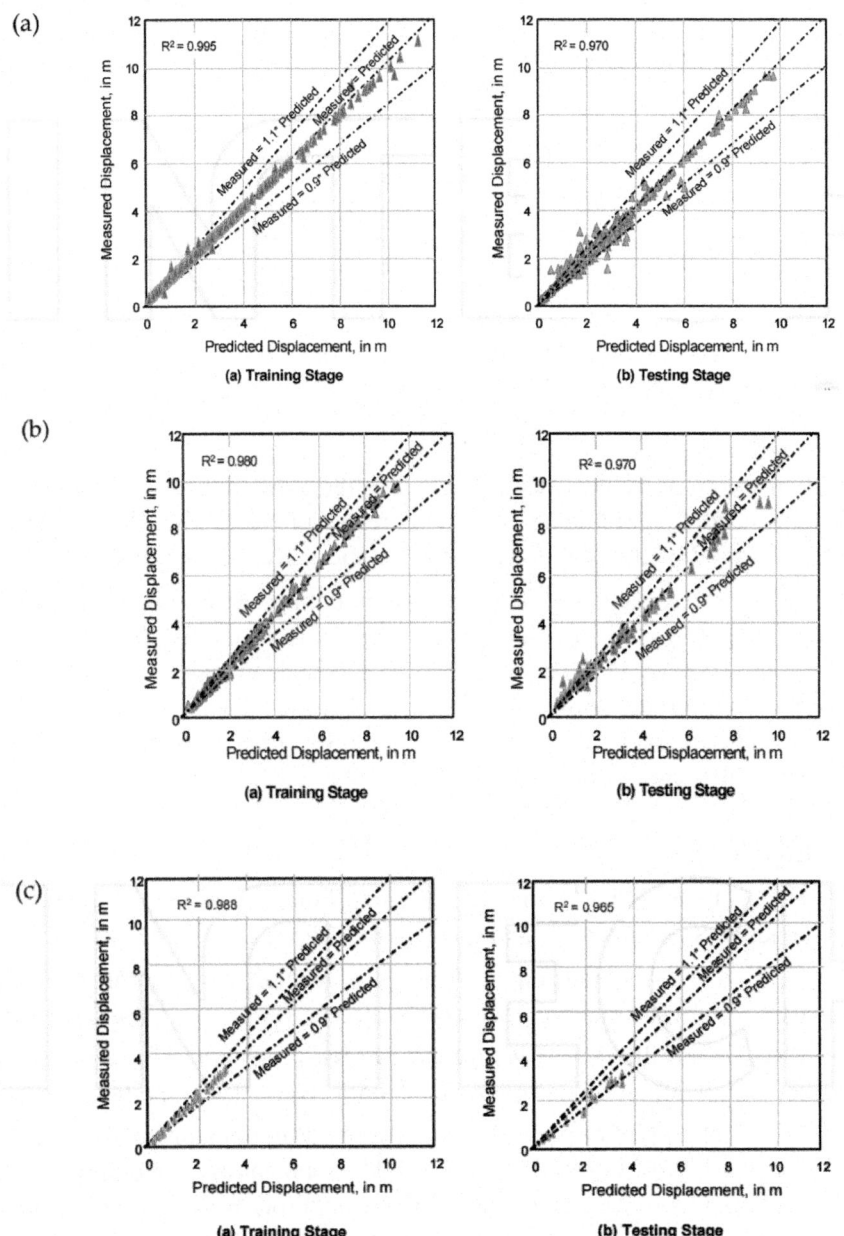

Figure 21: NN estimations vs measured displacements for a) the whole data set, b) Niigata Japan, c) San Francisco USA cases

CONCLUSIONS

Based on the results of the studies discussed in this paper, it is evident that cognitive techniques perform better than, or as well as, the conventional methods used for modeling complex and not well understood geotechnical earthquake problems. Cognitive tools are having an impact on many geotechnical and seismological operations, from predictive modeling to diagnosis and control.

The hybrid *soft* systems leverage the tolerance for imprecision, uncertainty, and incompleteness, which is intrinsic to the problems to be solved, and generate tractable, low-cost, robust solutions to such problems. The synergy derived from these hybrid systems stems from the relative ease with which we can translate problem domain knowledge into initial model structures whose parameters are further tuned by local or global search methods. This is a form of methods that do not try to solve the same problem in parallel but they do it in a mutually complementary fashion. The push for low-cost solutions combined with the need for intelligent tools will result in the deployment of hybrid systems that efficiently integrate reasoning and search techniques.

Traditional earthquake geotechnical modeling, as physically-based (or knowledge-driven) can be improved using soft technologies because the underlying systems will be explained also based on data (CC data-driven models). Through the applications depicted here it is sustained that cognitive tools are able to make abstractions and generalizations of the process and can play a complementary role to physically-based models.

REFERENCES

1. W.D.L.Finn State of the art of geotechnical earthquake engineering practice Soil Dynamics and Earthquake Engineering 20 (2000Elsevier

2. N. A. R. R. Abrahamson, A. Youngs, algorithm. stable, regression. for, using. analysis, random. the, model. effects, Bull Seismol Soc Am 1992

3. P. G. Somerville, T. Sato, Correlation of rise time with the style-offaulting factor in strong ground motions. Seismol Res Lett 1998pp (abstract).

4. Somerville PG, Greaves RL.Strong ground motions of the Kobe, Japan earthquake of January 17, 1995and development of a model of forward rupture directivity applicable in California. Proc. Western Regional Tech. Seminar on Earthquake Eng. for Dams. Assoc. of State Dam Safety Oficials, Sacramento, CA, April 11±12, 1996.

5. Abrahamson NA, Silva WJ.Empirical duration relations for shallow crustal earthquakes. Written communication, 1997

6. R. W. Graves, A. Pitarka, P. G. Somerville, Ground motion amplication in

the Santa Monica area: effects of shallow basin edge structure. Submitted for publication.

7. P. G. Somerville, art. Emerging, ground. earthquake, In. motion, P. Dakoulas, M. Yegian, R. D. Holtz, editors, Proc Geotechnical Earthquake Engineering in Soil Dynamics III, Geotechnical Special Publication 75 1ASCE, 1998 1

8. Abrahamson NA.Spatial variation of multiple support inputs. Proc. First US Symp. Seism. Eval. Retrofit Steel Bridges, UC, Berkeley, October 18, 1993

9. F. Naeim, M. Lew, On the use of design spectrum compatible motions. Earthquake Spec 1995II(1):111±28.

10. Boulanger R.W. and Idriss, I.M.2006Liquefaction Susceptibility Criteria for Silts and Clays," Journal of Geotechnical and Geoenvironmental Engineering, 132 (11), 1413 EOF1426 EOFpp.

11. Boulanger R.W. and Idriss, I.M.2007Evaluation of Cyclic Softening in Silts and Clays", Journal of Geotechnical and Geoenvironmental Engineering, 133 (6), 641 EOF652 EOFpp.

12. Youd and Idriss, NCEER. Proceedings, Workshop on Evaluation of Liquefaction Resistance of Soils.Technical Report NCCER-97970022National Center for Earthquake Engineering Research, University of Buffalo, Buffalo, New York, 1997

13. H. B. Seed, K. Tokimatsu, L. F. Harder, R. M. Chung, of. S. P. T. Influence, in. procedures, liquefaction. soil, evaluations. resistance, J Geotech Engng 19851425 EOF

14. Robertson PK, Fear CE.Liquefaction of sands and its evaluation. Proceedings, 1st Int. Conf. on Earthquake Geotechnical Engineering, Tokyo, Japan, 1995

15. Ambraseys NN. Engineering seismology.Earthquake Engng Struct Dynam 1988

16. Y. Wang, 2008aOn Contemporary Denotational Mathematics for Computational Intelligence. In Transactions of Computational Science (2629New York: Springer.

17. Y. Wang, 2009aOn Abstract Intelligence: Toward a Unified Theory of Natural, Artificial, Machinable, and Computational Intelligence. International Journal of Software Science and Computational Intelligence, 1 EOF17 EOF

18. Y. Wang, 2009bOn Cognitive Computing. International Journal of Software Science and Computational Intelligence, 1(3), 1-15.

19. M. Turing, 1950Computing Machinery and Intelligence. Mind, 59433460

20. Neumann. J. Von, 1946The Principles of Large-Scale Computing Machines. Reprinted in Annals of History of Computers, 3(3), 263-273.

21. Neumann. J. Von, 1958The Computer and the Brain. New Haven: Yale Univ. Press.

22. J. L. Gersting, 1982Mathematical Structures for Computer Science. San Francisco: W. H. Freeman & Co.

23. D. Mandrioli, C. Ghezzi, 1987Theoretical Foundations of Computer Science. New York: John Wiley & Sons.

24. H. R. Lewis, C. H. Papadimitriou, 1998Elements of the Theory of Computation, 2nd ed. Englewood Cliffs, NJ: Prentice Hall International.

25. J. Kephart, D. Chess, 2003The Vision of Autonomic Computing. IEEE Computer, 26(1), 41 EOF50 EOF

26. IBM2006Autonomic Computing White Paper. An Architectural Blueprint for Autonomic Computing, 4th ed., June, (137

27. Y. Wang, 2004Keynote: On Autonomic Computing and Cognitive Processes. Proc. 3rd IEEE International Conference on Cognitive Informatics (ICCI'04), Victoria, Canada, IEEE CS Press, (34

28. Y. Wang, 2007aJuly). Software Engineering Foundations: A Software Science Perspective. CRC Book Series in Software Engineering, Vol. II, Auerbach Publications, NY.

29. B. Bouchon-Meunier, R. Yager, L. Zadeh, 1995Fuzzy Logic and SoftComputing. World Scientific, Singapore.

30. J. C. Bezdek, is. a. What, intelligence?. computational, Zurada. J. M. In, I. I. R. J. Marks, C. J. Robinson, eds.) Computational Intelligence: Imitating Life, 112IEEE Press, Los Alamitos (1994

31. L. A. Zadeh, Sets. Fuzzy, Information and Control 8, 338-353 (1965

32. Zadeh L.A.The roles of fuzzy logic and soft computing in the conception, design and deployment of intelligent systems. BT Technol J. 14199443236

33. R. A. Aliev, R. R. Aliev, Computing. Soft, I. I. I. I. I. I. volumes, A. S. O. A. Baku, Press, 1997-1998in Russian).

34. R. Aliev, K. Bonfig, F. Aliew, Computing. Soft, Verlag. Berlin, Technic, 2000

35. Zadeh L.A. Foreword. In Proc.First European Congress on Intelligent Techniques and Soft Computing- EUFIT'95, page VII, 1995

36. L. A. Zadeh, Soft Computing and Fuzzy Logic. IEEE Software 11199464858

37. R. A. Aliev, Expert. Fuzzy, In. Systems, F. Aminzadeh, M. Jamshidi, S. O. F. T. C. O. M. P. U. T. I. N. G. (eds, Logic. Fuzzy, Networks. Neural, Artificial. Distributed, 9910 . Intelligence.pages, N. , NJ: PTR Prentice Hall, 1994

38. Zadeh L.A. Fuzzy logic, Neural Networks and Soft Computing .Comm of ACM 37199437784

39. Welstead S.T. (ed) Neural Networks and Fuzzy Logic Applications in C/ C++, Professional Computing.NY: John Wiley, 1994

40. Yager R.R. and Zadeh L.A. (eds) Fuzzy sets, neural networks and Soft Computing.NY: VAN Nostrand Reinhold , 1994

41. Nauck D., Klawonn F., and Kruse R., Foundations of Neuro-Fuzzy Systems.NY: John Wiley and Sons, 1997.

42. Mohamad H.Hassoun, Fundamentals of artificial neural networks. Cambridge: MIT Press, 1995

43. S. Haykin, Networks. A. Neural, Foundation. Comprehensive, Marmillau, I. E. E. E. Computer, Society, 1994

44. Goldberg D.E., Genetic algorithms in search, optimization and machine learning.Reading, MA: Addison-Wesley, 1989

45. Arciszewski, and De Jong K.A. (2001Evolutionary computation in civil engineering: research frontiers. Eight International Conference on Civil and Structural Engineering Computing, (Topping B. H. V., ed.), Saxe-Coburg Publications, Eisenstadt, Vienna, Austria.

46. K. Miettinen, P.and. Neittaanmaki, J. Periaux, 1999Evolutionary Algorithms in Engineering and Computer Science : Recent Advances in Genetic Algorithms, Evolution Strategies, Evolutionary Programming, John Wiley & Sons Ltd., pps. 483. 0-47199-902-4

47. R. R. Youngs, S. J. Chiou, W. J. Silva, J. R. Humphrey, 1997Strong ground motion attenuation relationships for subduction zone earthquakes. Seismol. Res. Lett., (68) 1, 58 EOF73 EOF

48. J. G. Anderson, 1997Nonparametric description of peak acceleration above a subduction thrust. Seismol. Res. Lett., (68) 1, 86 EOF93 EOF

49. C. B. Crouse, 1991Ground motion attenuation equations for earthquakes on the Cascadia subduction zone. Earth. Spectra, 7210236

50. S. K. Singh, M. Ordaz, M. Rodríguez, R. Quaas, V. Mena, M. Ottaviani, J. G. Anderson, D. Almora, 1989Analysis of near-source strong motion recordings along the Mexican subduction zone. Bull. Seism. Soc. Am., 7916971717

51. C. B. Crouse, Y. K. Vyas, B. A. Schell, 1988Ground motions from

subduction-zone earthquakes. Bull. Seism. Soc. Am.,78125

52. S. K. Singh, E. Mena, R. Castro, C. Carmona, 1987Empirical prediction of ground motion in Mexico City from coastal earthquakes. Bull. Seism. Soc. Am., 7718621867

53. K. Sadigh, 1979Ground motion characteristics for earthquakes originating in subduction zones and in the western United States. Proc. Sixth Pan Amer. Conf., Lima, Peru.

54. B. F. Tichelaar, L. J. Ruff, 1993Depth of seismic coupling along subduction zones. J. Geophys. Res., 9820172037

55. G. M. Atkinson, D. M. Boore, 2003Empirical ground-motion Relations for Subduction-Zone Earthquakes and Their Applications to Cascadia and other regions. Bull. Seism. Soc. Am., 93417031729

56. S. C. Gómez, M. Ordaz, C. Tena, 2005Leyes de atenuación en desplazamiento y aceleración para el diseño sísmico de estructuras con aislamiento en la costa del Pacífico. Memorias del XV Congreso Nacional de Ingeniería Sísmica, México, Nov. A-II-02

57. D. Gasparini, E. H. Vanmarcke, 1976SIMQKE: A Program for Artificial Motion Generation, Department of Civil Engineering, Massachusetts Institute of Technology, Cambridge, MA.

58. W. J. Silva, K. Lee, 1987WES RASCAL code for synthesizing earthquake ground motions." State-of-the-Art for Assessing Earthquake Hazards in the United States, Report 24, U.S. Army Engineers Waterways Experiment Station, Misc. Paper S-731

59. B. A. Bolt, N. J. Gregor, 1993Synthesized Strong Ground Motions for the Seismic Condition Assessment of the Eastern Portion of the San Francisco Bay Bridge", Report UCB /EERC-93/12, University of California, Earthquake Engineering Research Center, Berkeley, CA.

60. J. E. y. C. A. Carballo, Cornell, 2000Probabilistic seismic demand analysis: spectrum matching and design", Department of Civil and Environmental Engineering, Stanford University, Report RMS-41

61. C. Kircher, 1993Personal communication with Farzad Naeim and Marshall Lew.

62. F. Naeim, J. Kelly, 1999Design of Seismic Isolated Structures from Theory to Practice. New York, John Wiley & Sons. 289p.

63. N. E. Huang, S. Zheng, S. R. Long, M. C. Wu, H. H. Shih, Q. Zheng, N. Yen, C. , C. C. Tung, M. H. Liu, 1998The empirical mode decomposition and Hilbert spectrum for nonlinear and nonstationary time series analysis", Proc. R. Soc. London, Ser. A 454903995

64. Roulle, F. J. Chavez-Garcia, 2006The strong ground motion in Mexico City: Analysis of data recorded by a 3D array, Soil. Dyn. Eq. Eng. 2671

65. H. Kawase, K. Aki, 1989A study of the response of a soft bas in for incident S, P and Rayleigh waves with special reference to the long duration observed in Mexico City, Bull Seism. Soc. Am. 7913611382

66. N. Yoshida, S. Iai, 1998Nonlinear site response and its evaluation and prediction," IN Irikura, K., Kudo, K., Okada, K. & Sasatani, T. (Eds.) The Second International Symposium on the Effects of Surface Geology on Seismic Motion, Yokohama, Japan, A.A.Balkema, 7190

67. y. Herrera, E. . Rosenblueth, Response Spectra on Stratified Soil". Proc. 3rd. World Conference on Earthquake Engineering. Nueva Zelandia, 44561965

68. M. P. Romo, A. Jaime, 1986Dynamic characteristics of some clays of the Mexico Valley and seismic response of the ground". Technical Report, Apr., Instituto de Ingenieria, Mexico City, Mexico (in Spanish).

69. Jaime, M. P. Romo, D. Reséndiz, 1988Comportamiento de pilotes de fricción en arcilla del valle de México." Series of the Instituto de Ingeniería, Mexico City, Mexico

70. M. P. Romo, Seed, 1986Analytical modelling of dynamic soil response in the Mexico Earthquake of September 19, 1985". Proc. ASCE Int. Conf. on the Mexico Earthquakes-1985148162

71. S. R. García, M. P. Romo, J. Mayoral, 2007Estimation of Peak Ground Accelerations for Mexican Subduction Zone Earthquakes using Neural Networks", Geofísica Internacional, 46-15163enero-marzo

72. G. Castro, S. J. Poulos, J. W. France, J. L. Enos, 1982Liquefaction induced by cyclic loading. Winchester, Mass: Geotechnical Engineers Inc.

73. H. B. Seed, I. M. Idriss, I. Arango, . Evaluation, Liquefaction. of, Using. Potential, Performance. Field, Data, of. Journal, Geotechnical. the, Division. A. S. C. E. Engineering, Vol, GT31983Seed et al., 1983

74. H. B. Seed, I. Idriss, M.1971Simplified procedure for evaluation soil liquefaction potential. Journal of the Soil Mechanics and Foundations ASCE, 97 (9), 1249-1273.

75. T. L. Youd, I. M. . Idriss, R. D. Andrus, I. Arango, G. Castro, J. T. Christian, R. Dobry, F. Liam, L. F. Harder, M. E. Hynes, K. Ishihara, J. P. Koester, S. S. C. Liao, I. I. I. W. F. Marcuson, G. R. Martin, J. K. Mitchell, Y. Moriwaki, M. S. Power, P. K. Robertson, R. B. Seed, K. H. Stokoe, 200, Liquefaction resistance of soils. Summary report from the 1996NCEER and 1998 NCEER/NSF workshops on evaluation of

liquefaction resistance of soils. J. Geotech. Geoenviron. Eng., 127(10), 817-833.

76. R. Boulanger, I. M. Idriss, 2004State normalization of penetration resistance and the effect of overburden stress on liquefaction resistance. Proc. 11th International Conf. on Soil Dynamics and Earthquake Engineering and 3rd International Conference on Earthquake Geotechnical Engineering, Univ. of California, Berkeley, CA.

77. S. M. Fatemi-Agdha, M. Teshnehlab, A. Suzuki, T. Akiyoshi, Y. Kitazono, 1998Liquefaction potential assesment using multilayer artificial neural network. J. Sci. I.R. Iran, 9(3).

78. C. H. Juang, C. J. Chen, Y. M. Tien, 1999Appraising cone penetration test based liquefaction resistance evaluation methods: Artificial neural networks approach. Canadian Geotechnical Journal, 363443454

79. C. H. Juang, H. M. Yuan, D. H. Lee, Lin, 2003P. S., "Simplified cone penetration test-based method for evaluating liquefaction resistance of soils," Journal of Geotechnical and Geoenvironmental Engineering, 12916680

80. M. H. Baziar, N. Nilipour, 2003Evaluation of liquefaction potential using neural-networks and CPT results. Soil Dynamics and Earthquake Engineering, 237631636

81. S. Chern, C. Lee, 2009CPT-based simplified liquefaction assessment by using fuzzy-neural network. Journal of Marine Science and Technology, 174326331

82. García S.R. and Romo M.P.2007GENES: Genetic Generator of Signals, a Synthetic Accelerograms Application. , Proc. of the SEE5, SM-80.

83. S. F. Bartlett, T. L. Youd, 1992Empirical Analysis of Horizontal Ground Displacement Generated by Liquefaction Induced Lateral Spreads", Tech. Rept. NCEER 920021National Center for Earthquake Engineering Research, SUNY- Buffalo, Buffalo, NY.

84. Youd T.L., Hansen C.M. and Bartlett S.F.2002Revised Multilinear Regression Equations for Prediction of Lateral Spread Displacement' Journal of Geotechnical and Geoenvironmental Engineering, 1281210071017

Chapter 2

GEOTECHNICAL, MINERALOGICAL AND CHEMICAL CHARACTERIZATION OF THE MISSOLE II CLAYEY MATERIALS OF DOUALA SUB-BASIN (CAMEROON) FOR CONSTRUCTION MATERIALS

Elisabeth Olivia Logmo[1], Gilbert François Ngon Ngon[1], Williams Samba[1], Michel Bertrand Mbog[2], Jacques Etame[1]

[1]Laboratory of Subsurface, Department of Earth Sciences, Faculty of Science, University of Douala, Douala, Cameroon

[2]Laboratory of Environmental Geology, Department of Earth Sciences, Faculty of Dschang, Dschang, Cameroon

ABSTRACT

Geotechnical tests conducted on clayey materials of Missole II, Douala sub-basin of Cameroon showed that these materials present: fines particles (55 to 78 wt.%), sand (22 to 44 wt.%), and plasticity index of 13.8 to 21.6%. The X-ray diffraction (XRD) and the chemical analysis revealed a kaolinite amount of 46 to 56 wt.%, 19 to 27 wt.% of illite, 12 to 19 wt.% of quartz, 3 to 5 wt.% of goethite, 2 to 5 wt.% of hematite, 1.5 to 5 wt.% of anatase, 2 to 3 wt.% of feldspar-K with 52.87 to 63.11 wt.% of SiO_2, 18.08 to 24.31 wt.% of Al_2O_3, 3.28 to 11.45 wt.% of Fe_2O_3 and a small content of bases (<2 wt.%). The results of geotechnical tests combined to those of the XRD and the chemical analysis showed that the Missole II clayey materials are suitable for the manufacture of bricks, tiles and sandstones.

INTRODUCTION

Clay-rich materials are intensively used in the manufacturing of ceramics and as construction materials. However, the clayey deposit types are known as sedimentary, alluvial, and residual. A good knowledge of their occurrences, quantity and properties is required for their efficient exploitation.

In the Douala sedimentary sub-basin (South-Cameroon, Central Africa), some studies are done concerning the stratigraphic and tectonic evolution including [1-13]. Others studies are done to set up industrial units for manufacturing construction materials and ceramics [14] or concerning the mineralogical and chemical or thermal characteristics of the clay sediments [15-21].

In the Missole II area (Douala sub-basin), a geological study is carried out to locate and describe the clayey material outcrops and their provenance [22-24]. Also, some authors showed the possibility to obtain good ceramic building materials by mixing silica, feldspars, and kaolinitic and illitic clay of the Missole II area [25]. However, despite the preliminary works done on the field to describe clayey materials and to determine their sedimentation evolution, no further physical, mineralogical and chemical study is carried out to show their characteristics in order to their applications.

To that effect, the objective of this study is to associate the geotechnical characteristic to the mineralogical and chemical compositions of the clay occurrences of the Missole II deposit in order to evaluate its suitability for manufacturing of construction materials and ceramics.

GEOGRAPHIC AND GEOLOGICAL SETTING

Missole II is located on the Eastern part of the Douala sub-basin (Cameroon, Central Africa) between latitude 3°59' - 3°54'N and longitude 9°54' - 9°58'E. It is located within a humid equatorial climatic zone. Annual rainfall ranges between 3000 and 5000 mm, and the annual average temperature is 26°C. The vegetation is a dense rainforest transformed by human activities [26]. The geomorphology of the study area is a domain of the Cameroon coastal plain with low altitudes (40 - 120 m). The Missole II area shows hills with flat and sharp summits and is deeply dissected by V and U shaped valleys of MBongo, Bongougou, Missolo and Bongo the main rivers of the area. According to the geological map of SNH/UD report [12], the relative age of the Missole II sediments is Paleocene-Eocene corresponding to the N'Kapa Formation Figure 1.

The lithostratigraphy of Douala sub-basin is made of seven major Formations related to its geodynamic and sedimentary evolution [10-12]. 1) The synrift period represented by the Mundeck Formation (Aptian-Cenomanian) is discordant onto the Precambrian basement and consists of continental and fluvio-deltaic deposits, i.e., clays, coarse-grained sandstones, conglomerates. The postrift sequence includes; 2) the Logbadjeck Formation (Cenomanian-Campanian), discordant onto the Mundeck Formation and composed of micro conglomerates, sand, sandstone, limestone, and clay; 3) the Logbaba Formation (Maastrichtian), mainly composed of sandstone, sand and fossiliferous clay; 4) the N'kapa Formation (Paleocene-Eocene), rich in marl and clay with lenses of sand and fine to coarse-grained crumbly sandstone; 5) the Souellaba Formation (Oligocene) lying unconformably on N'kapa deposits and characterized by marl deposits with some interstratified lenses and sand channels; 6) the Matanda Formation (Miocene), dominated by deltaic facies interstratified with volcanoclasties layers; and 7) the Wouri Formation (Plio-Pleistocene) which consists of gravelly and sandy deposits with a clayey or kaolinic matrix.

RAW MATERIALS AND EXPERIMENTAL METHODS

The raw material used in this study comes from four representative clayey profiles of the Missole II area with three profiles of the interfluves along the Douala-Edéa road and one from the pit drilling on the lower slope of the valley. A geological survey shows different types of sediments with micro conglomerates, sandstones, fragments of ferruginous duricrusts and clays. Clayey layers are overlain upwards by sandstones or micro conglomerates, ferruginous duricrusts and sandy-clays, and occupy the lower part of the profiles. Four clayey facies identified with different mixed textures like sandy-clay, clayey- silt and silty-clay are mainly of sedimentary origin [23, 24].

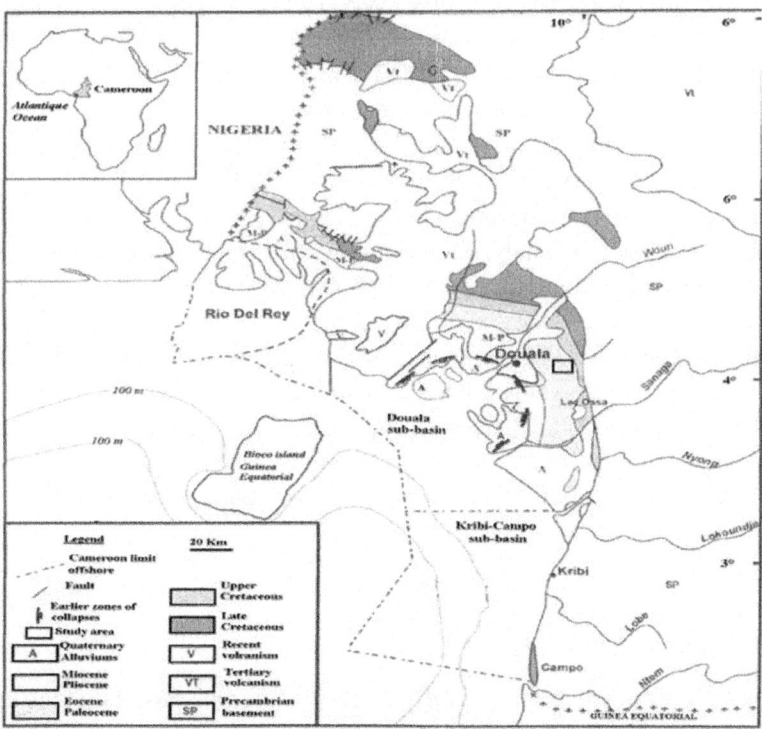

Figure 1: Geological sketch map of Cameroonian coastal basins (SNH/UD, 2005).

The average thickness of the exploitable layers is 2.5 m. Two clayey samples collected from clay layer of each representative profile for mineralogical and chemical data are mixed to obtain average sample which served to realize the geotechnical analyses. A sufficient quantity of the single mixture of 2 to 3 kg of sediments is collected from a meter-long groove. The sample is analyzed for physical, mineralogy and chemistry.

The particle size distribution has been achieved in two steps: 1) a conventional sieving for the 63 to 2000 µm fractions; 2) using a sedigraph 5000 in automatic procedure for clay and silt fractions. The liquid limit is measured by the method of the dish of Casagrande (wL) and the plastic limit by the method of the roller (wP). The blue methylene value (Vb) is determined on the total sample.

One hundred grams of each homogenized sample is grounded to -200 mesh (0.075 mm) in an agate mortar for chemical and XRD mineralogical study. Mineral identification is performed using a Setsys 2400 apparatus from SETARAM 85 equipped with a DSC 1500 heat system with Pt crucibles for thermal analysis, from room temperature up to 1100°C using a rate treatment of 10°C·min^{-1}and alumina heat treated at 1500°C serving as reference material. For XRD, a Brünker diffractometer D8 ADVANCE with a copper source ($\lambda = 1.5489$ Å) is used on bulk and fine (<2 µm) samples, working under 40 kV and 40 mA. The exposure time for qualitative analysis is 2 h. Mineralogical phases are identified (JCPDS, 1998). Semi-quantitative analysis is performed [27]. For microscopic analysis, clay samples are examined with a scanning electron microscope (SEM) (Cambridge stereos can 200) coupled with an energy dispersive spectra microprobe (EDS). Homogenized powder of sediment sample is chemically analyzed for major and trace elements by ICP-AES after dissolution using acid digestion procedure with HF, HNO_3, and $HClO_4$. Classification of the clayey materials is performed by Autret (1983) method [28,29]. This method differentiates lateritic materials to non lateritic materials. It is based on the ratio S/R where, S is the ratio between SiO_2 concentration and the molar mass and R the ratio between Al_2O_3 plus Fe_2O_3concentrations and their molar mass. In fact, for true lateritic material S/R is less than 1.33, for lateritic rocks it is 1.33 to 2, and for clay material it is more than 2.

RESULTS

Geotechnical Characteristics

The particle size distribution of the raw basic materials shows that the samples consist of 23 to 45 wt.% of sand, 17 to 33 wt.% of silts and 34 to 45 wt.% of clayey fractions. The geotechnical characteristics of the samples are reported in Table 1. These results are permitted to deduce that the class of these samples is in fine fractions [30]. The plasticity index (13.8% to 21.6%) is varied between 12 and 25, and shows that these materials are averagely plastic [31]. This plasticity is presumably a consequence of the average content of clayey minerals and quartz. The methylene blue value (0.4 to 0.91 g/100g) of the concerned materials, which is corroborated with the plasticity index, may suggest the absence of swelling clay minerals.

Mineralogical Composition

DTA analysis

Figure 2 represents the DTA analysis curves obtained for the raw materials M2P3 and M2A3 at room temperature up to 1100°C using a rate treatment of 10°C·min⁻¹. Differential scanning calorimetric (DSC) observed on these DTA analysis curves shows endothermic pheno- menon towards 110°C associated to the physisorbed water and another endothermic phenomenon towards 320°C due to the dehydroxylation of goethite.

Table 1: Geotechnical characteristics of the samples

Geotechnical characteristics	Results			
Average samples	M2A2	M2P3	M2P4	M2A3
Depth (cm)	600 - 750	700 - 800	700 - 800	50 - 150
Particles size distribution (wt.%)				
Sand (2000 - 63 µm)	45	31.3	44	22
Silts (63 - 2 µm)	21	25.5	17	33
Clays (<2 µm)	34	43.2	39	45
Atterberg limits				
Liquid limit, wL (%)	35.11	40.87	32.9	47.45
Plasticity limit, wP (%)	19.84	22.92	19.07	25.81
Plasticity index, PI (%)	15.27	17.95	13.83	21.64
Blue methylene value of the total sample (g/100g)	0.4	0.82	0.7	0.91

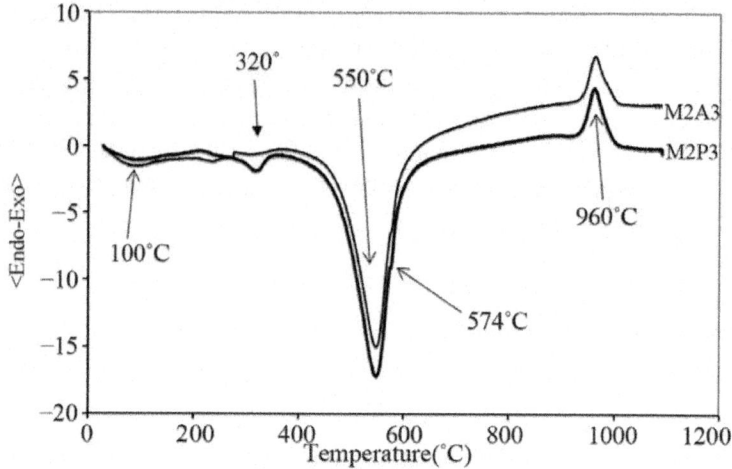

Figure 2: DTA curves of M2A3 and M2P3 materials for a heating rate of 10°C/min.

A specific endothermic transformation at 400°C - 600°C is due to the loss of structural hydroxyl groups of the kaolinite and an exothermic peak at 900°C - 1000°C due to the structural reorganization of the metakaolinite [32,33].

X-Ray Diffraction

The diffractograms of powder of the bulk samples ground until <80 μm revealed that clayey materials are kaolinitic and illitic and have permitted the detection of the following mineral phases: kaolinite, quartz, illite, goethite, hematite and accessory feldspar-K and anatase [24]. These mineral phases were usually present in the clayey materials of the Douala sub-basin [16,17,21].

The diffractograms of <2 μm fraction of Figure 3 also indicated the presence of the same mineral phases observed in the total samples with kaolinite, illite, quartz, goethite, hematite and accessory feldspar-K and anatase. Comparison of the minerals detected, these mineral phases are present in the fraction of the smaller and thicker granulometry.

The Table 2 gives the semi-quantitative mineralogical

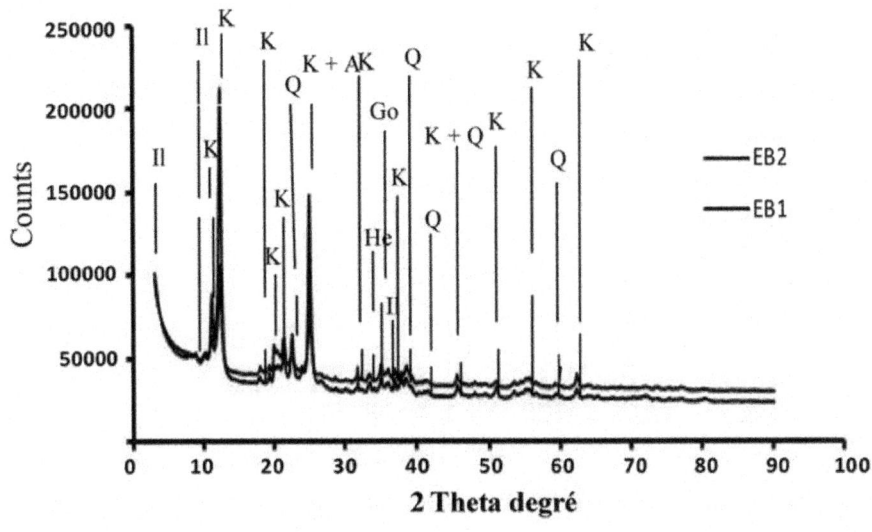

(a)

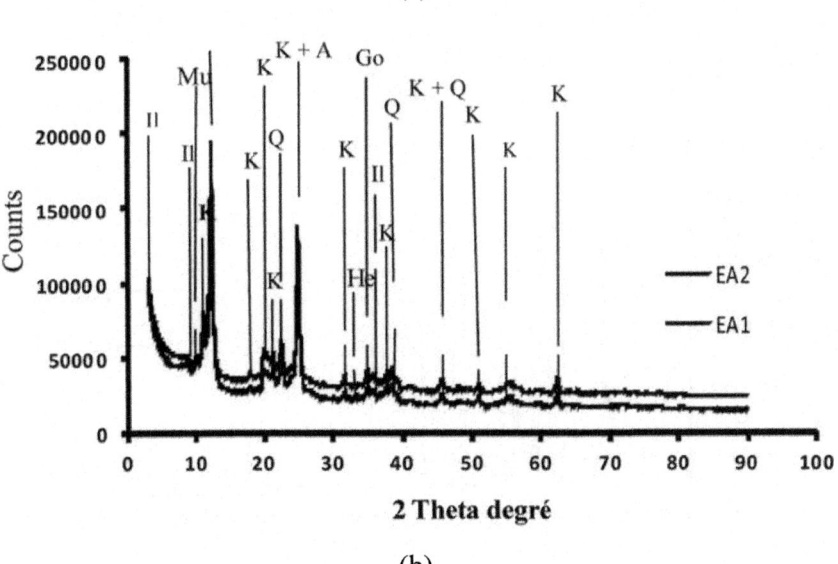

(b)

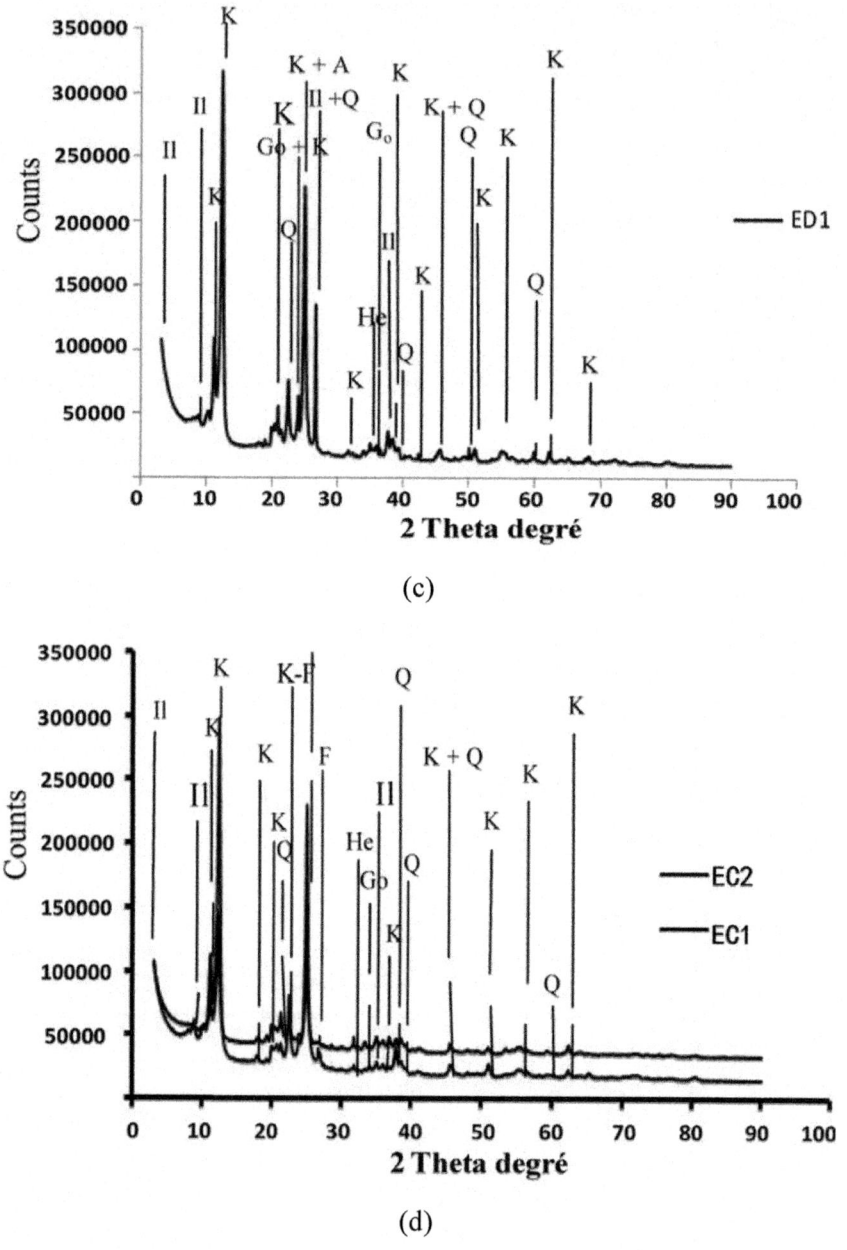

Figure 3: XRD patterns of <2 μm clay samples from Missolle II area of (a), (b), (c) and (d) profiles. A: Anatase; G: Goethite; He: Hematite; Il: Illite; K: Kaolinite; Q: Quartz; F: K-feldspar.

Table 2: Semi-quantitative mineralogical composition of the samples (wt .%)

Profiles										
			Profiles of the interfluves						Profiles of the pit	
Profiles	a			b			c		d	
Samples	EA1	EA2	M2A2	EB1	EB2	M2P3	EC1	EC2	M2P4	ED1	M2A3
Depth (cm)	600 - 700	700 - 750	600 - 750	700 - 750	700 - 750	750 - 800	700 - 750	750 - 800	700 - 800	50 - 100	50 - 150
Kaolinite	53.9	52.1	53.6	46.4	50.4	48.1	47.7	48.5	48.5	47.8	48.4
Illite	23.5	20.5	20.8	26.3	19.9	23.4	26.6	19.2	20.8	25.0	23.6
Quartz	16.1	15.0	17.1	18.8	17.7	18.2	12.5	17	16.9	13.7	16.1
Goethite	3.0	4.0	4.2	3.9	3.0	4.0	3.2	4.4	3.8	5.0	4.1
Hematite	-	2.2	1.6	-	3.0	2.1	3.2	4.4	3.6	2.9	3.2
Anatase	3.5	3.6	3.3	4.6	-	3.3	4.3	1.5	3.3	2.9	3.1
Feldspath-K	-	-	-	-	3.0	1.2	-	2.2	1.6	-	-

composition of the mineral phases of the samples. It reported an important proportion of kaolinite (46 to 56 wt.%), appreciable amounts of illite (19 to 27 wt.%) and quartz (12 to 19 wt.%), which are in relation with the value of the plasticity index (13.8 to 21.6%) and the methylene blue value (0.8 to 1.6 g/100g). The amounts of the other mineral phases are respectively goethite (3 to 5 wt.%), hematite (2 to 5 wt.%), anatase (1.5 to 5 wt.%) and feldspar-K (2 to 3 wt.%).

Microscopic Analysis

Figure 4 shows a SEM observation performed on samples of grey and mottled, with sandy-clay, silty-clay and clayey-silt texture. In clayey materials, piles of disordered kaolinites are observed with various sizes and irregular forms, which indicated that kaolinites of sedimentary clayey material of the Missole II area are poorly crystallized [34].

Chemical Composition

The chemical composition of the samples is given on Table 3. The results of the chemical analysis shows the presence of important amounts of SiO_2 (52.87 to 63.11 wt.%), associated with an appreciable amounts of Al_2O_3 (18.08 to 24.31 wt.%) and Fe_2O_3 (3.28 to 11.45 wt.%) and a small content of bases (<2 wt.%, $CaO + MgO + Na_2O + K_2O$).

These results show that quartz, alumino-silicates com- pounds and iron minerals are predominate in the samples.

M2P3

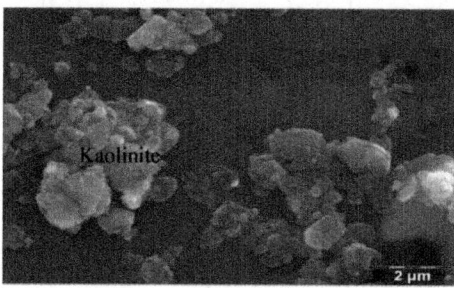

M2A3

Figure 4. SEM images of M2P3 and M2A3 clayey materials of Missole II.

Table 3: Chemical composition of the sample

						Profiles of the interfluves				Profiles of the pit	
Profiles		**a**			**b**		**c**			**d**	
Samples	**DL**	**EA1**	**EA2**	**M2A2**	**EB1**	**EB2**	**M2P3**	**EC1**	**EC2**	**M2P4**	**ED1**	**M2A3**
Depth (cm)		600 - 700	700 - 750	600 - 750	700 - 750	700 - 750	750 - 800	700 - 750	750 - 800	700 - 800	50 - 100	50 - 150
SiO_2	0.01	63.11	56.85	58.01	54.97	57.24	56.23	52.87	54.02	54.08	63.1	62.84
Al_2O_3	0.01	18.94	24.31	23.06	22.16	19.1	20.87	23.24	20.59	21.01	18.08	18.61
Fe_2O_3	0.01	3.28	3.45	3.39	5.96	7.54	8.02	3.51	11.45	8.89	4.0	3.54
CaO	0.01	0.04	0.02	0.02	0.02	0.2	0.02	0.02	0.02	0.02	0.02	0.02
MgO	0.01	0.1	0.1	0.1	0.13	0.08	0.1	0.13	0.08	0.12	0.1	0.1
Na_2O	0.01	0.06	0.06	0.06	0.1	0.03	0.1	0.06	0.06	0.06	0.06	0.06
K_2O	0.01	0.43	0.4	0.49	0.77	0.21	0.3	0.47	0.47	0.47	0.6	0.7
TiO_2	0.01	1.02	1.38	1.42	0.99	1.1	1.0	1.2	1.2	1.2	0.78	0.83
MnO	0.01	0.01	0.01	0.01	0.01	0.01	0.01	0.01	0.01	0.01	0.01	0.01
P_2O_5	0.01	0.02	0.04	0.04	0.04	0.04	0.03	0.02	0.05	0.05	0.03	0.03
Cr_2O_3	0.01	0.01	0.01	0.01	0.01	0.02	0.01	0.01	0.02	0.01	0.01	0.01
LOI		13.02	13.12	13.29	14.21	13.85	13.21	17.38	11.81	14.01	14.4	12.34
Total		99.94	99.75	99.9	99.37	99.22	99.9	99.29	99.79	99.93	99.25	99.34

DL: Detection limit; LOI: Loss on ignition.

The weak content of oxide calcium or potassium has proven that the calcitic and potassic minerals like calcite or feldspar are absent or are in a weak proportion. With S/R more than 2, the analyzed samples are clays [28].

DISCUSSION

The particle size in this study revealed texture classes: silty-clay, clayey-silt and sandy-clay described in many deposits of the Douala sub-basin [17,21]. These classes characterized very sticky and plastic materials in accordance with some field characteristics such as smoothness, stickiness and plasticity [35]. Indeed, the fine particles and especially the very fine clay minerals determine the plasticity of an earthly material [32]. The plasticity index (13.8% to 21.6%) of the above clayey samples is supported by the high amounts of fine particles (silt + clay) and fine clay minerals. The amount of clay fraction is used to determine the manufactured products as is shown in Table 4[36]. However, particle size and plasticity index are amongst the main geotechnical characteristics used to determine a convenient choice of construction material. In this case, taking into account of the amounts of sand (22 to 44 wt.%), fine particles notably the clay fractions (34 to 45 wt.%), and plasticity index, the Missole II clayey materials are convenient to the manufacturing of bricks, tiles and sandstones [37-39].

The mineralogical composition and geotechnical characteristics of a construction material must be known simultaneously in order to make a suitable choice for this raw basic material. It is well known that clay minerals have detrimental effects on the geotechnical properties; since some swelling clay minerals such as smectites absorb more water than the other (e.g. kaolinite). The volumetric shrinkage which increases with swelling clay minerals such as smectites constitutes an important parameter to take into account for civil public engineering [29,40]. The amount of clay minerals and especially swelling clay minerals contribute to the increasing of volumetric shrinkage and the swelling phenomena. Some authors joined the swelling potential and/ or volumetric shrinkage to two parameters [36]. A first ordering is based on the plasticity index and the percentage of fine clayey fraction (<2 μm size).

Table 4: Relation between particles size and type of ceramic product to realize (after Cere and Mazel, 1993)

ϕ < 2 μm (wt.%)	Ceramic products to realize
5 - 25	Full bricks
25 - 35	Perforated bricks
35 - 45	Hollow bricks, drains
40 - 50	Tiles, hordes

A second method proposes to use the liquid limit and the percentage of fine particles which size is less than 74 μm [41]. In this case, the studied materials, which have plasticity index between 12% and 25%, 34 and 45 wt.% of clayey fraction (<2 μm size), 54 and 78 wt% of fine fraction (<80 μm size) and liquid limit between 32% and 47% are very low to moderate swelling materials as is indicated in **Table 5**. This observation is corroborated with the mineralogical composition which shows that fine fraction is mainly composed of kaolinite and few illites, as well as smectites are absent.

The geotechnical and mechanical characteristics of clayey materials are essentially depending on their chemical and mineralogical compositions as well as of the minerals distribution [37,39]. In this study, the clayey samples mainly constituted of poorly crystallized kaolinite (46 to 56 wt.%), which is in agreement with the loss on ignition (12% to 17%), illite (19 to 27 wt.%) and quartz (12 to 19 wt.%), are adequate for the composition of ceramic pastry [42] and for the manufacturing of bricks and tiles [14,37]. Also, clayey materials are siliceous (SiO_2 > 60 wt.%), with their alumina (Al_2O_3) content less than 35 wt.% and appreciable amount of iron (<10 wt.% with the exception of EC2 sample) is suitable for the manufacturing of tiles and bricks [41]. However, the weakest amounts of bases <2 wt.% (oxides of calcium, magnesium, sodium and potassium) which are the smelting or cimentitious compound corroborated with the flexural strength less than 10 MPa of the Missole II clayey material [25].

CONCLUSIONS

The following conclusions can be drawn out from the present work:

1) The geotechnical, mineralogical and chemical results show that the clayey materials of the Missole II, Douala sub-basin of Cameroon are mainly constituted of sand (23 to 45 wt.%), silts (17 to 33 wt.%) and clays (34 to 45 wt.%) with plasticity index of 13.8% to 21.6%.

2) Kaolinite, illite and quartz are the main minerals.

3) Clayey materials are also siliceous and aluminous, and have appreciable amount of iron in most clayey samples (Fe_2O_3 less than 10 wt%).

Table 5: Swelling potential of soils according to their plasticity index, liquid limit and respectively their <2 and <74 μm fractions.

PI (%)	% <2 μm	% <74 μm	wL (%)	Swelling potential
>35	>95	>95	>60	Very high
22 - 35	60 - 95	60 - 95	40 - 60	High
18 - 22	30 - 60	30 - 60	30 - 40	Moderate
<18	<30	<30	<30	Low

4) Based on the obtained geotechnical, mineralogical and chemical results, and the literature data, the studied materials are suitable for the manufacturing of bricks, tiles and sandstones.

5) These clayey materials present poor mechanical characteristics (feeble flexural strength and great fragility) partly linked to the chemical and mineralogical nature fundamentally complex of natural clays.

ACKNOWLEDGEMENTS

The authors acknowledge the financial support of the University of Douala and the geochemical analysis of Geo Labs (Geoscience Laboratories) in Canada. Many thanks are also given to Professor David Smith, Director of the GEMH laboratory at ENSCI and Professor Gisèle Lecomte of Centre Européen de la Céramique, Limoges, France, and to all of their collaborators.

REFERENCES

1. Y. C. Belmonte, "Stratigraphie du bassin sédimentaire du Cameroun," Proceeding on 2nd West African Micropaleontology Colloquium, Ibadan, 1-5 September 1966, pp. 7–24.

2. M. E. Brownfield and R. R. Charpentier, "Geology and Total Petroleum Systems of the West-Central Coastal Province (7203) West Africa," USGS: Geological Survey Bulletin, 2006, 2207-B 52 p.

3. E. Dartevelle and P. Brebion, "Mollusques fossiles du Crétacé de la Côte occidentale d'Afrique du Cameroun à l'Angola,"Annales du Musée Royal Congo Belge, Sciences Géologiques de Tervuren I-Gastéropodes, Vol. 8, No. 20, 1956, pp. 1-128.

4. E. Dartevelle, S. Freinex and J. Sornay, "Mollusques fossiles du Crétacé de la Côte occidentale d'Afrique du Cameroun à l'Angola,"Annales du Musée Royal Congo Belge, Sciences Géologiques de Tervuren II-Lamellibranches, Vol. 8, No. 20, 1957, pp. 1-271.

5. J. F. Dumort, "Identification par la Telédétection de l'Accident de la Sanaga (Cameroun)," Géodynamique, Vol. 1, No. 3, 1968, pp. 13-19.

6. J. B. Meyers, B. R. Rosendahl and H. Groschel-Becker, "Deep Penetrating MCS Imaging of the Rift-to-Drift Transition Offshore Douala and North Gabon Basins West Africa," Marine Petrology Geology, Vol. 13, No. 7, 1996, pp. 791-835. doi:10.1016/0264-8172(96)00030-X

7. D. Reyre, "Histoire géologique du bassin de Douala," In: D. Reyre, Ed., Symposium sur les Bassins Sédimentaires du Littoral Africain, Association du Service Géologique d'Afrique, IUGS, 1966, pp. 143-161.

8. P. R. N. Ngaha, "Contribution à l'Etude Géologique, Stratigraphique et Structurale de la Bordure du Bassin Atlantique du Cameroun," Thèse 3e Cycle, Université de Yaoundé, 1984.

9. P. Maurizot, A. Abessolo, J. L. Feybesse, V. Johana and P. Lecomte, "Synthèse des Travaux de 1978 a` 1986," Rapport 85CM066, 1986.

10. J. M. Regnoult, "Synthèse Géologique du Cameroun," D.M.G., Yaoundé, Cameroun, 1986.

11. F. R. Nguene, S. Tamfu, J. P. Loule and C. Ngassa, "Paleoenvironments of the Douala and Kribi/Campo Subbasins in Cameroon, West African," Colloque de Géologie Africaine, Libreville, Recueil des Communications, Géologie Africaine, 6-8 May 1991, 1992, pp. 129-139.

12. SNH/UD, "Stratigraphie Séquentielle et Tectonique des Dépôts Mésozoïques Synrifts du Bassin de Kribi/Campo," Rapport Non Publié, 2005.

13. C. S. Manga, "Stratigraphy, Structure and Prospectivity of the Southern Onshore Douala Basin Cameroon—Central Africa," In: M. J. Ntamak-Nida, G. E. Ekodeck and M. Guiraud, Eds., Cameroon and Neighboring Basins in the Gulf of Guinea (Petroleum Geology tectonics Geophysics Paleontology and Hydrogeology), African Geoscience, 2008, pp. 13-37.

14. P. M. Thibaut and P. Le Berre, "Recherche d'Argiles Pour Briques Dans la Région de Yaoundé, Douala et Edéa," Rapport 85CM065, 1985.

15. D. Njopwouo, "Minéralogie et Physico-Chimie des Argiles de Bomkoul et de Balengou (Cameroun). Utilisation Dans la Polymérisation du Styrène et Dans le Renforcement du Caoutchouc Naturel," Thèse d'Etat, Faculté des Sciences, Université de Yaoundé, 1984.

16. D. Njopwouo and R. Wandji, "Minéralogie de l'Argile Kaolinique de Bomkoul (Cameroun)," Revue de Sciences et Technique, Série des Sciences de la Terre, I 3-4, 1985, pp. 71-81.

17. D. Njopwouo and S. Kong, "Minéralogie de la Fraction Fine des Matériaux Argileux de Bomkoul et de Balengou (Cameroun)," Annales de la Faculté des Sciences, Série des Sciences Chimiques, I 1-2, 1986, pp. 17-31.

18. Elimbi and D. Njopwouo, "Firing Characteristics of Ceramics from the Bomkoul Kaolinite Clay Deposit (Cameroon)," Tile and Brick International, Vol. 18, No. 6, 2002, pp. 364-369.

19. M. M. Mpondo, "Cartographie des Affleurements de la Localité de Bomkoul (Sous-Bassin de Douala)," Université de Douala, Mémoire, 2010.

20. M. B. Mbog, "Etude Morphologique, Physico-Chimique et Minéralogique des Argiles de Bomkoul Dans le Sous-Bassin Sédimentaire de Douala-Cameroun," Université de Douala, Douala, 2010.

21. G. F. Ngon Ngon, J. Etame, M. J. Ntamak-Nida, M. B. Mbog, A. M. Maliengoue Mpondo, M. Gérard, R. Yongue-Fouateu and P. Bilong, "Geological Study of Sedimentary Clayey Materials of the Bomkoul Area in the Douala Region (Douala Sub-Basin, Cameroon) for the Ceramic Industry," Comptes Rendus Geoscience, Vol. 344, No. 6, 2012, pp. 366-376.doi:10.1016/j.crte.2012.05.004

22. W. Samba, "Etude Morphologique, Géotechnique et Minéralogique des Argiles de Missole 2 Dans le SousBassin de Douala-Cameroun," Université de Douala, Douala, 2010.

23. E. O. Logmo, "Etude Géologique, Minéralogique et Géochimique des Argiles de Missole II Dans le Sous-Bassin de Douala (Cameroun)," Université de Douala, Douala, 2012.

24. G. F. Ngon Ngon, E. Bayiga, M. J. Ntamak-Nida, J. Etame, S. Noa Tang and R. Yongeu-Fouateu, "Trace Elements Geochemistry of Clay Deposits of Missole II from the Douala Sub-Basin in Cameroon (Central Africa): A Provenance Study," Sciences, Technologie et Développement, Vol. 13, No. 1, 2012, pp. 20-35.

25. G. F. Ngon Ngon, G. L. Lecomte Nana, R. Yongue Fouateu, G. Lecomte and P. Bilong, "Physicochemical and Mechanical Characterisation of Ceramic Materials Obtained from a Mixture of Silica, Feldspars and Clay Material of the Douala Region in Cameroon (Central Africa)," Advances in Ceramic Science and Engineering (ACSE), Vol., 2, No. 1, 2013, pp. 23-31.

26. R. Letouzey, "Atlas du Cameroun, Phytogéographie Camerounaise," Imprimerie Nationale Yaoundé, 1968.

27. K. Chakravorty and D. K. Ghosh, "Kaolinite-Mullite Reaction Series: The Development and Significance of a Binary Aluminosilicate Phase," Journal of the American Ceramic Society, Vol. 74, No. 6, 1991, pp. 1401-1406. doi:10.1111/j.1151-2916.1991.tb04119.x

28. P. Autret, "Latérites et Graveleux Latéritiques," Laboratoire Central des Ponts et Chaussées, 1983.

29. M. Younoussa, "Etude Géotechnique, Chimique et Miné- ralogique de Matières Premières Argileuses et Latéritiques du Burkina Faso Améliorées aux Liants Hydrauliques: Application au Génie Civil (Bâtiment et Route)," Thèse, Université de Ouagadougou, 2008.

30. NF P94-057, "Analyse Granulométrique des Sols. Méthode par

Sédimentation," AFNOR, 1992.

31. NF P94-051, "Détermination des Limites d'Atterberg," AFNOR, 1993.

32. A. Jouenne, "Traité de Céramiques et Matériaux Mineraux," Septima, Paris, 1990.

33. S. Lee, Y. J. Kim and H. S. Moon, "Phase Transformation Sequence from Kaolinite to Mullite Investigated by an Energy-Filtering Transmission Electron Microscope," Journal of the American Ceramic Society, Vol. 82, No. 10, 1999, pp. 2841-2848.doi:10.1111/j.1151-2916.1999.tb02165.x

34. M. W. Carty, "The Colloidal Nature of Kaolinite," The Bulletin of the American Ceramic Society, Vol. 78, 1999, pp. 72-76.

35. E. A. Fitzpatrick, "Soils-Their Formation, Classification and Distribution," Longman, Berlin, 1983.

36. L. Cere and F. Mazel, "Caractérisation d'Argiles," ENSCI Limoges, 1993.

37. Y. Beron and P. Le Berre, "Guide de Prospection des Matériaux," Manuels et Methodes du BRGM, No. 5, 1983.

38. V. Rigassi, "Bloc de Terre Comprimée," Manuel de Prospection, Vol. 1, Craterre EAG, 1995.

39. N. Française, "Sols: Reconnaissance et Essais. Description-Identification-Dénomination des Sols," XP P94-011, 1999.

40. Al-Rawas and M. Qamarouddin, "Construction Problems of Engineering Structures Founded on Expansive Soils and Rocks in Northern Oman," Building and Environment, Vol. 33, No. 2-3, 1998, pp. 159-171.

41. Djedid, A. Bekkouche and A. M. Aissa Mamoune, "Identification and Prediction of the Swelling Behavior of Some Soils from the Tlemcen Region of Algeria," Bulletin des Laboratoires des Ponts et Chaussées, Vol. 233, 2001, pp. 69-77.

42. O. Castelein, "Influence de la Vitesse du Traitement Thermique sur le Comportement d'un Kaolin: Application au Frottage Rapide," Thèse, Université de Limoges, 2000.

Chapter 3

EVOLUTION OF LATERITIC SOILS GEOTECHNICAL PARAMETERS DURING A MULTI-CYCLIC OPM COMPACTION AND CORRELATION WITH ROAD TRAFFIC

Meissa Fall[1], Déthiè Sarr[1], Makhaly Ba[1], Etienne Berbinau[2], Jean-Louis Borel[2], Mapathé Ndiaye[1], Cheikh H. Kane[1]

[1]Laboratoire de Mécanique et Modélisation, UFR Sciences de l'Ingénieur, University of Thies, Thies, Senegal

[2]RAZEL sa, Christ de Sarclay, 3 rue René razel, Orsay cedex, Orsay, France

ABSTRACT

Gravel lateritic soils are intensively used in road geotechnical engineering. This material is largely representative of engineering soil all around the tropical African Countries [1,2]. Gravel lateritic soils from parts of Burkina Faso and Senegal (West Africa) are used to determine the evolution of the geotechnical parameters from one to ten cycles of modified Proctor compaction. This test procedure is non-common for geotechnical purposes and it was found suitable and finally adopted to describe how these problematic soils behave when submitted to a multi-cyclic set of Modified Proctor compactions (OPM) [3,4]. On another hand, we propose a correlation between the traffic and the cycles of compaction considered as the repeated load. From that, this work shows the generation of active fine particles, the decrease of the CBR index and also the mechanical characteristics (mainly the Young Modulus, E) that contribute at least to the main deformation of the road structure.

INTRODUCTION

This paper is primarily intended to demonstrate that under unpredicted traffic and repeated loading, properties of gravel lateritic soils used as pavement layer can significantly change. According to [5-10], gravel lateritic soils are very sensitive to an exceptional variation of stresses under which they are subjected in a pavement structural fill. Thus, it is expected that most of the physical and mechanical properties of gravel lateritic soils evolves during the design life.

It is then important to find an adequate method of testing that can deal with such behavior already known in the literature. It is then necessary to perform usual characterization tests on these kinds of materials by studying the evolution of their main properties under traffic such as gradation, plasticity, CBR (Californian Bearing ration), Los Angeles loss, Shear strength (UCT), etc.

To do this, tests are conducted so that they can simulate multi-cyclic axial loading generated by traffic loads. The first cycle of OPM compaction (cycle 1) corresponds to the specifications that are led to the initial design of pavement:

- Compaction at the Optimum Modified Proctor (OPM) and determination of the initial CBR value of the material that will have to support traffic.

- Determination during the same initial state of all physical and mechanical characteristics of materials, as reference values such as gradation, Atterberg limits, CBR, Los Angeles loss, Shear strength as Unconfined Compression Test characteristics (UCT), etc.

- And finally, perform multi-cyclic compaction procedure to determine soil characteristics at each cycle of compaction.

TEST PROCEDURE AND MATERIAL PROPERTIES

After complete characterization of a gravel lateritic specimen from Burkina Faso (between Boromo and Bobo Dioulasso mainly used for the design of this West African International Road) and Senegal (in the western part of the country, as Yenne and Thiès), (sieve and hydrometer analysis, Atterberg limits, methylene blue, etc.), soils are compacted and subjected to mechanical tests at the Optimum Modified Proctor (OPM). Theses mechanical tests are essentially CBR tests, unconfined compression test and resistance to degradation by abrasion and impact in the Los Angeles machine. After the first cycle, the remaining material is used to perform exactly the same tests during the subsequent cycles (2nd, 3rd, ..., 10th cycles) (Table 1). The purpose of these tests is to compare the evolution of main properties (particle size distribution, CBR, Young modulus, etc.) with repeated cycles of compaction. Tables 2(a) and (b) below summarize the overall results:

Table 1: Values of material properties at cycle 0 (raw material), (The main Lateritic Soils used in this paper are sampled from Burkina Faso between Boromo and Bobo Dioulasso)

	(fines) % < 80 m	*PI (%)*
Pk 247	14	15
Pk 272 + 600	16	17
Pk 284	15	10
Pk 288	18	15
Pk 342	20	12

Table 2a: Summary of the test results depending on the soil provenance and the cycles of compaction

Soil provenance (Pk)	Cycles of Compaction	Compaction Modified Proctor γd max (kN/m³)	W opt (%)	Grain size distribution (f) %<5 mm	(m) %<2 mm	(f) %<0.08 mm	Atterberg Limits LL (%)	LP (%)	PI (%)	Other Soil Characteristics γr (kN/m³)	VBS	CBR (%)	Evolution (%)	E (MPa)	Rc (UCT) (MPa)
Pk 247	1	21.6	8.7	40.4	30.4	17.3	33.5	13.4	20.5	2.75	2.15	88	100	26.90	0.15
	2	22.8	7.7	43.7	14.4	23.1	35.9	14.2	21.7			93	106	16.96	0.12
	3	23.3	8.2	60.5	31.4	25.0	39.4	15.4	24.0			95	108	59.50	0.35
	4	22.6	10.1	55.4	31.5	27.2	41.4	16.6	24.8		1.7	101	115	31.50	0.16
	5	23.2	10.3	60.0	24.8	29.9	43.3	17.4	26.0		1.65	99	113	44.90	0.35
	6	23.2	8.4	58.0	20.0	33.3	46.1	19.4	26.7			87	99	61.10	0.34
	7	22.2	9.4	69.5	24.4	37.1	48.0	20.3	27.6			54	61	57.56	0.31
	8	21.5	9.8	45.9	22.1	44.1	52.6	21.8	30.8			49	56	45.98	0.28
	9	21.2	9.3	42.9	17.2	45.2	56.0	24.7	31.3			30	34	34.25	0.15
	10	22.3	7.6	31.9	16.3	46.2	57.1	26.6	30.6			29	33	29.06	1.10
Pk 272+600	1	23.5	8.1	23.7	10.0	20.3	43.5	14.2	29.3	2.81	2.15	96	100	19.45	0.23
	2	23.4	9.3	33.8	12.1	23.4	49.5	15.4	34.1	2.47	1.95	103	107	23.43	0.25
	3	23.5	8.6	72.7	34.4	25.5	46.8	16.6	30.2	2.45	1.8	110	115	26.76	0.36
	4	23.9	8.7	51.0	20.0	25.0	49.6	19.1	30.5		1.75	102	106	45.87	0.39
	5	23.8	8.0	57.7	19.0	29.7	52.3	18.1	34.2		1.7	113	118	76.98	0.45
	6	24.4	7.8	57.3	21.8	31.5	52.7	18.1	34.6			76	79	40.50	0.37
	7	23.1	8.4	59.8	18.8	39.6	53.3	18.9	34.4			48	50	15.90	0.22
	8	23.4	8.7	61.0	26.5	40.0	57.0	22.5	34.5			47	49	16.98	0.18
	9	22.5	9.8	44.7	22.6	45.2	62.3	21.3	41.0			35	36	14.00	0.16
	10	22.3	7.6	34.4	19.6	47.0	64.0	24.0	40.0			36	38	14.50	0.16
Pk 284	1	22.5	8.7	63.8	40.5	28.5	33.9	13.4	20.5	2.87	1.55	75	100	35.71	0.15
	2	22.7	7.7	71.4	54.7	40.0	41.8	16.0	25.8	2.32	1.25	84	112	54.17	0.28
	3	23.3	7.1	58.3	38.6	45.1	52.4	22.2	30.2	2.12	1.1	54	72		
	4	23.8	6.7			47.5						50.7	68		
	5	23.7	6.7									56	75		
	6											54	72		
	7											48	64		
	8														
	9														
	10														

Table 2b: Summary of the test results depending on the soil provenance and the cycles of compaction. * (Empty cells indicate insufficient quantity of materials for further testing. Multi-cyclic compaction uses a large amount of material per cycle. In this case, several samples were compacted at the same water content in order to provide enough amount of material for each cycle)

Soil provenance (Pk)	Cycles of Compaction	Compaction Modified Proctor γd max (kN/m²)	Wopt (%)	Grain size distribution (f) % < 5 mm	(m) % < 2 mm	(f) % < 0.08 mm	Atterberg Limits LL (%)	LP (%)	PI (%)	γs (kN/m²)	VBS	CBR (%)	Evolution (%)	E (MPa)	Rc (UCT) (MPa)
Pk 288	1	23.0	7.1	58.4	15.9	26.2	34.7	11.8	22.9	2.73	2.25	81	100	15.60	0.18
	2	22.6	9.3	43.4	19.7	30.3	42.8	16.4	26.4	2.5	1.9			23.15	0.20
	3	23.6	8.1	44.5	22.1	31.0	43.8	17.4	26.5	2.39	1.8			15.60	0.18
	4	24.2	7.6	44.0	21.0		47.4	18.1	29.3		1.75			32.30	0.38
	5	24.0	8.6	60.1	25.4		49.6	19.4	30.2					32.40	0.32
	6	22.5	11.8	57.0	19.8									28.70	0.32
	7	22.2	9.4	58.9	18.1									25.32	0.35
	8	21.5	9.8	44.7	20.4									20.50	0.25
	9	21.2	9.3	39.7	14.6									22.30	0.11
	10	20.4	9.3	33.6	18.6									18.15	0.12
Pk 342	1	23.4	7.3	43.7	33.7	28.2	31.3	15.5	15.8	2.87	1	84	100	33.33	0.11
	2	23.4	7.4	50.7	39.7	35.3	47.4	19.1	28.3	2.71	1.25	90	107	25.00	0.11
	3	23.3	5.5	50.4	38.8		61.8	21.1	40.6	2.33	0.9	97	115		
	4	23.7	5.7									105	125		
	5	23.6	6.8									95	113		
	6														
	7														
	8														
	9														
	10														

Lateritic Soils sampled between Boromo and Bobo Dioulasso (Burkina Faso)

INTERPRETATION OF RESULTS

Generation of Fine particles and Changing in Characteristics of Consistency

As shown by figures below, the transition between first to 10th cycles contributes to a strong generation of fine particles, as well as a gradual increase of plasticity (Figure 1). The amount of fines particles (% < 80 μm) increases from 17% (which is the limit generally accepted for such materials) for the first cycle and reaches 46% for the 10th cycle. From the first to the 10th cycle, plasticity of materials also changes from 21% to 31% for the sample of Pk 247 and from 29% to 40% for the sample of Pk 272 + 600.

The Figure 2 gives the results of Los Angeles tests performed on gravel lateritic soils samples. The test was conducted in a particular procedure that is "unconventional". In the case of the strict application of the standard, the test is performed in the fraction 10/14 with a mass of test sample of 5 kg. For our purposes, we took care to fill the hollow steel cylinder with the total fraction of the material without any selection. This procedure allows testing the total mass of the initial material without any selection and therefore allows completing Figure 3 showing the generation of fine particles and changes in plasticity. Since the test measures the resistance to degradation by abrasion and impact of the material in a rotating steel drum containing a specified number of steel balls, results show a strong increase of percent loss by abrasion and impact as the number of cycle increases. In this sense, both coarse and fine aggregates fragment extensively during the test. This further demonstrates the problematic behavior of all gravel lateritic soils related in the literature [10].

Comparison with the Specifications in the Western African Area (West African Standards—WAS)

From Figure 3 we can remark that, at the end of compaction cycles, materials tested are outside of specifications for the plasticity index and the amount of fine particles (<80 mm) as required by specifications.

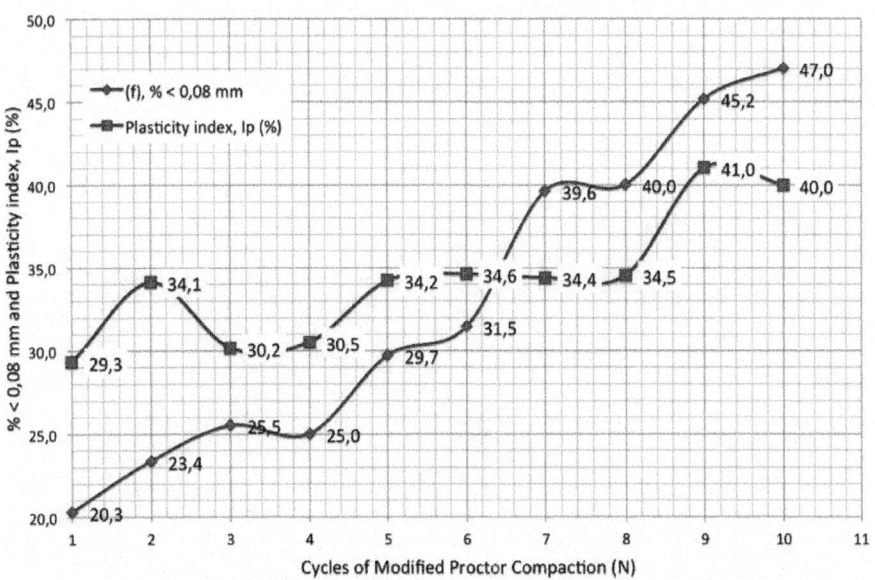

Figure 1: Evolution des fines (% < 80 m) et de la plasti-cité (PI) (Pk 272 + 600).

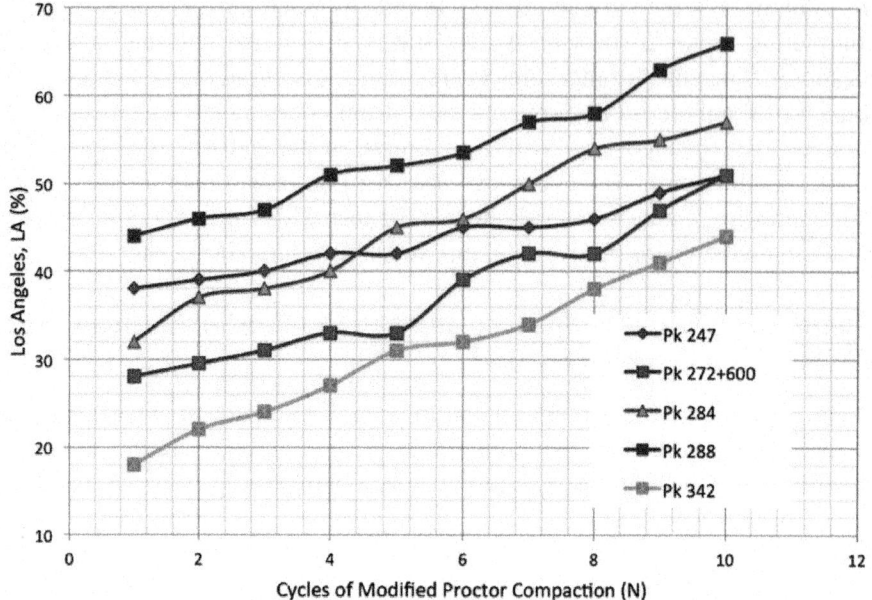

Figure 2: Evolution of percentage loss by abrasion.

PK 247

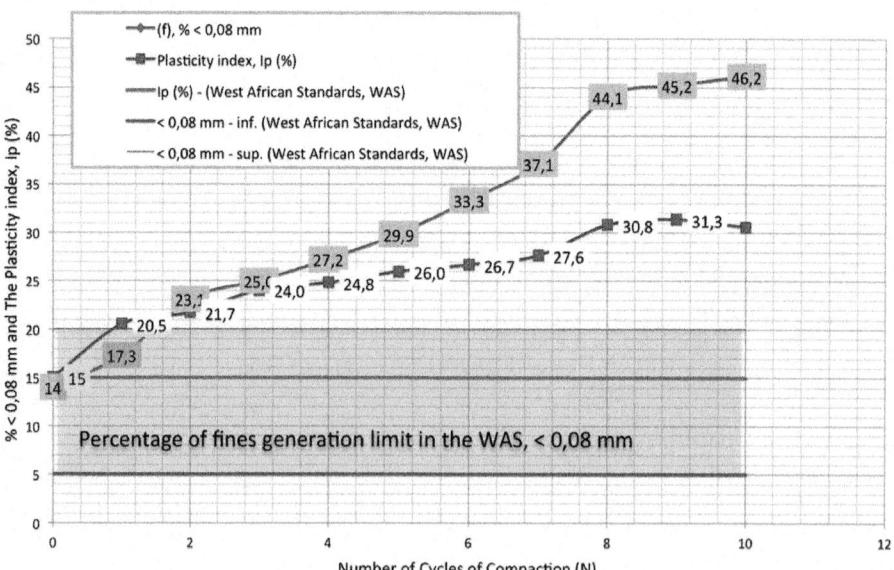

Figure 3: Comparison between results (<0.08 mm et PI (%)) and specification of the WAS.

Figure 4 shows the variation of CBR values with the cycles of modified Proctor compaction. Table 3below reminds technical recommendations contained in current textbooks approved by the CEBTP, the BCEOM and the LCPC [11] for the use of gravel lateritic soils as base courses and in the case of a T_1 to T_2 traffic level:

Evolution of the CBR Values

CBR is analyzed in several ways (Figure 7):

- In gross value, the CBR is changing slightly for all materials up to the 5th cycle. This trend towards material stiffening is well known. Fall et al. [10] underlines that behavior and attributes it to the fact that the soil is becoming denser during the first cy cles. It gradually changes from a loose state to a dense state. Air void between coarse grains tends to be reduced and filled by fine particles generated by the breaks of the material.

- The trend to the fifth cycle is to increase the CBR, which passes from a reference value of 100% and goes up to 118% or 113%. In gross value, the CBR increases from 88% to 101% and from 96 to 113%.

- After the fifth cycle, the CBR begins to drop strongly and eventually reaches extremely low values such as 29% and 36% (sometimes approaching 67%) for the gravel lateritic base course.

Trends explained in figures 5 and 6 are much clearer in Figure 7 where the material stiffening is more perceptible. The stiffness increases from 0 to 5 cycles and then decreases considerably after the fifth cycle.

Note:

Whatever the type of correlations made on the basis of CBR, we should have, in all cases, moduli that drop significantly when the number of cycles increases. In these cases, the design of pavement base courses should lead to a significant increase in thicknesses.

Table 3: Specification for a base layer for traffic $T_1 - T_2$

	$CBR_{4d\ imbibition}$ at 95% OPM	PI (%)	% inf. at 80 μm (%)
CEBTP	80	<15	4 à 20
CEBTP-LCPC	80	<15	<15
CEBTP-BCEOM	80	<15	<15

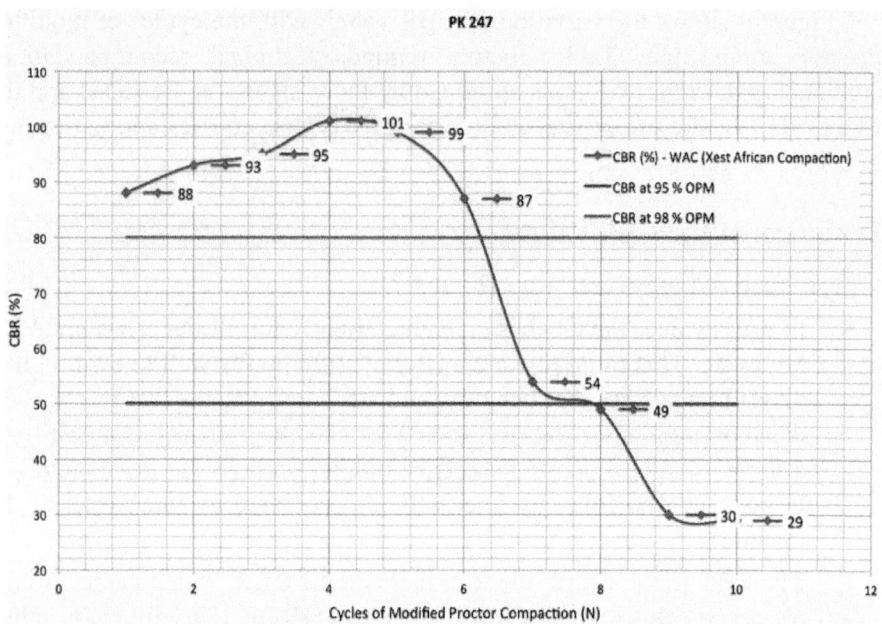

Figure 4: Comparison of CBR values with the requirements of the WAS.

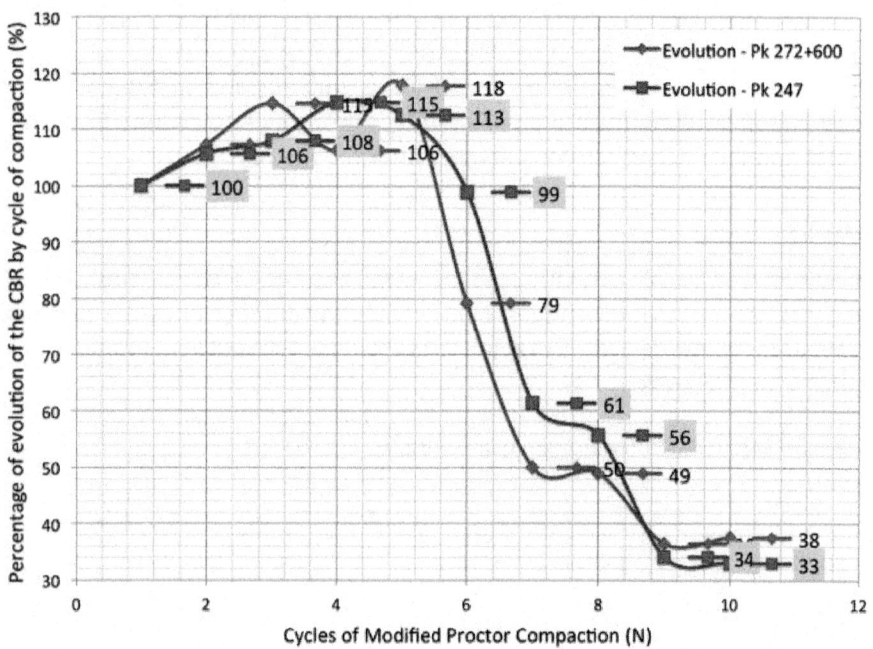

Figure 5: Percentage of evolution of the CBR values with cycles of compaction.

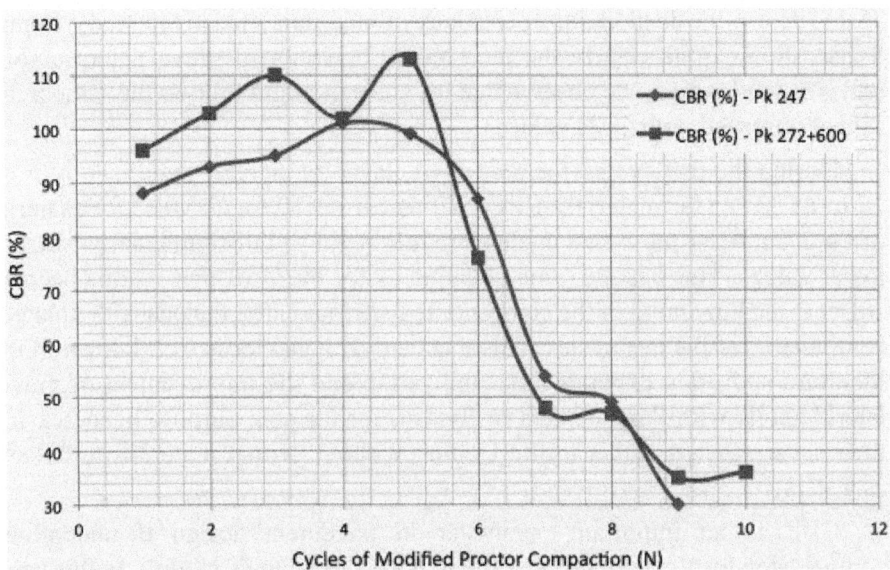

Figure 6: Evolution CBR value with cycle of compaction.

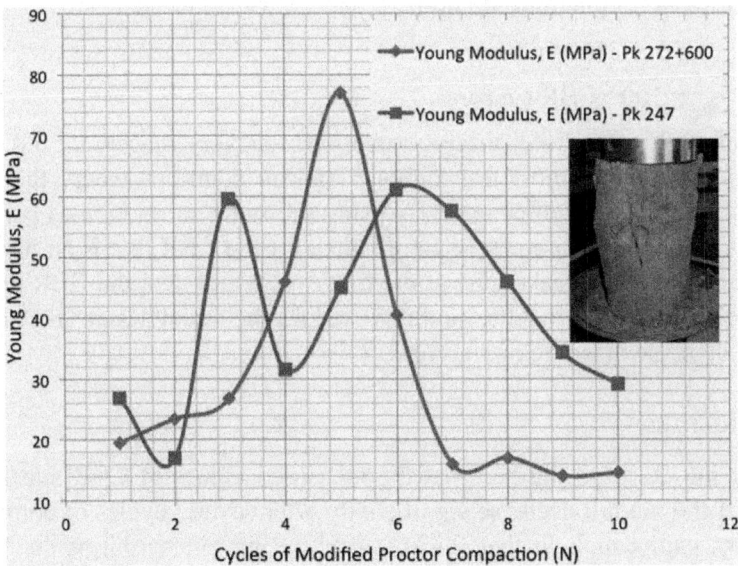

Figure 7: Evolution of the young modulus.

Evolution of Young Moduli (E: Obtained from UCT Tests)

Figure 7 give an illustration of the samples during the Unconfined Compression

Test to determine de modulus of elasticity of materials used in this study. Figure 8 also shows quite clearly the mechanical behavior of gravel lateritic soils under cyclic loading. We observe that the shear strength follows the same trend as that observed with CBR values.

During the first cycles, the moduli of elasticity increase significantly from 17 to 61 MPa (for material of Pk 247) and from 19 to 77 MPa (for material of Pk 272). After these first cycles, moduli begin to fall significantly towards lower values. This has the same meaning as for the CBR that is soils become stiffer at the beginning of the compaction cycles and after the behavior changes completely for the last cycles. This mechanical behavior is well known in the literature and often explains the stabilization and the improvement of gravel lateritic soils with lime, cement or fly ash, for the sole purpose of increasing their shear strength under traffic loading without getting materials to behave as a slab.

CBR is an important parameter in pavement design if unconfined compression tests cannot be performed to get the Young's moduli. In this case, it is often used in empirical correlations to obtain static and dynamic moduli. Thus, for most tropical countries and according to textbook used as reference, the following correlations are used:

√ $E_{static} = 50 \times CBR$ (in bars)

√ $E_{dynamic} = 100 \times CBR$ (in bars).

Although often used, these correlations are very inaccurate but still are references today in most francophone African countries where the state of the research is still rudimentary. By using the same correlation as part of this project, we get of course the same trend as for the CBR that is an increase of modulus towards a peak value at the first cycles and then the CBR decreases beyond. This implies that the modulus used for the initial design (E_0) decreases due to increase in traffic on the road.

Conclusions

Taking into account the fact that the measured values of CBR and likewise those of the moduli decrease significantly after several cycles of compaction, we may well conclude that thickness of the pavement during its life will also differ significantly from the initial designed thickness. The immediate conclusion to this is that:

- The design life of the pavement is significantly reduced and lead to premature ruin of the structurel Initial thicknesses should be higher if the designer was well aware of these behaviors.

CORRELATION BETWEEN ENERGY OF COMPACTION AND ENERGY OF TRAFFIC

Energy of Compaction

The energy of compaction is given by:

$$E_C = \frac{N \times m \times g \times h}{V_m}$$

N: number of blows; m: mass of the hammer; g: acceleration due to gravity; h: height of drop of the hammer and Vm: volume of the Proctor or CBR mold.

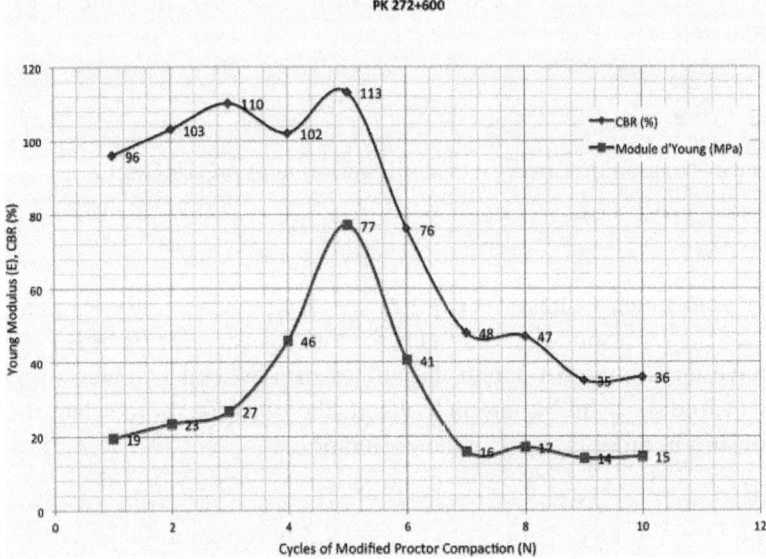

Figure 8: Evolution of design parameters (young's modulus and CBR).

Energy Due to Traffic

By analogy with the energy of compaction (Table 4), the energy of traffic can be expressed as below:

$$E_t = \frac{K \times g \times h}{V}$$

K is then defined by:

$$K = Q \times TJM$$

Q is the standard axle load converted to kgG is the acceleration due to gravity.

h (m) is the thickness of the pavement, $V = h \times \pi r^2$ is the volume of materials involved under the standard axle (in m³).

After simplification E_t becomes:

$$E_t = \frac{Q \times TJM \times h}{V}$$

So in that relationship, the only variable is TJM and the energies of traffic are determined for TC_0, TC_1, TC_2, TC_3, TC_4, TC_5 (Table 5 and Figure 9» target=»_self»> Figure 9).

For given values of x and y, we can directly calculate the TJM by the equation below:

$$TJM = \frac{y \times log_{10} \, y \times S}{Q \times g}$$

The corresponding energy of compaction is expressed as:

$$E_C = \frac{y \times log_{10} \, x \times S'}{V_m}$$

For four given values of x ($x_1 = 2$, $x_2 = 5$, $x_3 = 8$, $x_4 = 10$), we identify the values of the corresponding ordinates for the two energies (E_t et E_c). Applying the above equations, we obtain the values of TJM and E_c given in the table below (Table 6). k_t and k_c are calculated; the objective is to relate them in a relationship in order to achieve the correlation.

Let:

1 kt(1) = 5.4, kt(2) = 12.6 et kt(3) = 19.2 progression factors of Etl kc(1) = 23, kc(2) = 50; kc(3) = 66 progression factors of the energy of compaction.

These ratios are calculated as below:

$$\frac{K_C(1)}{K_t(1)} = \frac{K_C(2)}{K_t(2)} = \frac{K_C(3)}{K_t(3)}$$

$K_c = 4k_t$, let 1 $TJM_0 = 10$ and $TJMi$ (i = 1, 2, 3...n)

1 $E_0 = 8\,761{,}5$

Table 4: Summary of the parameter of the curve E_c

Cycles	1	2	3	4	5	6	7	8	9	10
N	275	550	825	1100	1375	1650	1925	2200	2475	2750
E_n (KJ)	2635	5276	7904	10,539	13,174	15,809	18,443	21,078	23,713	26,348
p	1	2	3	4	5	6	7	8	9	10

Table 5: Summary of the parameters for drawing the curve E_t

Classes TCi	TC_0	TC_1	TC_2	TC_3	TC_4	TC_5
TJM (in heavy trucks)	2	14	27	68	164	342
E_t (kJ)	51,908	363,354	700,754	1,764,862	4,256,431	8,876,217
Progression factor	1	7	13.7	34	82	171

Table 6: Increase in the number of heavy load vehicles and energy of compaction

TJM	10	54	126	192
k_t	1	5.4	12.6	19.2
Ec (kJ)	8761	203,435	438,073	639,587
k_c	1	23	50	73

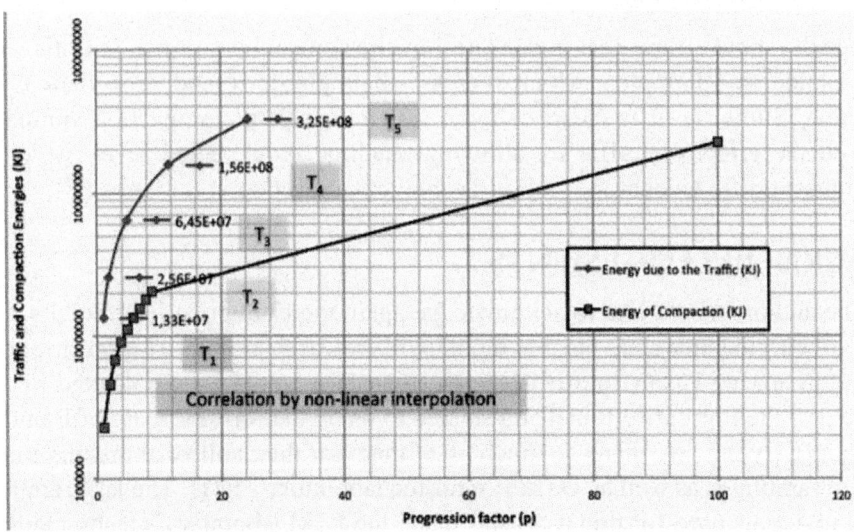

Figure 9: Curves of traffic energy and compaction energy vs progression factor.

We can write :

$$\frac{E_C(i)}{E_0} = K_C = 4Kt = 4.\frac{TJM_i}{TJM_0}$$

Hence,

$$E_C(i) = \frac{4E_0}{TJM_0}.TJM_i$$

In this formula, the energy is expressed in k_j. The formula reflects a geometric series with a common ratio expressed as below:

$$q = \frac{4E_0}{10}$$

This result allows the designer to assume the desired traffic and then deduce the corresponding energy of compaction.

Conclusions

We note well that the multi-cyclic compaction simulate exactly the effect of traffic loading. In this sense, observe that increase in traffic can be simulated by an increase in the compacting cycle. At the end of compaction, the traffic reaches very high level (T_4 to T_5).

CONCLUSIONS

Results show clearly that under multi-cyclic compaction, gravel lateritic soils generate fine particles, which increase their plasticity and drop their CBR value. Similarly, it is clearly shown that multi-cyclic compaction simulates well the effect of traffic by allowing reaching its expected level, which is highest traffic level at the end of the cyclic compaction.

ACKNOWLEDGEMENTS

The authors would like to acknowledge again the Entreprise Jean Lefebvre sa for their guidance and valuable input in this research project. The lateritic soils from Burkina Faso (from Boromo to Bobo Dioulasso) were collected by the help of the Entreprise Razel and sended to us by the Razel's Team of Bamoko (Mali). By this, we thank so much all of them for their ability of making the in situ samplings as well as we desire for the laboratory's tests. The large amount of materials used for this work has made the L2M laboratory's technicians as busy as they never be, the authors would like to thank them for their important contributions.

REFERENCES

1.   Remillon, "Les recherches Routières Entreprises en Afrique D'expression Française, Application à la Conception et au Renforcement des Chaussées économiques, " ITBTP, 1967.

2.   M. D. Gidigasu, "Laterite Soil Engineering-Pedogenesis and Engineering Principles," Elsevier Scientific Publishing Company, Amsterdam, 1976.

3.   T. Aashto, "Standard Method of Test for MoistureDensity Relations of Soils Using a 2.5-kg (5.5-lb) Rammer and a 305-mm (12-in.) Drop. In Standard Specifications for Transportation Materials and Methods of Sampling and Testing," American Association of State Highway and Transportation Officials (AASHTO), Washington DC, 1999.

4.   T. Aashto, "Standard Method of Test for the California Bearing Ratio," In Standard Specifications for Transportation Materials and Methods of Sampling and Testing, American Association of State Highway and Transportation Officials (AASHTO), Washington DC, 1993.

5.   H. R. Sreekantiah, "Laterites and Lateritic Soils of West Coast in India," Southeast Asian Geotechnical Conference, Bangkok, Vol. 1, 27 December 1987, pp. 159-170.

6.   Sridhan, "General Report: Engineering Properties of Tropical Soils," International Conference on Geomechn in Tropical Soils, Singapore, 12 December 1988.

7.   G. Sweere, P. Galjaard and T. Tjong, "Engineering Behaviour of Laterites in Road Constructions," International Conference on Geomechn in Tropical Soils, Singapore, Vol. 1, 1988, pp. 421-427.

8.   A. Vallerga, et al., "Engineering Study of Laterites and Lateritic Soils in Connection with Construction of Road, Highways and Airfields," Country Report, AID/CSD- 1810, 1969, p. 165 and appendices.

9.   L. W. Ackroyd, "Engineering Classification of Some Western Nigerian Soils and Their Qualities in Road Building," Road Research Laboratory, British, 1959.

10.   M. Fall, "Caractérisation et Identification mé- canique de Trois Graveleux Latéritiques du Sénégal Occidental: Application au domaine routier," Thèse de Doctorat de l'INPL en Génie Civil et Minier de l'INPL-Nancy France, 1993, p. 277.

11.   Secrétariat d'Etat aux Affaires Etrangères – CEBTPBCEOM (1984)

- Guide pratique de dimensionnement des chaussées dans les pays tropicaux, CEBTP ; 2 ème édition: tome 1: Politique et économie routière ; 2 ème édition, 1991; tome 2: Etudes techniques et construction ; 2 ème édition, 1991; tome 3: Entretien et gestion des routes ; 2 ème édition, 1991.

Chapter 4

ENVIRONMENTAL AND GEOTECHNICAL ASSESSMENT OF THE STEEL SLAGS AS A MATERIAL FOR ROAD STRUCTURE

Wojciech Sas[1], Andrzej Głuchowski[2], Maja Radziemska[3], Justyna Dzięcioł[1] and Alojzy Szymański[2]

[1]Water Centre Laboratory, Faculty of Civil and Environmental Engineering, Warsaw University of Life Sciences, 02-787 Warsaw, Poland

[2]Department of Geotechnical Engineering, Faculty of Civil and Environmental Engineering, Warsaw University of Life Sciences, 02-787 Warsaw, Poland

[3]Department of Environmental Improvement, Faculty of Civil and Environmental Engineering, Warsaw University of Life Sciences, 02-787 Warsaw, Poland

ABSTRACT

Slags are the final solid wastes from the steel industry. Their production from waste and associated materials is a proper implementation of the basic objectives and principles of the waste management. This study aims to investigate the chemical and selected significant geotechnical parameters of steel slag as the alternative materials used in road construction. These investigations are strongly desired for successful application in engineering. Young's modules E, and resilient modules M_r showed that their values corresponding with requirements for subbase (principal or auxiliary) and riding surface as well. Tested mechanical properties were conducted in soaked and un-soaked (optimal moisture content) conditions. The designated high content of chromium and zinc are strongly associated with the internal crystal structure of steel slag. The results do not lead to threats when they are applied in roads' structures. Mechanical characterization was obtained by performing California bearing ratio (CBR) tests for steel slag in fixed compaction and moisture content conditions. Moreover, cyclic loading of steel slag was conducted with the application of cyclic California bearing ratio (cCBR) apparatus to characterization of this material as a controlled low-strength material. Finally, field studies that consist of static load plate VSS tests were presented.

INTRODUCTION

Construction engineering, as well as the building materials industry, makes a significant contribution to the production of industrial wastes and environmental pollution. On the flip side, this industry utilizes a large amount of waste, using the industrial waste as raw materials or ingredients of building materials, an example of which can be steel slags [1]. These are the final waste products from the steel industry and make up a proportion of approximately 15% by mass of the steel output [2]. Steel slag is produced from Basic Oxygen Furnace and Electric Arc Furnace in steel making and the main chemical compositions include CaO, SiO_2, Al_2O_3, Fe_2O_3, MgO and FeO [3,4,5].

Huawei [6] reports that the production of each of these three tons of stainless steel will create one tonne of waste. According to European Slag Association (EUROSLAG), in Europe, a production of steelmaking slag is 21.8 Mt in 2010 [7]. The refining process of the so-called secondary steel making operation produces stainless steel reducing slag (SSRS). SSRS includes argon oxygen decarburization and ladle metallurgy slag [8]. When SSRS compares with ground granulated blast furnace slag (GGBFS) from iron making, steel slag contains toxic ingredients such as nickel, cadmium, chromium and strontium. These compounds could be harmful not only for environment but also for human health [9,10]. Steel slag recycling issues have gained importance in research areas and environmentally friendly processes.

Artificial and recycled aggregates are alternative building materials [11,12]. Their production from waste and associated materials is a proper implementation of the basic objectives and principles of waste management such as: reduction of the amount of waste and its negative impact on the environment, recovery of materials, disposal of waste and production of artificial aggregates. The application of such materials requires preliminary strength property estimation. These properties depend on many variants, which were summarized in ACI 229R-99 [13]. Many materials, such as plastic-soil elements and slurry material, were proposed as controlled-low strength materials (CLSM). CLSM could consist of many various materials—for example, recycled concrete aggregates that could be deposited on construction site in unbound or bound form, with addition of lime in the last case [14,15,16,17]. Many low strength clayey soils could also be a base for CLSM [18,19]. CLSM can compound from various chemically active substances but also can contain another materials which improve physical properties as rubber waste from tires to decrease the weight of CLSM [20].

In the case of steel slag, this material could be deposited in its raw state as a subgrade or subbase material. The application of steel slag in asphalt concrete also gives good performance [21,22]; however, the use of steel slag as a construction material has been a problem, because free CaO abundant in steel slag in contact with water transforms into $Ca(OH)_2$ [23]. Moreover, steel slag could be subjected to various treatments. The activation of hydraulic properties during crushing is one of them [24]. A large amount of steel slag is employed, as was mentioned above, to road construction as an aggregate in concrete production or fertilizer production [25] and after relevant treatments as a hydraulic binder [26,27], to replace river sand for preparing the waterproof mortar [28]. Steel slag could be used as a source of reactive elements for the purpose of CO_2 isolation [5].

CLSMs under their own weight consolidate themselves. This property allows the replacement of natural soil by those materials [13]. An important feature of CLSM as opposed to virgin soils, is that, after the compaction and hardening process, CLSMs do not stop to settle. Hardening obtained by controlled-low strength materials (CLMS) is greatly dependent on the quantity of cementitious material which it contains. Hardening time can be as short as one hour but usually takes three to five hours to reach proper bearing capacity conditions. Requirements for such materials as CLSM are to achieve low compressive strength of 0.5–2.0 MPa for possible re-excavation in the future [29]. ACI-229R requirements for compressive strength are of 8.3 MPa or less, for steel slags (7.23 MPa), this requirement is fulfilled [30]. During CBR tests, steel slag also exhibited the abovementioned properties and is discussed in this paper.

The utilization method of steel slag is closely related to the chemical and physical characteristics and should be given a priority from environmental considerations. Nowadays, almost 100% of steel slag is utilised in many areas, such as cement production, raw material for brick production, landfill daily cover, road construction, civil engineering work, filtering media for waste water treatment and fertilizer production [31,32].

Nevertheless, for road construction, the possibility of using unbound material with the potential to a self-cementing feature in time after the mechanical stabilization is desired. Possible settlement due to consolidation and repeated loading could be limited by a volume expansion of steel slag when it reacts with free lime or another alkali compounds [33]. Steel slag could be also milled to powder form which exhibits cementitious properties [9,34,35] and after compaction of unbound material could be filled with slag powder to create a mortar.

In recent years, there has been a significant increase of the usage of these aggregates in road construction in Poland. Very good quality of road surface, together with the pro-environmental and economic reasons, support their wider utilization. The technical requirements for unbound aggregates in the road construction in Poland for natural aggregates are presented in the appropriate document [36].

Mechanical characterization of materials such as steel slag are indicated in road engineering by performing the CBR test. The CBR values are key factors of estimating the roadbed thickness. The evaluated strength of subgrade filling give the indices of different highway grade and different road base requirements. This procedure is an element of empirical road design. The right choice of materials, which was characterized by this value, ensure the quality of road construction which nowadays play huge role [37].

The aim of this study was to determine the chemical and selected significant geotechnical parameters of steel slag as the alternative materials used in roads construction, as well as to confirm their environmental and technical quality.

MATERIALS AND METHODS

Material and Sample Preparation

In order to evaluate the chemical composition and obtain geotechnical parameters, the steel slag samples were collected from a metallurgical landfill, which provided the examined aggregate to improve the municipal low frequented village road. The test site was a real proving ground, the municipal road with the asphalt surface. It was the main road of the L class, traffic categories KR5 in polish standards and by the ARRB AR354 [38] is represented by class 50, which is the local throughout route. The subbase of this road was constructed from steel slag. Steel slag was cut off by construction of the drainage layer under the subbase to prevent capillary actions. Over the subbase, an impermeable asphalt layer was structured. Such construction leads to averting any outflows and inflows of water in a subbase layer.

The commercial aggregate samples had the size fraction of: 0–10 mm; 0–31 mm, and 0–63 mm. After compaction of a steel slag layer, field tests were performed. Four static load plate VSS tests in distant points were conducted until the first loading caused 0.8 mm of surface displacement. During these tests, stress and displacement were noted.

Laboratory and Chemical Analysis

Based on averaging collected material, melted below 80 μm by using a centrifugal mill (Retsch ZM100, Haan, Germany) and powdered samples were prepared. The chemical composition was determined by wavelength dispersive X-ray fluorescence (XRF) spectrometry analysis on Philips PW 2400 (Eindhoven, The Netherlands), quantified using the Super Q software.

The heavy metal content was analyzed with inductively coupled plasma optical emission spectrometry (ICP-OES, Varian 720-ES, Mulgrave, Australia). The operating conditions were: argon gas used as plasma gas flow at the rate of 15.0 L·min⁻¹, auxiliary gas flow rate 1.50 L·min⁻¹, nebulizer gas flow rate 0.75 L·min⁻¹. The room temperature was fixed at 25 °C for the analyses. Analyses were performed three times. Samples were digested with nitric acid (Merck, Darmstadt, Germany, 69% m/v) on a microwave oven (Milestone, Italy). All reagents were of analytical reagent grade unless otherwise stated. Stock solutions of metals (1000 mg/L) were prepared from their nitrate salts. Ultra-pure (UP) water (Millipore System, Bedford, MA, USA) 0.055 µS·cm⁻¹ resistivity was used for preparing the solutions and dilutions for all dilutions.

Geotechnical Laboratory Analysis

Geotechnical laboratory analysis consisted of a number of bearing capacity tests according to PN-S02205: 1988 [39] so-called CBR tests [40]. The test relates the bearing capacity of the material in CBR test conditions to that of a standard crushed gravel represented as % of standard material bearing capacity. Representative specimens were prepared from large samples of slag material, with respect to Proctor's method, preliminary tests lead to estimate optimal moisture content equal 14.9% at dry density equal 2.02 g/cm³ . Four special mixture aggregates: 2–10 mm; 2–25 mm; 2–10 mm and 2–25 mm with 30% addition fraction 1–2 mm were tested. The reason of the addition of 1–2 mm fractions was to simulate real field conditions connected with the seal of surface subgrade and to prevent them from loose contact between plungers and specimens in laboratory tests. Each compound was studied ten times.

Cyclic Loading Tests

Repeated loading on soil samples prepared in CBR mould was performed with the use of cyclic CBR test (cCBR) procedure. cCBR method was based on a common CBR test. The main idea behind using this equipment came from its popularity. The long existence of the CBR method and its usefulness in road design, resulted in its worldwide spread.

By using CBR, mould and repeated loading apparatus, the cCBR test method was established. The main principle of this test approach is to use standard CBR test procedures as a reference in order to study the later cyclic loading stage.

As was mentioned above, the first step is standard CBR test loading to 2.54 mm. After reaching the desired displacement with the use of a plunger, the unloading procedure was attempted, with the use of up to 10% of force obtained at 2.54 mm. Loading and unloading is treated as the first cycle of the cCBR test. The next cycles are determined by the maximal and minimal force from the first loading. The test was carried out with standard 1.27 mm/min. velocity. The number of cycles is determined by the percentage of plastic strain in one cycle. cCBR method assumes that the test can be stopped, when 1% or less of plastic displacement in one cycle will occur. The amount of the cycles to obtain this condition usually oscillates around 50 [41,42].

RESULTS AND DISCUSSION

Chemical and Mineralogical Characterisation of the Slag

The chemical composition of steel slag varies with the furnace type, steel grades and pre-treatment method. The chemical composition, by XRF, of the steel slag is given in Table 1.

Table 1: Chemical composition of steel slag samples, mass percent (%)

CaO free	SiO₂	TiO₂	Al₂O₃	Fe₂O₃	MnO	MgO free	Na₂O	K₂O	P₂O₅	SO₃	Cl	F
27.46	16.69	0.43	6.64	33.82	3.87	6.68	0.68	0.10	0.30	0.45	0.003	2.89

The composition of steel slag is variable and may depend on, among others, the size of fraction. The main primary solid phases consist of a Fe_2O_3. The main mineral compounds are metallic iron (dicalciumsilicate, dicalciumferrite) [4]. The two major mineral phases present in the steel slag samples were hematite Fe_2O_3 and lime (CaO), content of ~34 wt % and ~27 wt %, respectively. Other minor phases identified were quartz (SiO_2) and Al_2O_3. The content of CaO, MgO, SiO_2 and Al_2O_3 that may principally substitute raw materials for cement production [43,44]. Because of the disintegration process, the residue values of free lime (CaO), are most harmful for unbound steel slag composition. Concentrations of lime (CaO) are dominant in steel slag [45]. The content of free CaO and free MgO is the most important factor for the disposal of slag and their use in the building industry because of their durability volume. Steel slag with CaO content above 50% can be used as sinter ore fluxing agent, partially replacing the commercial lime.

In contact with water, these minerals react with hydroxide, depending on the level of free lime and free MgO reaction causes an increase in the volume of slag, which is mainly linked to the breakdown of the particles and a decrease in the strength. The stability of the volume is a key criterion by using slag as a construction material. The analyzed material was characterized by a relatively low content of free CaO and MgO, so that significant changes in its volume as well as a decrease of its strength in time can be excluded. Experience in Germany has found that steel slags with a free lime content up to 4% in asphaltic layers and up to 7% may be used in unbound layers [4]. Some steel slags contain a higher amount of P_2O_5; this may affect the direct recycling of the steel slags to the iron and steel making process. Steel slag was identified as a material which causes 10% higher P retention capacity in columns for phosphorus removal than columns without steel slag [46]. This component might lead to decomposition of C_3S, which reduced activity of steel slag. The silica (SiO_2) content in slag samples was 16.69%, the Al_2O_3 and MnO contents are in 6.64% and 3.87%, respectively. The silicate glass at the level 42.38–78.23 wt % SiO_2 and 7.48–38.87 wt % Al_2O_3 were found in slag from the former Hegeler Zn-smelting facility in Illinois (USA) [47]. Puziewicz *et al.* [48] reported for a Zn-smelter waste dump in Upper Silesia (Poland) having 12.43–41.27 wt % SiO_2, 3.38–20.69 wt % Al_2O_3, 9.86–29.68 wt % Fe_2O_3 and 4.97–23.53 wt % CaO.

Heavy metal toxicity and mobility in the natural environment depends on their chemical speciation. In steel, slags are present in relatively high values [49].

Steel slag contains trace amounts of elements potentially mobile and toxic to the environment [50]. To get information about the effect on the soil and ground water, it is interesting to know that the concentrations of those environmentally relevant components can be leached out. Some steel slags contain higher amounts of toxic metals, such as chromium, nickel, manganese, vanadium and molybdenum [51]. Lottermoser [52] identified in slag from Río Tinto (Spain) contain elevated concentrations of potentially toxic trace elements such as As, Cd, Co, Pb, Sb and Zn.

Steel slag from analysing samples also consists of several different types of heavy metals in various concentrations. The heavy metal composition of the steel slag samples is given in Table 2. Results of the analysis of samples confirm a high content of chromium and zinc. The analysis of the chemical composition of steel slag shows the content of Cr at the level of 2 915 mg·kg^{-1}. Chromium exists in slags as magnesiochromite ($MgO·Cr_2O_3$) or solid chromium oxide (Cr_2O_3). In contact with water the tri- and hexavalent chromium (highly soluble in water) can be released from the slag by means of a leaching process [53]. The hexavalent cation can be produced by oxidation with atmospheric oxygen and contact with CaO. Leaching of Cr increases as soon as the iron (Fe^{2+}) is oxidized into Fe^{3+}. Zinc has a high affinity for mineral colloids, characterized by a high mobility in the soil and the high bioavailability of the plants due to the rate of dissolution of the compounds in which they occur, in particular in an acidic environment [54,55]. In the tested sample of slag, zinc was the second in terms of the content of the element (1 084 mg·kg^{-1}). This value is similar in composition to Zn-rich slags from Poland with 2600–27,200 mg·kg^{-1} [45] and 212 to 14,900 mg·kg^{-1} in slag from the former Hegeler Zn-smelting facility in Illinois (USA) [47]. In sandy soils, zinc can be toxic at concentrations of 6.9–12.8 g·kg^{-1} of soil, in clay soils it is revealed at the higher concentrations of 16.2–21.5 g·kg^{-1} soil. In the presented study, concentrations of the barium in steel slag samples were 380 mg·kg^{-1}. According to Piatak and Seal [47], the content of Ba in slag was between 788 and 1170 mg·kg^{-1}. The tested steel slags were characterized by less than half as much Cu as niobum, lead, and nickel. Studies of authors [47] examined the As contained in slags from Hegeler and the value of As was 1–45 mg·kg^{-1}; in samples presented in this study, the concentration of arsenium value was 10 mg·kg^{-1}.

Table 2: Total composition of the steel slag samples, mg·kg^{-1}

Element	Value	Element	Value
Chromium (Cr)	2915	Rubidium (Rb)	11
Zinc (Zn)	1084	Arsenium (As)	10
Barium (Ba)	380	Cadmium (Cd)	8
Strontium (Sr)	266	Uranium (U)	4
Cupper (Cu)	175	Bromine (Br)	5
Circonio (Zr)	109	Cerium (Ce)	<5
Vanadium (V)	92	Cobalt (Co)	<5
Niobum (Nb)	62	Lanthanum (La)	<5
Lead (Pb)	59	Yttirum (Y)	<3
Nickel (Ni)	26	Thorium (Th)	<3
Tin (Sn)	15	Bismuth (Bi)	<3
Molybdenum (Mo)	11	Gallium (Ga)	<3

Other elements that may pose a potential risk of soil degradation occur in samples of slag at levels exceeding the limit values for soil category A: cobalt (Co) and nickel (Ni); exceeding the limit values for soil category A, but allowing free use of the soil category B and C: molibdenium (Mo), lead (Pb) and cadmium (Cd). To limit the effects of toxic elements, leaching from slag should be minimized to reduce long term leaching and minimized contact with water e.g., cut of water inflow and outflow in layers constructed with use of steel slag to prevent toxic effects of chromium from slag to environment. In the tested sample of slag, zinc was second in terms of the content of the element. Other elements that may pose a potential risk of soil degradation, occur in samples of slag at levels exceeding the limit values for soil category A: arsenium (As), cobalt (Co) and nickel (Ni); exceeding the limit values for soil category A, but allowing free use of the soils category B and C: cupper (Cu), molibdenium (Mo), lead (Pb) and cadmium (Cd).

Bearing Capacity Tests

Laboratory experimentation of the bearing capacity of steel slag is one of basic tests for classification of unbound material for road structure such as subbase (principal or auxiliary) and subgrade (compacted or natural). For each of those layers, the minimum value of bearing capacity ratio (CBR) is required [36]. It is also a very important parameter used for the design of roads. The results of the tests done on mixture aggregates with grain size range 1–25 mm and 2–25 mm (Figure 1) show that obtained results of CBR are in a range of 35%–42%.

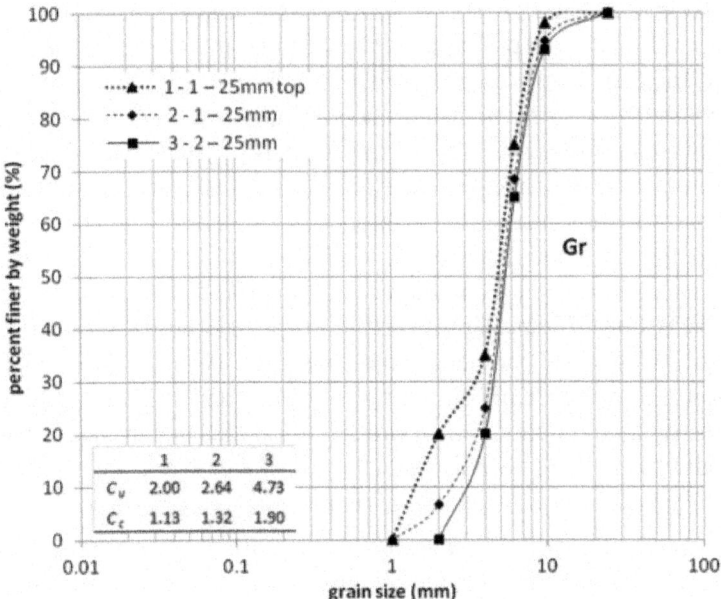

Figure 1: Grain size distribution for steel slag.

The addition of 1–2 mm fractions to both of the aforementioned mixtures and located in the upper part of samples, allows the obtaining of the bearing capacity ratio on higher levels of more than 60%. The addition was performed by mixing the top layer of the CBR sample with a fraction, whose size was equal to 1–2 mm. Curve 3 (Figure 1) represents sieve analysis results for the top layer. Curve 2 presents the result of sieve analysis for all samples consisting of a mixed top layer. Steel slag was sieved and a gradation curve was estimated. For both mixes, the coefficient of uniformity (Cu) and coefficient of curvature (Cc) was calculated. Fractions with the size of 2–25 mm was taken to studies due to possibility of the lack of proper material or the quality control process on the construction area.

The received values of CBR allows for the conclusion that the tested material meets the requirements [36] for the subgrade layer of the road (for traffic category KR1-KR6) where the minimum was specified on 35%. Improved by 1–2 mm grain size fraction samples achieved requirements for auxilary subbase (for traffic category KR1-KR6) were 60% CBR values are needed. The quality of CBR results predispose this steel slag material as unbound riding surface where there is no recommendation of minimum value for traffic category KR1-KR2 [36]. Using the analogy with other artificial and recycled (anthropogenic) materials with well-graded curves of grain size distribution in the range of 0–31.5 mm or 0–61 mm, it can be said that slag material is sufficient for principal subbase. Results are presented on Figure 2 where CBR test results are detailed for each test. Figure 3 and Figure 4, present detailed views of CBR values dependence with density and void ratio. Figure 3 presents impact of dry density on CBR value, rise of density has a bigger impact on the mix presented on Curve 3 (Figure 1). In the case of soaked-unsoaked conditions, the difference is also clear. The change of CBR value corresponds with material size grain composition rather than with saturation state. Nevertheless, in both cases, unsoaked samples perform better during the CBR test. Figure 4 presents the impact of void ratio on CBR value. On this plot, the same phenomena can be observed. An increase of CBR value depends on grain size composition. On Figure 3 and Figure 4, the trend of closing interpolated function is also worth noting. This occurrence is caused by the fact that void volume decreasese in the soil skeleton and, because of that, a smaller impact of saturation and bigger contact between particles.

After compaction of this layer, the proposition of adding well graded material (Curve 3) was conducted in studies. These conditions, when two layers of the same material but with other grading would be part of road construction as subbases, for example, were taken into consideration in further studies. Curve 1 (1–25 mm top) is characterized by Cu 2.00 and Cc 1.13, which means poor graded material. Curve 3 (2–25 mm) is characterized by Cu 4.73 and Cc 1.90, which is stated for well graded material [15].

The compaction of gravely materials is problematic and the occurrence of poorly graded material in construction layers could be present. Curve 3 was created on the basis of Curve 1 by adding to 2–25 mm fraction 20% of its weight fraction 1–2 mm. Curve 2 represents average grain size distribution after CBR tests as a control of properly designing the layers in CBR mould (Curve 1 was placed in 1/3 of CBR mould at the top).

Statistical analysis of obtained test results (Figure 5) was conducted in order to find correlation between physical properties and CBR value of steel slag. Table 3 presents those results. The average error of the presented equation in comparison to test results was between 2% and 5%.

Nevertheless, physical properties calculated with CBR values which have their engineering applications should be extended to more sophisticated relationships. Therefore, the dependence estimation of CBR values and density of samples to degree of saturation during the tests was conducted. This method connects the soaked and unsoaked state of material and clearly, on Figure 6 and Figure 7, presents the decrease of CBR bearing capacity when saturated conditions occur.

In other words, Figure 6 and Figure 7 concern the impact of saturation. Soaked samples assumed to have 100% moisture and representing the results for soaked and un-soaked specimens with various moisture content were used to find formulas describing this phenomena.

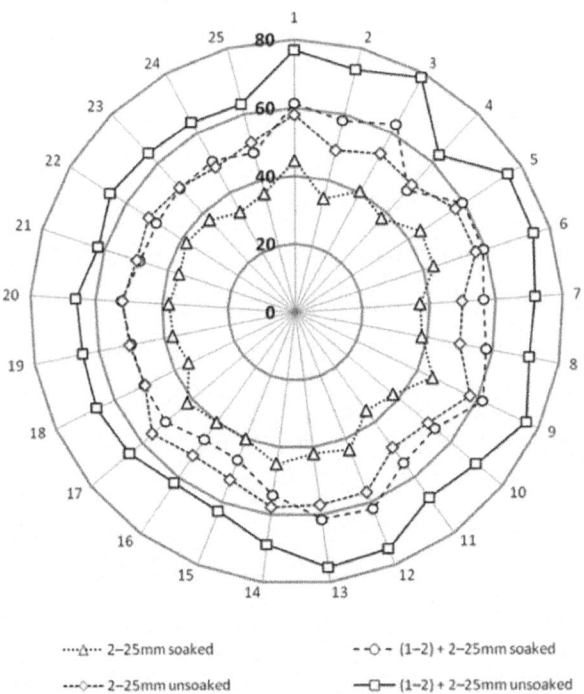

Figure 2: Bearing capacity tests for steel slag mixture aggregate with various grain size and curing conditions.

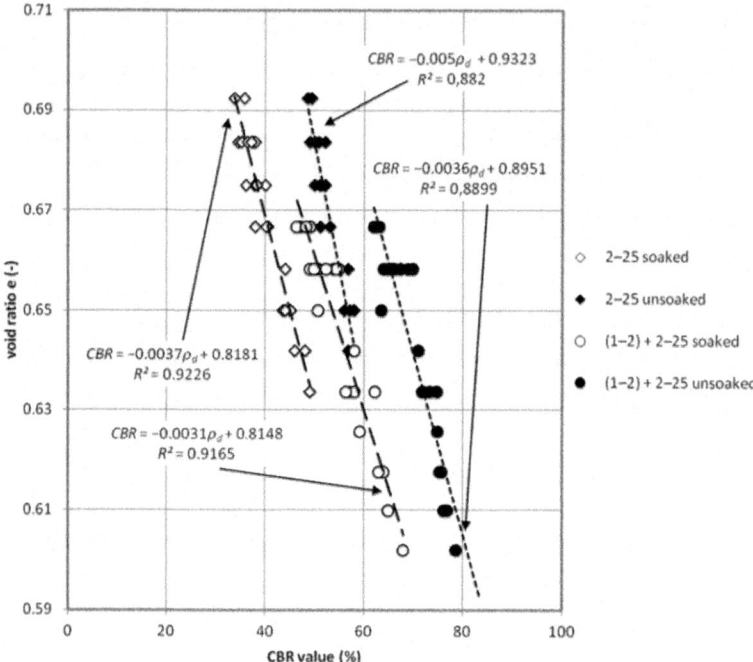

Figure 3: Plot of California bearing ratio (CBR) test results against density of tested samples.

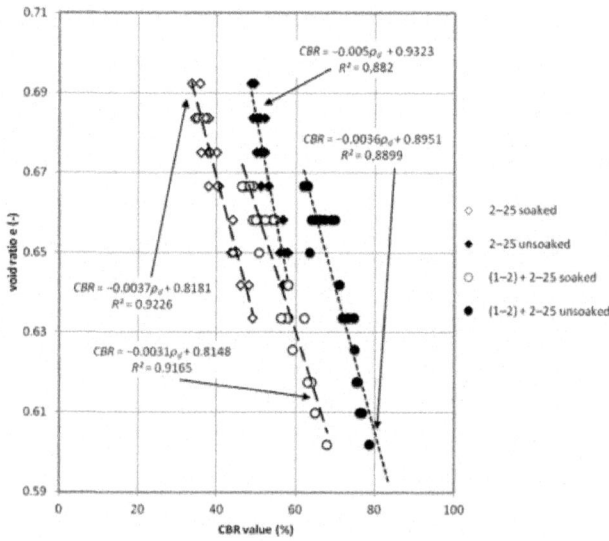

Figure 4: Plot of void ratio diversity for obtained CBR values tested samples.

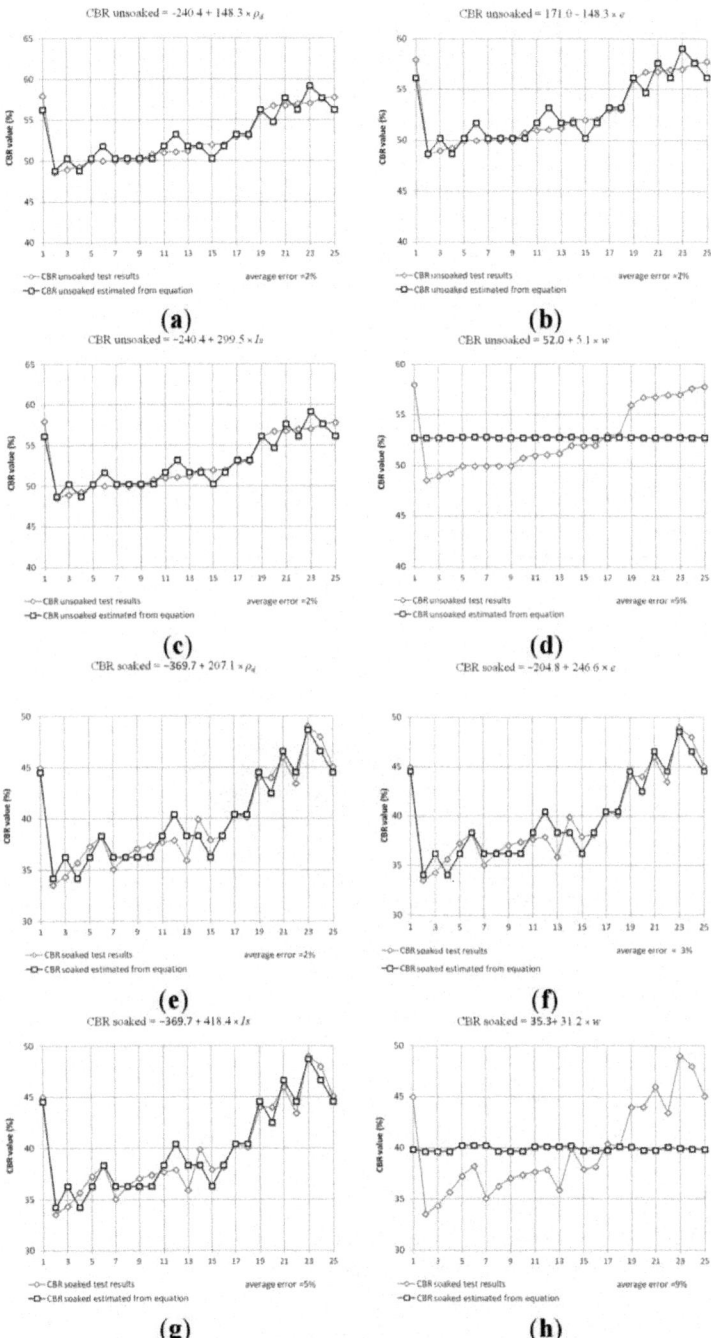

Figure 5: Plot of estimated equations for CBR values and test results for unsoaked test conditions, fraction 2–25 mm. (**a**) CBR unsoaked calculated on the base density.

(b) CBR unsoaked calculated on the base void ratio. **(c)** CBR unsoaked calculated on the base degree of compaction. **(d)** CBR unsoaked calculated on the base moisture content. **(e)** CBR soaked calculated on the base density. **(f)** CBR soaked calculated on the base void ratio. **(g)** CBR soaked calculated on the base degree of compaction. **(h)** CBR soaked calculated on the base moisture content.

Table 3: Results of statistical correlation estimation from California bearing ratio (CBR) test results. ρ_d: dry density; e: void ratio; w: moisture Is: relative compaction

CBR Test Condition	2–25 mm Fraction	(1–2) + 2–25 mm Fraction
CBR unsoaked	$240.4 + 148.3 \cdot \rho_d$	$330.7 + 139.2 \cdot \rho_d$
	$171.0 - 176.7 \cdot e$	$227.3 - 245.5 \cdot e$
	$52.0 + 5.1 \cdot w$	$74.3 + 36.8 \cdot w$
	$240.4 + 299.5 \cdot I_s$	$330.7 + 402.4 \cdot I_s$
CBR soaked	$369.7 + 207.1 \cdot \rho_d$	$429.4 + 241.5 \cdot \rho_d$
	$204.8 - 207.1 \cdot e$	$46.8 - 2297.3 \cdot e$
	$35.3 + 31.2 \cdot w$	$63.6 - 58.6 \cdot w$
	$369.7 + 418.4 \cdot I_s$	$429.4 + 487.8 \cdot I_s$

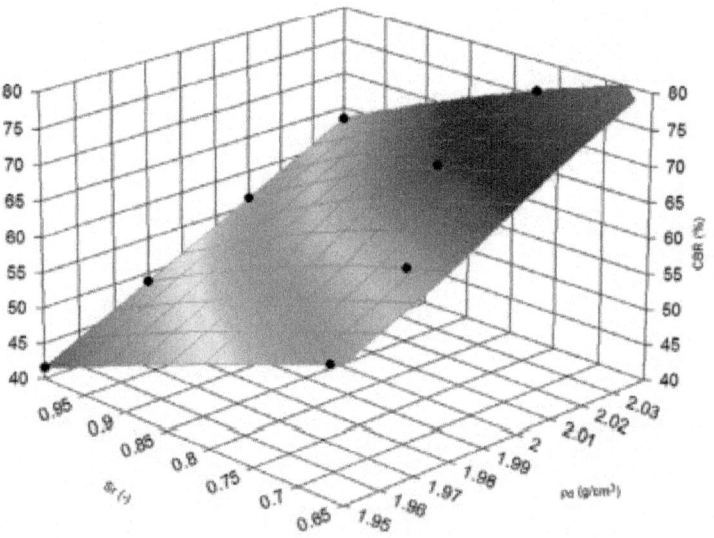

Figure 6: Three-dimensional (3D) view of CBR value dependence from density and saturation ratio (Sr), fraction (1–2) + 2–25 mm.

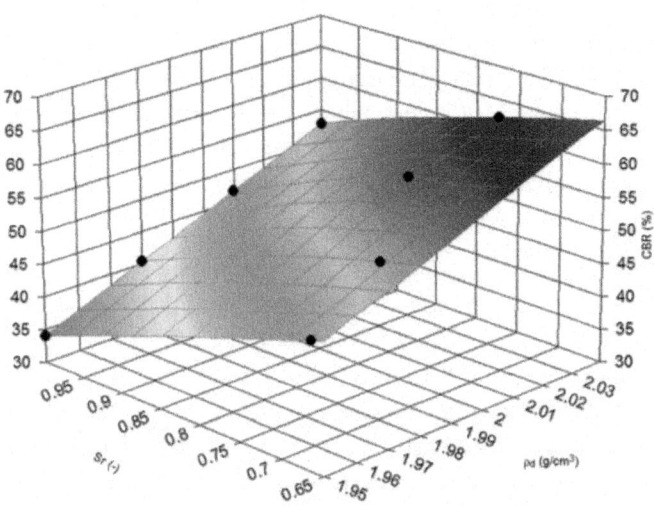

Figure 7:   3D view of CBR value dependence from density and saturation ratio, fraction 2–25 mm.

Equation (1) presents a formula for calculating the compressive CBR value (%) of steel slag (z) with varying degrees of saturation (-) (y) and density (x), expressed in g/cm³. For fraction (1–2) + 2–25 mm:

$$z = a + \frac{b}{x} + cy + \frac{d}{x^2} + ey^2 + f\frac{x}{y}$$

(1)

where letters from a to f are constants: a = 560.1161825; b = −2308.7165; c = 738.5252195; d = 2144.540662; e = −225.272043; f = −803.121929. For this equation the R^2 value is 0.999.

Equation (2) presents a formula for calculating the compressive CBR value of steel slag (%) with varying degree of saturation (-) and density (g/cm³), expressed in g/cm³. For fraction 2–25 mm:

$$z = a + \frac{b}{x} + cy + \frac{d}{x^2} + ey^2 + f\frac{x}{y}$$

(2)

where letters from a to f are constants: a = 124.4164903; b = −880.826622; c = 788.1567559; d = 1064.132286; e = −188.72681; f = −1009.85394. For this equation, the R^2 value is 0.999.

Saturation ratio clearly impacted the results of the CBR tests and the decreasing of bearing capacity from CBR tests although both gradations have the same surface form represented by Equations (1) and (2). Further studies could lead to useful results for engineers' correlations between soaked and unsoaked conditions. Moreover, when this dependence characterizes the saturation ratio, adding to fraction 2–25 mm, the fraction 1–2 mm, increases the CBR bearing capacity in soaked and unsoaked test conditions.

Cyclic Loading Tests

Tests were performed on steel slag that contains 2–25 mm grains and density equal 2.02 g/cm^3. Results are presented onFigure 8, Figure 9, Figure 10, Figure 11 and Figure 12. Figure 8 and Figure 9 presents a detailed view of displacement variation during loading and unloading phases (soaked and unsoaked conditions, respectively). Figure 10 presents a detailed view of stress variation during the loading and unloading phases. Total displacement after 50 repetitions was equal to 3.57 mm and consists of a 96% elastic response of material to repeated loads. The material was subjected to a stress equal to 3.84 MPa and unloaded to about 10% of maximal stress. The detailed view of this process is presented in Figure 10. Figure 11 and Figure 12 presents views of cCBR tests in axial stress-displacement configuration.

The resilient modulus cannot be calculated directly from stress-displacement plots and needs to be calculated in another manner. In literature, such recalculation was presented by Arraya [56]. Resilient modulus Mr from repeated loading CBR can be obtained as follows:

$$M_r = \frac{1.513 \cdot \left(1 - \upsilon^{1.104}\right) \cdot \Delta\sigma_p \cdot r}{\Delta u^{1.012}}$$

(3)

where: υ—Poisson's ratio (-) (in this study 0.35 for granular materials), $\Delta\sigma_p$—change between maximum and minimum axial stress in 50th cycle (MPa), r—radius of plunger (mm), Δu—recoverable displacement in one cycle (mm).

Resilient modulus for steel slag calculated in this manner is equal to M_r = 331 MPa which is reasonable result [57].

Figure 8 and Figure 11 presents cCBR tests for fraction 2–25 mm in soaked conditions. Axial stress reached 3.05 MPa and total displacement was 4.26 mm. Important to note is the fact that a huge plastic displacement occurred in the first 3 cycles. It could be explained by delayed pore pressure distribution over the sample and no negative pore pressure in pores which additionally increases the strength of soil mass.

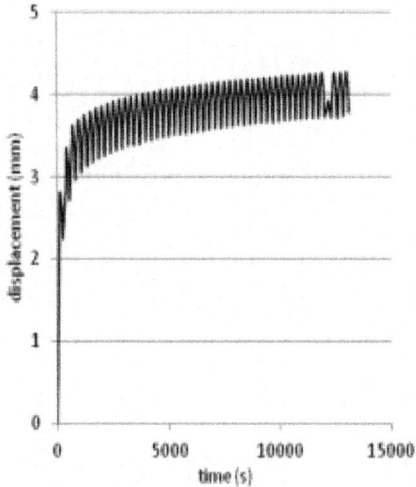

Figure 8: Plot of displacement over time from cCBR tests (soaked conditions).

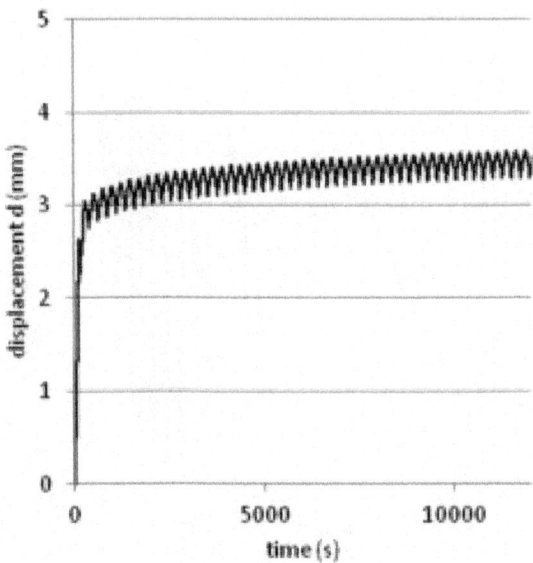

Figure 9: Plot of displacement over time from cCBR tests (unsoaked conditions).

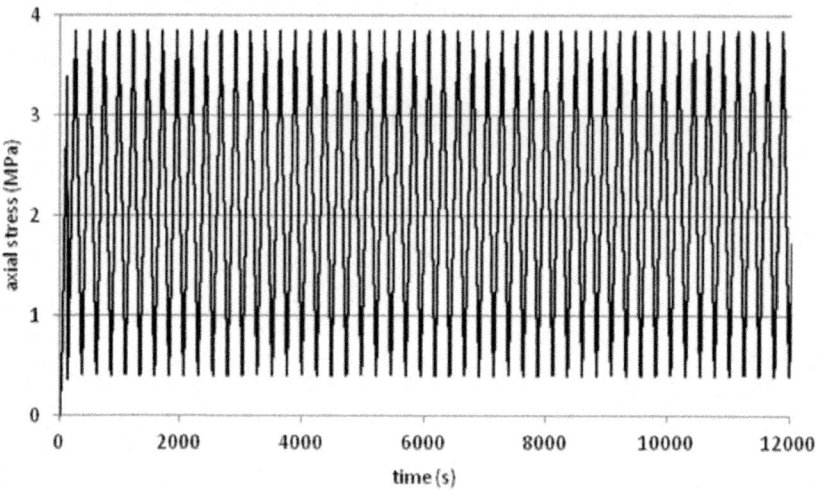

Figure 10: Plot of axial stress over time from cCBR tests (unsoaked conditions).

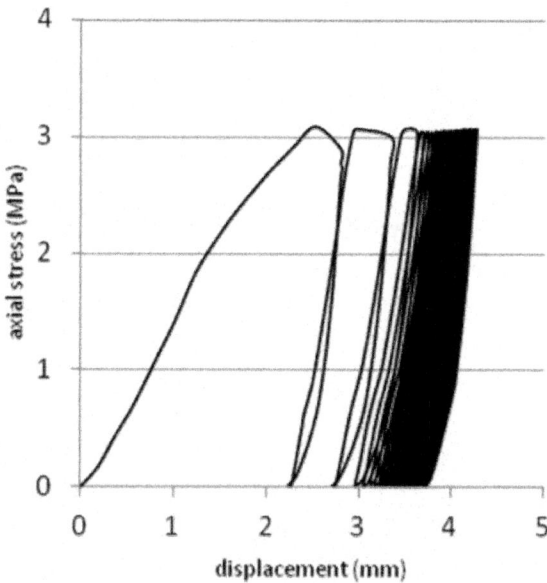

Figure 11: Plot of cCBR test results for steel slag after 50 cycles of loading (soaked conditions).

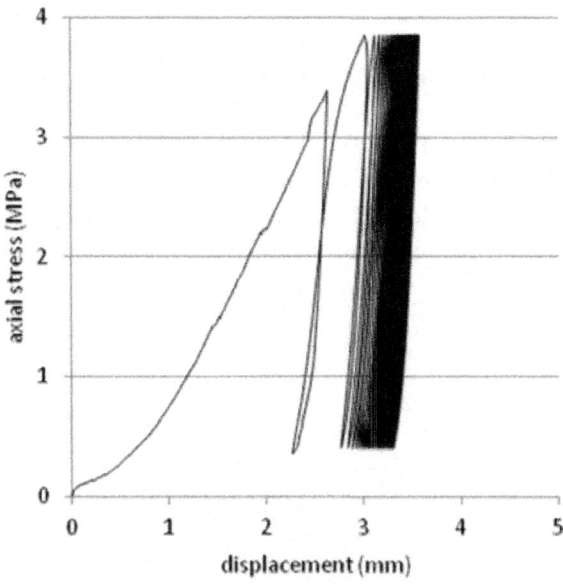

Figure 12: Plot of cCBR test results for steel slag after 50 cycles of loading (unsoaked conditions).

Field Tests

In situ static load plate tests (VSS) performed on roads improved by steel slag showed that Young's modulus (modus of elasticity) E_1 (determined from the first loading of subbase) reached values of 54–72 MPa. The values of Young's modus (modus of elasticity) E_2 (determined from the second loading of subbase) were 103–126 MPa. The associate soil deformability ratio I_0 reached 1.8–2.0 and fulfilled the requirements $I_0 < 2.2$. These results show that the tested layer of steel slag fulfilled the requirements for auxiliary subbase ($E_1 > 50$ MPa).

Static plate load testing done on layers from the second section of road proved that the steel slag mixture fulfilled the requirements for principal subbase (($E_1 > 100$ MPa, $E_2 > 140$–170 MPa) and riding surface. Young's moduli E_1 were 103–126 MPa and E_2 reached 158–209 MPa. The associate soil deformability ratio I_0 reached 1.5–1.7 and fulfilled the requirements $I_0 < 2.2$ as well. The value of soil deformability ratio I_0 ($I_0 < 2.2$) can be also recognized as the compaction ratio I_S with value equal to 1.0. It also means that steel slag layers have been well compacted.

Young modulus plot and Deformability ratio I_0 for each test are presented in Figure 13.

Results of cCBR were compared with VSS test data and plotted on a stress-displacement chart (Figure 14). The first loading on a cCBR test overlaps with the VSS test results. Lower displacement obtained during the VSS test leads to greater resilient strain occurrence during the unloading phase. This phenomenon can be utilised to support field studies in laboratories by performing numerous cyclic loading tests and evaluating more reliable parameters for road designers.

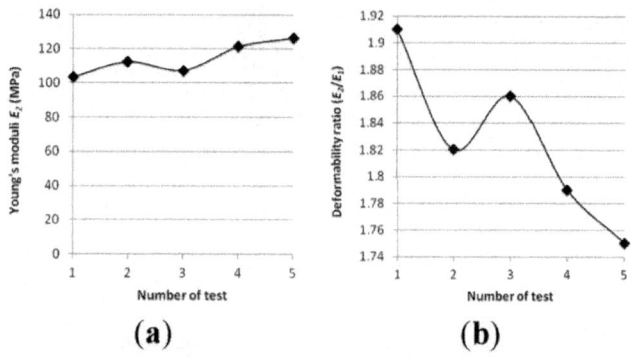

(a)   (b)

Figure 13: Results of Young's moduli E_2 and deformability ratio I_0 for ground road improved by steel slag. (**a**) Young's moduli *versus* number of static plate VSS tests. (**b**) Deformability ratio values *versus* number of static plate VSS tests.

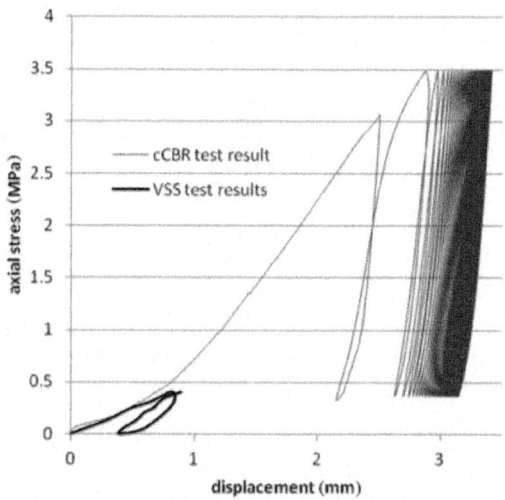

Figure 14: Plot of stress-displacement tests results for cCBR and static load plate VSS on steel slag in duplicate conditions.

CONCLUSIONS

Aggregates of steel slag have chemical and mechanical properties comparable to similar natural aggregates. An important fact is that varying results are common to virgin soils, so there is no uncertainty regarding performance under working stress. However, due to their heterogeneity, it is important to verify them before using.

Their parameters fill requests for road construction independently from their origin. The designated high content of chromium and zinc are strongly associated with the internal crystal structure of steel slag. Consequently, they can be safely applied in roads' structure.

The aforementioned results suggest a possibility to recycle steel slag as a material for subbases. Test results concerning lack of fines and soaked conditions were mixed. Those circumstances are negative and mostly affect the performance of subgrade constructed from steel slag. CBR tests show a drop of bearing capacity of this material when the aforementioned phenomena occur. Therefore, the constructed subbase is cut off from any water inflows or outflows. This treatment makes obtaining steady conditions possible. Well-graded steel slag reached the CBR value exceeding 60% in each of the 25 repetitions. On the other hand, poorly graded soaked samples mostly did not cross 40% of CBR value. Upon comparing the impact of the previously mentioned properties on bearing capacity gradation and saturation conditions, steel slag is more sensitive to poor gradations or simply to a lack of fine graded grains.

Saturation ratio decreases bearing capacity from CBR tests. Equations (1) and (2) could in further studies lead to the estimation of a new equation, which can take into account change in gradation. This is made possible by the fact that for steel slag with varying gradation, a change of CBR value with density and saturation ratios has the same surface represented by Equations (1) and (2).

Moreover, cyclic loading shows good performance of steel slag and plastic displacement was 1 mm greater after the 50th load repetition than after the first loading. Eurocode 7 EN 13286-7:2004 [58] classifies steel slag by its mechanical performance by the resilient modulus M_r parameter, and this material reached the C2 class.

Field tests including the static plate load test have proved that steel slag mixture fulfilled the requirements for principal subbase (($E_1 >$ 100 MPa, $E_2 >$ 140–170 MPa) and riding surface. It is interesting to note that VSS tests and cCBR tests seem to be interrelated. This would be important for supporting field tests with laboratory studies. Mechanistic-empirical pavement design states that, for every layer designed, the resilient modulus M_r should be taken in the calculation. The cCBR, therefore, could be a response to the lack of simple methods for estimating such a parameter.

Taking into consideration the above test results, it can be concluded that steel slag can be used in base courses in road structures for motorways and roads with medium traffic loads. The most important selected geotechnical parameters, bearing capacity ratio (obtained from laboratory tests) and resilient modulus M_r showed that their values correspond with requirements for subbase (principal or auxiliary) and riding surface as well. The use of steel slag in road structure courses would be desired from both the economic and environmental point of view: great quantities of waste material would thus be used, reducing the amount of slag deposited in landfills.

AUTHOR CONTRIBUTIONS

Wojciech Sas and Andrzej Głuchowski managed the research in this paper and also performed geotechnical laboratory and field studies. Wojciech Sas prepared the manuscript. Maja Radziemska performed chemical studies. Justyna Dzięcioł performed statistical analysis. Alojzy Szymański consulted on the research program and checked the manuscript.

CONFLICTS OF INTEREST

The authors declare no conflict of interest.

REFERENCES

1.   Faullman, A.M. Leaching of chromium and barium from steel slag in laboratory and field tests—a solubility controlled process? *Waste Manag.* 2000, *20*, 149–154

2.   Furlani, E.; Tonello, G.; Maschio, S. Recycling of steel slag and glass cullet from energy saving lamps by fast firing production of ceramics. *Waste Manag.* 2010, *30*, 1714–1719

3.   Shi, C.J.; Qian, J.S. High performance cementing materials from industrial stags—a review. *Resour. Conserv. Recycl.*2000, *29*, 195–207

4.   Motz, H.; Geiseler, J. Products of steel slags an opportunity to save natural resources. *Waste Manag.* 2001, *21*, 285–293

5.   Baciocchi, R.; Costa, G.; Di Bartolomeo, E.; Polettini, A.; Pomi, R. The effects of accelerated carbonation on CO_2 uptake and metal release from incineration APC residues. *Waste Manag.* 2009, *29*, 2994–3003

6.   Zhang, H.W.; Hong, X. An overview for the utilization of wastes from stainless steel industries. *Resour. Conserv. Recycl.* 2011, *55*, 745–754.

7. Directive, W.F. Position paper on the status of ferrous slag. *Regulation* 2012. Available online: http://www.euroslag.com/fileadmin/_media/images/ Status_of_slag/Position_Paper_April_2012.pdf (accessed on 8 July 2015).

8. Shen, Y.N.; Huang, L.J.; Wang, H.Y.; Le, D.H. Experimental study and strength formulation of soil-based controlled low-strength material containing stainless steel reducing slag. *Constr. Build. Mater.* 2014, *54*, 1–9

9. Sheen, Y.N.; Wang, H.Y.; Sun, T.H. A study of engineering properties of cement mortar with stainless steel oxidizing slag and reducing slag resource materials. *Constr. Build. Mater.* 2013, *40*, 239–245

10. Yildirim, I.Z.; Prezzi, M. Chemical, mineralogical, and morphological properties of steel slag. *Adv. Civ. Eng.* 2011, *201*, 463638

11. Polish Committee for Standardization. *Mixtures of Unbound Materials–Requirements*; PN-EN 13285:2004; Polish Committee for Standardization: Warsaw, Poland, 2004.

12. Polish Committee for Standardization. *Aggregates for Unbound and Hydraulically Bound Materials for Use in Building and Road Construction*; PN-EN 13242:2004; Polish Committee for Standardization: Warsaw, Poland, 2004.

13. American Concrete Institute. *Controled-Low Strength Materials*; ACI 229R-99; American Concrete Institute: Detroit, MI, USA, 2005.

14. Achtemichuk, S.; Hubbard, J.; Sluce, R.; Shehata, M.H. The utilization of recycled concrete aggregate to produce controlled low-strength materials without using Portland cement. *Cem. Concr. Compos.* 2009, *31*, 564–569

15. Chen, Y.; Chen, Z.; Huang, J.; Xue, J.; Zhou, C. Mechanical properties of recycled pebble aggregate concrete. *J. Build. Mater.* 2014, *17*, 465–469. (In Chinese).

16. Sas, W.; Głuchowski, A.; Szymański, A. Impact of the stabilization of compacted cohesive soil–sandy clay on field criterion improvement. *Ann. Wars. Univ. Life Sci. Land Reclam.* 2014, *46*, 139–151.

17. Colangelo, F.; Cioffi, R. Use of cement klin dust, blast furnace slag and marble sluge in the manufacture of sustainable artificial aggregates by means of bonding pelletization. *Materials* 2013, *6*, 3139–3159

18. Wu, J.Y.; Lee, M.Z. Beneficial reuse of construction surplus clay in CLSM. *Int. J. Pavement Res. Technol.* 2011, *4*, 293–300.

19. Sas, W.; Głuchowski, A. Effects of stabilization with cement on mechanical properties of cohesive soil–sandy-silty clay. *Int. J. Pavement*

Res. Technol. 2013, *45*, 193–205.

20. Wang, H.Y.; Chen, B.T.; Wu, Y.W. A study of the fresh properties of controlled low-strength rubber lightweight aggregate concrete (CLSRLC). *Constr. Build. Mater.* 2013, *41*, 526–531

21. Ivana, S.D.; Barišić, I.N.; Dimter, S. Possibilities of application of slag in road construction. *Tech. Gaz.* 2010, *17*, 523–528.

22. Dunster, A.M. The use of blastfurnace slag and steel slag as aggregates. In Proceedings of the 4th European Symposium on Performance of Bituminous and Hydraulic Materials in Pavements, Nottingham, UK, 11–12 April 2002; pp. 257–260.

23. Hwang, E.H.; Koa, Y.S.; Kim, J.M.; Hwang, T.S. Mechanical/physical characteristics of polymer mortar recycled from rapid-chilled steel slag. *J. Ind. Eng. Chem.* 2009, *15*, 628–634

24. Kriskova, L.; Pontikes, Y.; Cizer, Ö.; Mertens, G.; Veulemans, W.; Geysen, D.; Jones, P.T.; Vandewalle, L.; Van Balen, K.; Blanpain, B. Effect of mechanical activation on the hydraulic properties of stainless steel slags. *Cem. Concr. Res.* 2012, *42*, 778–788

25. Pellegrino, C.; Cavagnis, P.; Faleschini, F.; Brunelli, K. Properties of concretes with black/oxidizing electric arc furnace slag aggregate. *Cem. Concr. Compos.* 2013, *37*, 232–240

26. Adolfsson, D.; Robinson, R.; Engström, F.; Björkman, B. Influence of mineralogy on the hydraulic properties of ladle slag. *Cem. Concr. Res.* 2011, *41*, 865–871

27. Pasetto, M.; Baldo, N. Experimental analysis of hydraulically bound mixtures made with waste foundry sand and steel slag. *Mater. Struct.* 2014

28. Liu, S.J.; Hu, Q.Q.; Zhao, F.Q.; Chu, X.M. Utilization of steel slag, iron tailings and fly ash as aggregates to prepare a polymer-modified waterproof mortar with a core–shell styrene–acrylic copolymer as the modifier. *Constr. Build. Mater.* 2014, *72*, 15–22

29. Katz, A.; Klover, K. Utilization of industrial by-products for the production of controlled low-strength materials (CLSM). *Waste Manag.* 2004, *24*, 501–512

30. Pasetto, M.; Baldo, N. Cement bound mixtures with metallurgical slags for road constructions: Mix design and mechanical characterization. *IM In . Miner.* 2013, *14*, 15–20.

31. Sakai, S.I.; Hiraoka, M. Municipal solid waste incinerator residua recycling by thermal processes. *Waste Manag.* 2000, *20*, 249–258.

32. Zhang, W.S.; Shi, D.; Shao, Z.J.; Ye, J.Y.; Wang, Y. The properties of

steel slag bricks prepared by both alkali activation and accelerated carbonation. *Key Eng. Mater.* 2012, *509*, 113–118

33. Wang, G.; Wang, Y.; Gao, Z. Use of steel slag as a granular material: Volume expansion prediction and useability criteria. *J. Hazard. Mater.* 2010, *184*, 555–560

34. Kourounis, S.; Tsivilis, S.; Tsakiridis, P.E.; Papadimitriou, G.D.; Tsibouki, Z. Properties and hydration of blended cements with steelmaking slag. *Cem. Concr. Res.* 2007, *37*, 815–822

35. Setién, J.; Hernández, D.; González, J.J. Characterization of ladle furnace basic slag for use as a construction material.*Constr. Build. Mater.* 2009, *23*, 1788–1794

36. WT-4. Unbound mixtures for national roads. (In Polish). Available online: http://www.gddkia.gov.pl/userfiles/articles/d/Dokumenty_techniczne/WT4.pdf.

37. Zhao, W.; Fan, C. Discussion about technical problems in CBR test. *Appl. Mech. Mater.* 2014, *587–589*, 1286–1290

38. Giummarra, G. *Road Classifications, Geometric Designs and Maintenance Standards for Low Volume Roads*; ARRB Transport Research Ltd.: Vermont South, VIC, Australia, 2001.

39. Polish Committee for Standardization. *Motor roads. Earthworks. Requirements and Tests*; PN-EN 13242:2004; Polish Committee for Standardization: Warsaw, Poland, 1988. (In Polish)

40. American Society for Testing Materials. *Standard Test Method for CBR (California Bearing Ratio) of Laboratory-Compacted Soils*; ASTM D1883-87; ASTM International: West Conshohocken, PA, USA, 1987.

41. Sas, W.; Głuchowski, A.; Szymański, A. Determination of the resilient modulus M_R for lime stabilized Clay obtained from repeated loading CBR tests. *Ann. Wars. Univ. Life Sci. Land Reclam.* 2012, *44*, 143–153

42. Sas, W.; Głuchowski, A. Methods of determination of the modulus of elasticity (E and Mr) from the repeated loading tests CBR. *Sci. Rev. Eng. Env. Sci.* 2012, *57*, 171–181.

43. Kumar, S.; Kumar, R.; Bandopadhyay, A. Innovative methodologies for the utilization of wastes from metallurgical and allied industries. *Resour. Conserv. Recycl.* 2006, *48*, 301–314

44. Kumar, D.S.; Umadevi, T.; Paliwal, H.K.; Prasad, G.; Mahapatra, P.C.; Ranjan, M. Recycling steelmaking slags in cement. *Worldcement* 2010, *11*, 1–6.

45. Das, B.; Mohanty, J.K.; Reddy, P.S.R.; Ansari, M.I. Characterization and

beneficiation studies of charge chrome slag.*Scand. J. Metall.* 1997, *26*, 153–157.

46. Chazarenc, F.; Filiatrault, M.; Brisson, J.; Comeau, Y. Combination of slag, limestone and sedimentary apatite in columns for phosphorus removal from sludge fish farm effluents. *Water* 2010, *2*, 500–509

47. Piatak, N.M.; Seal, R.R., II. Mineralogy and the release of trace elements from slag from the Hegeler Zinc smelter, Illinois (USA). *Appl. Geochem.* 2010, *25*, 302–320

48. Puziewicz, J.; Zainoun, K.; Bril, H. Primary phases in pyrometallurgical slags from a zinc-smelting waste dump, Świętochłowice, Upper Silesia, Poland. *Can. Mineral.* 2007, *45*, 1189–1200

49. Proctor, D.M.; Fehling, K.A.; Shay, E.C.; Wittenborn, J.L.; Green, J.J.; Avent, C.; Bigham, R.D.; Connolly, M.; Lee, B.; Shepker, T.O.; Zak, M.A. Physical and chemical characteristics of blast furnace, basic oxygen furnace, and electric arc furnace steel industry slags. *Env. Sci. Technol.* 2000, *34*, 1576–1582

50. Geiseler, J. Use of steelworks slag in Europe. *Waste Manag.* 1996, *16*, 59–63

51. Mombelli, D.; Mapelli, C.; Barella, S.; Gruttadauria, A.; Le Saout, G.; Garcia-Diaz, E. The efficiency of quartz addition on electric arc furnace (EAF) carbon steel slag stability. *J. Hazard. Mater.* 2014, *279*, 586–596

52. Lottermoser, B.G. Evaporative mineral precipitates from the historical smelting slag dump, Río Tinto, Spain. *Neues Jahrb. Miner. Abh. J. Miner. Geochem.* 2005, *181*, 183–190

53. Hill, A. Understanding the leaching behaviour of slags, testing and interpretation. In Proceeding of the 3th European Slag Conference—Manufacturing and Processing of Iron and Steel Slags, Keyworth, UK, 16–18 October 2002.

54. Ansari, A.; Mehrabian, M.A.; Hashemipour, H. Zinc ion adsorption on carbon nanotubes in an aqueous solution. *Pol. J. Chem. Technol.* 2012, *14*, 29–37.

55. Thamilarasu, P.; Kumar, G.; Tamilarasan, R.; Sivakumar, V.; Karunakaran, K. Kinetic, equilibrium and thermodynamic studies on the removal of Cr(VI) by activated carbon prepared from Cajanus Cajan (L) Milsp seed Shell. *Pol. J. Chem. Technol.* 2011, *13*, 1–7

56. Araya, A.A. Characterization of Unbound Granular Materials for Pavements. Ph.D. Thesis, Technical University Delft, Delft, The Netherlands, March 2011.

57. Hainin, M.R.; Yusoff, N.I.M.; Mohammad Sabri, M.F.; Abdul Aziz, M.A.; Sahul Hameed, M.A.; Farooq Reshi, W. Steel slag an aggregate replacement in malaysian hot mix asphalt. *ISRN Civ. Eng.* 2012, 459016

58. BSI Standards Publication. *Unbound and Hydraulically Bound Mixtures. Cyclic Load Triaxial Test for Unbound Mixtures*; BS EN 13286-7:2004; British Standards Institute: London, UK, 2004.

Chapter 5

EARTH RESOURCES EXPLOITATION AND SUSTAINABLE DEVELOPMENT: GEOLOGICAL AND ENGINEERING PERSPECTIVES

C. C. Iwuji, O. C. Okeke, B. C. Ezenwoke, C. C. Amadi, H. Nwachukwu

Department of Geology, Federal University of Technology, Owerri, Nigeria

ABSTRACT

Earth resources are the essential basis for an economy of a country and well-being of her citizens. Their exploitation is a key factor in economic growth and development of the nation, but one that can have serious negative environmental and socio-economic impacts. Therefore it is important to understand their origin/occurrence and exploitation in terms of environmental sustainability. The human society has profited from exploitation of earth resources, precisely when energy use became much more efficient during the industrial era. Some of these earth resources (renewable and non-renewable) are oil and gas, coal, water, metal ore, wind, air etc. and their methods of exploitation are surface and subsurface mining methods [1] . Thus the geological perspectives of earth resources refer to origin and occurrence of these earth resources while the engineering perspectives refer to the processes involved in the extraction (mining), processing and utilization of the resources. To fulfill the requirement of sustainable development, the efficiency with which resources are utilized has to be improved. The main concerns associated with earth resources, therefore, are generally the costs and environmental impacts of extracting, transporting, and refining them. The economics of earth resources deals with the supply, demand and allocation of the resources.

INTRODUCTION

The Earth resources are referred to as natural resources and that are useful raw materials for manufacturing/ production of goods. They occur naturally, meaning that man cannot make them but instead use and modify the resources in ways that are beneficial. These resources include minerals such as fossil fuels, solid minerals, air, water, soil, wind, geothermal, tidal, and solar energy. Hence, earth resources are grouped into: 1) Renewable resources: this implies that they are replaced or regenerated naturally for example tides, solar power and wind energy, water etc. 2) Non-renewable resources: these are earth resources that cannot be replaced or regenerated naturally or otherwise for example coal, petroleum, metals etc.

The word "sustainable" means able to be maintained. Sustainability is defined as the process applied to aid the quality of human life within the limitations of the environment. The United Nations World Commission on Environment and Development coined a definition of sustainable development, which is probably the most well known in all of the sustainability literature: "development that meets the needs of the present without compromising the ability of future generations to meet their own needs" [2] . However, sustainable development implies a dynamic balance between maintenance/conservation (sustainability) and development [3] , both directed towards human needs. The concept of sustainable development therefore involves the successful integration of environmental considerations into development management [4] . Sustainability is in turn defined by Lawrence as meeting the ecological, social and economical needs and aspirations of human and other species [5] .

In practical, non-renewable resources are not fully exhausted or exploit all the renewable resources. Hence, the most valuable deposits and sites are discovered and exploited first, followed by other less quality sources as demand increases. As demand increases and scarcity of resources set in, its price increases. Thus reduction in demand gives explorers incentives to look out for sources that are of lower quality and/or more expensive to exploit, and to improve methods for locating, extracting, and processing the resources. Increasing prices also encourage the development of substitutes that were less demanded when then original resource was cheap.

Sustainable development requires the maintenance, rational use and enhancement of natural resources, as well as a balanced consideration of ecology, economy and social justice. More so, it involves solutions for improving human welfare that does not result in degrading the environment on impinging in well-being other people [6] [7] . Industrialized countries have to transfer their advanced technologies to developing countries in order to avoid undesirable development in the mining industry and use of natural resources in those regions. Hence, people who study energy resources seek to know the type of resources that are available and where they are located, and to create new methods for locating, extracting, and exploiting the resources. Discovery of new sources of supplies and using more energy resources is one way to derive more benefits [8] . But we can also use these resources more efficiently, so that we can obtain a rising amount of service from a constant level of inputs.

OVERVIEW OF EARTH RESOURCES AND SUSTAINABLE DEVELOPMENT

Earth Resources

The earth is rich in natural resources that we use every day. These resources are very economical materials of geologic origin that can be extracted from the earth. However, earth resources are classified into: 1) non-re- newable resources: suggest that they are natural resources that are not replaced or re- growth naturally. These resources are exhaustible and are extracted faster than the rate at which they formed [9] [10] . Some common examples are fossils fuels (coal, oil, and natural gas), precious gems and metals and ores etc. 2) Renewable resources are replaced through natural processes at a rate that is equal to or greater than the rate at which they are exploited, and depletion is usually not a threat. Some common examples include; air (wind), water, soil, trees, sun-light etc. Though many renewable resources do not have such a rapid recovery rate, these resources are susceptible to depletion by over-use. Resources from a human use perspective are classified as renewable only so long as rate of replacement exceeds that of the rate of consumption [11] .

The Notion of Sustainable Development

Sustainable development is a normative term—according to the philosopher Immanuel Kant—like liberty and equality. In the UN report "Our Common Future", commonly called the Brundtland Report, sustainable development is suggested to be development that meets the needs of the present without compromising the ability of the future to meet their own needs [2] . However, this definition was expanded by the United Nations Environment Programme (UNEP), which has added that this notion also requires the maintenance, rational use, and enhancement of the natural resources base that underpins ecological resilience and economic growth, and that it implies progress towards international equity [12] .

The next level was the Rio declaration at the UN conference on Environment Development in Rio de Janeiro in 1992 and Agenda 21, which stresses the three elements of sustainable development—economy, ecology, and social justice. To conserve the basic needs of life and also to enable all people to achieve economic success, and to strive towards social justice. All three objectives should initially be considered to have the same priority. While the definition above emphases intergeneration fairness. The first real guidelines for sustainable development came from the German forestry administration.

The man credited with inventing sustainable development was a miner, Oberberghauptmann Johnann-karl Von carlowitz. He was the head of the mining administration of the famous silver-mining district in freiberg, Saxony. Meanwhile, he was not only responsible for the mining and smelting operations in this district, but also for forestry, because timber was needed in the underground mining operations, and the smelting of the silver ores required vast amounts of wood for charcoal. He realized that unchecked deforestation for these purposes would lead to the collapse of timber production. He was the first person to spell out what sustainable development meant in forestry, that the amount of wood cut should not exceed the regeneration rate.

Sustainable Use of Earth Resources

The issue of depletion plays an important role in the use of non-renewable and renewable earth resources. In the renewable resources, depletion occurs when extraction exceeds renewal rate. Flow and environmental resources are not depleted and always exist. However, environmental resources can be degraded by pollution, and contamination, rendered useless. The way development takes place at the moment is extremely wasteful of natural resources.

More so, proper resources uses are the backbone of every economy and provide two basic functions; raw material for production of goods and services, and environmental service.

Management of Earth Resources

The main concerns associated with earth resources, therefore, are generally the costs and environmental impacts of extracting, transporting, and refining them. Scientists who study energy and material resources seek to understand what types of resources are available and where they can be found, and to develop new technologies for locating, extracting and exploiting them. Discovering new supplies and using more energy and materials is one way to derive more benefits. But these resources can be used more efficiently to obtain a rising amount of service from a constant level of inputs [13] [14] . Over the longer term, scientific and technological advances may enable societies to substitute new energy source and material stocks for old ones.

This typically happens when new resources perform as well as or better than current options and produce fewer negative impacts, such as pollution or health and safety risks. Energy resources and other mineral resources is the centre of development in developed countries. In situation of high cost and scarce supply of resources every part of the economy will be affected. In such situation, political leaders and major consumers want know which way forward.

However, alternating from resources to another involves more than exploration and exploitation of these resources. It also means altering the systems that produce process and distribute these resources. For example, major commercial energy fuels like coal, natural gas, and uranium are mined, cleaned, processed, refined and delivered through complex, multi-stage systems that represent billions of dollars in infrastructure investments and complicated logistical interconnections.

Facilities built for energy resources averagely operate between 30 to 50 years, hence cannot change to different resource or technology mixes over a short period of time. Retiring them prematurely to replace them with something better, is very costly even if the new plants are not more expensive than the old one [15] . Thus, as there is increase in population in developing countries and their citizens demand better service, global energy use will also continue to rise, with developing nations accounting for a growing share of total world demand.

GEOLOGICAL AND ENGINEERING ASPECTS OF EARTH RESOURCES EXPLOITATION

Geological Aspects: Formation and Occurrence of Earth Resources

This is concerned with the origin/formation and occurrence of non-renewable resources such as crude oil, coal, minerals, rock resources, and processing and utilization.

Oil and Natural Gas

Petroleum is not a single chemical compound. Liquid petroleum or oil comprises a variety of liquid hydrocarbon compounds (carbon and hydrogen). There are also gaseous hydrocarbons (natural gas), of which the compound Methane (CH_4) is the most common.

The production of a large deposit of any fossil fuel requires a large initial accumulation of organic matter, which is rich in carbon and hydrocarbon. Another requirement is that the organic debris be buried quickly to protect it from the air so that decay by biological means or reaction with oxygen will not destroy it. Microscopic life is abundant over much of the oceans. When these organisms die, their remains can settle to the sea floor. There are also underwater areas near shore, such as on many continental shelves, where sediments derived from continental erosion accumulated rapidly. In such a setting, the starting requirements for the formation of oil are satisfied. There is an abundance of organic matter rapidly buried by sediments. Oil and much natural gas are believed to form from such accumulated marine microorganisms. Continental oil fields often reflect the presence of marine sedimentary rocks below the surface. Additional natural gas deposits not associated with oil may form from deposits of plant material in sediment on land.

As burial continues, the organic matter begins to change. Pressures increase with weight of the overlying sediment or rock; temperatures increase with depth in the earth; and slowly, over a long period of time, chemical reactions take place. These reactions break down the large, complex organic molecules into simpler, smaller hydrocarbon molecules. The nature of the hydrocarbons changes with time and continued heat and pressure [16] .

The early stages of petroleum formation in a marine deposit, the deposit may consists mainly of larger hydrocarbon molecules (heavy hydrocarbon), which have the thick, nearly solid consistency of asphalt. As the petroleum matures, and as the breakdown of large molecules continues, successively "lighter" hydrocarbons are produced. Thick liquids give way to thinner ones, from which are derived lubricating oils, heating oil, and gasoline. In the final stages, most or all of the petroleum is further broken down into very simple, light, gaseous molecules; natural gas. Most of the maturation process occurs in the temperature range of 50°C to 100°C.

The hydrocarbons can migrate out of the rocks in which they formed. Such migration is important if the oil or gas is to occur in an economically valuable and practically usable deposit. The majority of petroleum source rocks are fine grained clastic sedimentary rocks of low permeability, from which it would be difficult to extract large quantities of oil or gas quickly. Despite the low permeabilities, oil and gas are able to migrate out of their source rocks and through more permeable rocks over the long spans of geologic time [1] [16] . The pores, holes, and cracks in rocks in which fluids can be trapped are commonly full of water. Mostly oils and all natural gases are less dense than water, so they tend to rise as well as to migrate laterally through the water filled pores of permeable rocks.

The most economical deposits are those in which large quantity of petroleum has been concentrated and trapped by an impermeable rock in geologic structures.

Coal

This is formed not from marine organisms, but from the remains of land plants. A swampy setting, in which plant growth is lush and where there is water to cover fallen trees, dead leaves, and other debris, is especially favourable to the initial stage of coal formation. The process requires anaerobic conditions, in which oxygen is absent or nearly so, since reaction with oxygen destroys the organic matter.

The first combustible product formed under suitable condition is peat. Peat can form at the earth surface, and there are places on earth where peat can be seen forming today. Further burial, with more heat, pressure, and time , gradually dehydrates the organic matter and transforms the spongy peat into soft brown coal (lignite) and then to the harder coals (Bituminous and Anthracite) [16] .

As the coals become harder, their carbon content increases, and so does the amount of heat released by burning a given weight of coal.

Solid Minerals

Basically, by cooling down of magma, atoms are linked into crystalline patterns and subsequently different minerals are formed. When the formation takes place in the depths of the earth's crust (approx. 33 km deep) quite large rocks may be formed (for instance granite).

Igneous rocks are formed and created by magmatic processes in the earth. To form very large crystals of rare minerals, exceptional conditions are needed. For instance, a rock called pegmatite is formed by the crystallization of magma enriched with water in the veins of other rocks, and may contain berly, tourmaline and topaz [17] .

Engineering Aspects of Earth Resources Exploitation

This is concerned with the exploitation methods of non-renewable resources and renewable resources.

Drilling of Oil and Gas Wells

Once the site has been selected, scientists survey the area to determine its boundaries, and conduct environmental impact studies if necessary. The oil company may need lease agreements, titles and right of way accesses before drilling the land. For off-shore sites, legal jurisdiction must be determined.

After the legal issues are settled, the crew goes about preparing the land: 1) The land must be cleared and leveled and access roads may be built. 2)

Because water is used in drilling, there must be a source of water nearby. If there is no natural source, the crew drills water well. 3) The crew digs a reserve pit, which is used to dispose of rock cuttings and drilling mud during drilling process, and lines it with plastic to protect the environment. If the site is an ecologically sensitive area, such as a marsh or wilderness, then the cuttings and mud must be disposed of offsite—trucked away instead of placed in a pit.

Once the land has been prepared, the crew digs several holes to make way for the rig and the main hole. A rectangular pit called a cellar is dug around the location of the actual drilling hole. The cellar provides a work space around the hole for the workers and drilling accessories. The crew then begins drilling the main hole, often with a small drill truck rather than the main rig, see Figure 1 below. The first part of the hole is larger and shallower than the main portion, and is lined with a large-diameter conductor pipe. The crew digs additional holes off to the side to temporarily store equipment. When these holes are finished, the rig equipment can be brought in and set up. Depending upon the remoteness of the drill site and its access, it may be necessary to bring in equipment by truck, helicopter or barge. Some rigs are built on ships or barges for work on inland water where there is no foundation to support a rig [18] .

Surface and Underground Mining Methods

Coal mining is the extraction of deposits from the surface of the earth and from the underground, see Figure 2. Coal mining is the most abundant fossil fuel on earth. Its predominant use has always been for producing heat energy. It was the basic energy source that fueled the industrial revolution of the 18th and 19th centuries, and the industrial growth of that era in turn supported the large-scale exploitation of coal deposits [19] .

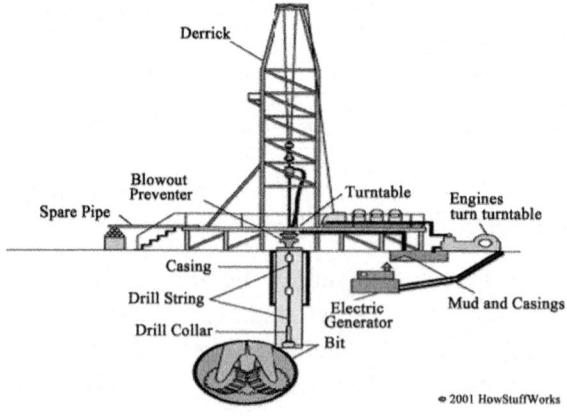

Figure 1: Rotary drilling rig (modified from Craig and Jonathan, 2001).

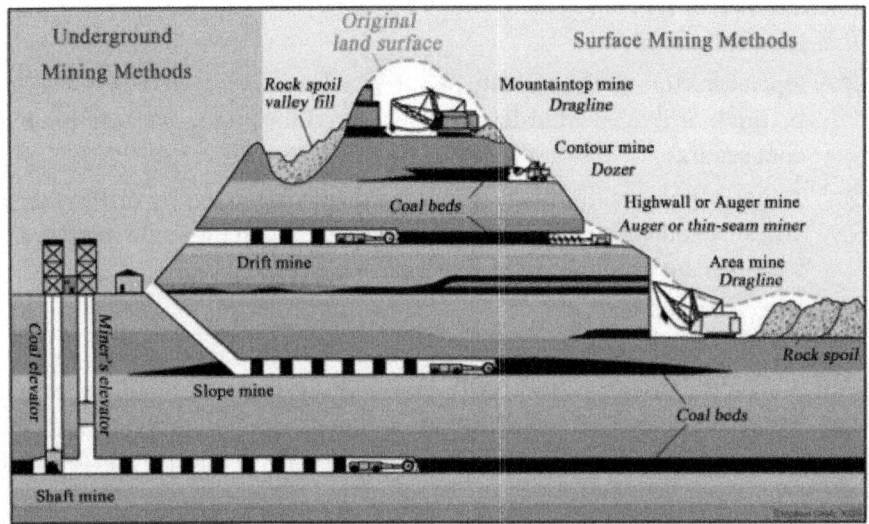

Figure 2: Different coal mining methods. Source:www.worldcoal.org/coal/coal-mining/.

The mining of coal from surface and underground deposits today is a highly productive and mechanized operation.

1)   Surface Mining

Surface coal mining generally involves the following sequence of unit operations:

a)   Clearing the land of trees and Vegetation.

b)   Removing and storing the top layers of the unconsolidated soil.

c)   Drilling the hard strata over the coal seam.

d)   Fragmenting or blasting the hard strata with explosives.

e)   Removing the blasted material, exposing the coal seam, and cleaning the top of the coal seam.

f)   Fragmenting the coal seam, as required by drilling and blasting.

g)   Loading the loose coal onto haulage conveyances.

h)   Transporting the coal from the mine to the plant, and

i)   Reclaiming lands affected by the mining activity.

   Surface techniques can be broadly classified into:

a)   Contour Strip Mining: Is commonly practiced where a coal seam outcrops in rolling or hilly terrain.

b)   Area Strip Mining: Is applied where the terrain is flat, commences with a trench or box cut made through the overburden to expose a portion of the coal seam.

c)   Open Pit Mining: is generally practiced where thick coal seam s overlain by thick or thin overburden. It is also used for mining steeply pitching coal seams.

d)   Auger Mining: with this method, the coal is removed by drilling auger holes from the last contour cut and extracting it in the same manner that shavings are produced by a carpenter's bit.

2)   Underground Mining

Here, the working environment is completely enclosed by the geologic medium which consists of the coal seam and the overlying and underlying strata. Modern underground coal mining methods can be classified into four distinct categories: (a) Room and Pillar (b) Longwall (c) Short-wall (d) Thick-seam.

Room and Pillar: Here, coal deposits are mined by cutting a network of "rooms" into the coal seam and leaving behind "pillars" of coal to support the roof of the mine.

Longwall Mining: This involves the full extraction of coal from a section of the seam or face using mechanical shearers. A long-wall face requires careful planning to ensure favourable geology exists throughout the section before development work begins.

Shortwall: method of mining coal is best described as a method similar to longwall mining with two exceptions. The blocks of panels are smaller, usually ranging from 100 to 150 feet wide and 300 feet long and the coal is cut with a continuous miner and is loaded into shuttle cars.

Processing and Utilization of Earth Resources

Here, the occurrences and principal applications of mineral and rock resources are discussed.

1)   Metals

The term mineral resource usually brings metals to mind first. The most heavily used metal is iron. And it is also one of the most common metals. Iron ore is mined from ancient sedimentary deposits, laterites and from concentrations of magnetite in some igneous bodies [20] .

Aluminum is the second most widely used. Its light weight, couple with its strength, make it particularly useful in the transportation and construction industries; it is also widely used in packaging especially for beverage cans. It is the third most common element in the earth's crust, but there, it is most often found in silicates, from which it is extremely difficult to extract. Bauxite is an aluminum-rich laterite in which the aluminum is found as a hydroxide and is commercially mined for the production of aluminum [21] .

Other important metals include copper, zinc, Lead, Nickel, Cobalt etc are found in sulfide ore deposits. Sulfides occur frequently in hydrothermal deposits and may also be concentrated in igneous rocks. Copper, Lead, and Zinc may also be found in sedimentary ores; some laterites are moderately rich in Nickel Cobalt [20] .

Metal refining consists of purifying an impure material. It is to be distinguished from other processes such as smelting and calcining in that those two involve a chemical change to the raw material, whereas in refining, the final material is usually identical chemically to the original ore, only it is purer. The processes used are of many types including pyrometallurgical and hydrometallurgical techniques [17] [21] .

2)   Nonmetallic minerals

Another by-product of mining sulfides is the nonmetal sulfur. Sulfur may also be recovered from petroleum during refining, from volcanic deposits (sulfur is sometimes precipitated as pure native sulfur from fumes escaping from volcanic vents).

Several important minerals are recovered from evaporate deposits. The most abundant is halite, or rock salt, used principally as a source of the sodium and chlorine of which it is composed, and secondarily for road salt, either directly or through the production of other salts from it. Gypsum, essential to the manufacture of plaster, Portland cement, and wallboard for construction, is another evaporate mineral. Others include phosphate rock and potassium-rich potash, key ingredients of the synthetic fertilizers.

However, clay is not a single mineral, but a group of layered hydrous silicates that are formed at low temperature, commonly by weathering, and they are abundant in sedimentary deposits.

Rocks such as granite, granodiorite, diorite and other basement complex rocks are usually exploited by excavation and quarried into aggregates for engineering purposes like construction of roads, buildings, dams, and many other important structures, see Figure 3. Aggregates are used in construction to provide drainage, fill voids, protect pipes, and to provide hard surfaces. They are also used in water filtration and sewage treatment processes. Water will percolate through a trench filled with aggregate more quickly than it will through the surrounding soil, thus enabling an area to be drained of surface water. This is frequently used alongside roads in order to disperse water collected from the asphalt surfacing. The three main types of rock used to produce crushed rock aggregates are igneous, sedimentary and metamorphic [17] .

Renewable Resources Exploitation

Hydropower

Hydropower makes use of the kinetic energy water gains when it drops in elevation; is a renewable energy source. Typically, water dammed in a lake or reservoir is released through turbines and generators to produce electricity, as shown in Figure 4. Hydropower has been a staple of electricity since the beginnings of the electric age. However, very little of this potential is currently slated for development. Significant legal and regulatory impediments, such as land acquisition and environmental protection, will be a part of any major hydro project. Additionally, reservoirs are typically built and managed as municipal water supply and flood control systems and secondarily for power production [22] [23] .

Figure 3: Rock aggregates and quarrying (modified from Adekoya, 2003).

Figure 4: Kainji-dam. Sources: www.nigeriaalternativeenergy.org.

Solar Energy

Solar energy is energy in the form of electromagnetic radiation emitted from the sun and harnessed using sophisticated technologies such as solar heating, solar photo-voltaics, solar thermal energy, solar architecture and artificial photosynthesis. It is one of the most important sources of renewable energy and its technologies are categorized as either passive solar or active solar depending on the way they capture and distribute solar energy or convert it into solar electricity (power). Active solar methods include the use of photovoltaic systems, concentrated solar power and solar water heating to tap the energy. Passive solar methods include orienting a building to the Sun, selecting materials with advantageous thermal mass or light dispersing properties, and designing spaces that naturally circulate air. Businesses and industry also use these methods to diversify their energy sources, improve efficiency, and save cost. Solar photovoltaic and concentrating solar power methods are also being used by developers and utilities to produce electricity to power houses both residential and commercial [23] [24] .

Geothermal Energy

Geothermal energy is the heat from the Earth. It is clean and sustainable. Resources of geothermal energy range from the shallow ground to hot water and hot rock found a few miles beneath the Earth's surface, and down even deeper to the extremely high temperatures of molten rock called magma. Of all the elements of a geothermal system, the heat source is the only one that need be natural. Providing conditions are favourable, the other two elements could be "artificial". For example, the geothermal fluids extracted from the reservoir to drive the turbine in a geothermal power-plant could, after their utilization, be injected back into the reservoir through specific injection wells [22] . In this way the natural recharge of the reservoir is integrated by an artificial recharge. For many years now re-injection has also been adopted in various parts of the world as a means of drastically reducing the impact on the environment of geothermal plant operations [24] .

Ocean Energy (Waves and Tides)

Three distinct types of ocean resource are commonly mentioned as possible energy sources: tides, waves, and ocean temperature differentials (ocean thermal energy conversion, or OTEC). For example, tidal energy schemes capture water at high tide and release it at low tide. Wave energy generation devices fall into two general classifications, fixed and floating. In both cases, the oscillating motion of an incoming and outgoing wave is used to drive turbines that generate electricity. Tide energy systems traps high tides in a reservoir. When the tide drops, the water behind the reservoir flows through a power turbine, generating electricity. Ocean thermal energy conversion uses the difference in temperature between warm surface water and cold deep ocean water to make electricity [21] [24] .

Surface and Groundwater

Water is usually contained naturally in aquifers or as surface water both of which are deposited by geological process known as water cycle. Exploitation is usually by drilling boreholes or wells to access their aquifers or direct access to surface sources such as rivers, lakes and so on. Water wells are excavations or structures created in the ground by digging, driving, boring, or drilling to access groundwater in underground aquifers. They are drawn by pumps, or by using containers, such as buckets, that are raised mechanically or by hand. Water may flow to the surface naturally after excavation of the hole or shaft. Such a well is known as a flowing artesian well. In numerous recently published state-of-the-environment reports, water turns out to be the factor most limiting sustainable development. In the next decade the availability of fresh water will dictate all agendas of development, not only in developing countries, and might cause political instability in larger dry regions on Earth. .The availability of fresh potable water is seriously restricted by:

1) Depletion of aquifers; and

2) Pollution of both surface and groundwater resources. The problem of aquifer depletion is partly caused by inadequate aquifer management (over-exploitation, poor recharge, etc.) but mainly by the fact that ever-

smaller quantities of fresh groundwater will have to supply more and more people on earth [21] [25] .

Environmental Problems of Earth Resources Exploitation and Their Mitigation

The most obvious and severe disruptions of the natural environment have come from the exploitation of resources. Mining, quarrying, dredging, and the drilling and extraction of oil and gas from wells are all activities that have made irreversible impacts on the landscape and therefore on the environment. Directly linked to the extraction activities are issues concerned with the disposal of the waste product that accompany extraction [1] .

Further environmental problems may be caused at the site of exploitation when various extraction or concentration processes are employed. For example, most metal mines remove an ore that contains only a very small proportion of the metal being extracted. Various physical and chemical processes, often culminating in smelting, are then needed to extract the metal from the mined ore. For metal extraction, at least three aspects of the process create potential environmental problems: the mining operation, the disposal of very large quantities of waste rock, and the processing, smelting, and refining of the ore [26] [27] .

Environmental Impact of Oil and Gas Exploitation

The exploitation of Oil and gas drilling can pose adverse health and environmental impacts, from contamination of surface and underground water system with drilling mud and fluids. Offshore and onshore drilling can cause blowout and put the life of oil workers at risk. Spills and leakages may occur and pollute ocean waters or close land areas, either as a result of industrial accidents or through storm damage to drilling rigs or Formation pressure upsurge. Oil spills from wells are not uncommon and can pollute vast areas both offshore and onshore, generating clear and measurable environmental impacts. In Nigeria, an estimated 260,000 barrels of oil (41,000 m^3) spill each year into the Niger Delta and surrounding areas, with devastating impacts on people, plants, and wildlife. Marine spills such as these can lead to freshwater system contamination when the oil hits the shoreline and drifts up through estuaries into streams. From extraction to end use, petroleum products affect surface water and groundwater, impairing water quality with hydrocarbons, salts, nutrients, a host of organic compounds, and various heavy metals. Movement of oil and gas from wells to processors needs enormous facilities and such creates environmental risks. Pipelines and tankers are used in the world today to convey oil and gas, both of which are prone to spills. Producers convey crude oil from the well via pipelines or ships to refineries, where crude oil is distilled into a variety of petroleum products, including gasoline, kerosene, fuel oils, liquefied petroleum gas, various lubricants, asphalt, and precursors to plastic and pharmaceutical products, among others. These petroleum products are then distributed via various modes to other manufacturers and to end users. Each of these steps can affect surface water sources negatively. After production, crude oil is refined through a series of water-intensive processes: water is used for steam, as part of the refining process, as wash water, and for cooling. Process water typically becomes contaminated with sulphur and ammonia, requiring treatment [13] [28] .

Impacts of Mining and Quarrying

Most of the impacts arising from mining and quarrying are obvious. They include the disruption of land otherwise suitable for agricultural, urban, or recreational use; the deterioration of the immediate environment through noise and airborne dust; and the most dangerous for its workers and is potentially also hazardous for the public.

The generation of sulfuric acid can occur when hazardous minerals are exposed to air in underground mines or in open pits, or in the dumps of waste materials left by mining operations. Water passing through the mines or dumps becomes acidified and then can find its way into rivers and streams or into the groundwater system [16] .

Acid Rain

Rainfall is generally thought of as beneficial, washing and watering Earth's surface. While this is still true in many places, the image of rainfall has become tarnished because we now know that rainfall is not always pure water and its effects are not always beneficial. As early as 1872, the British chemist Robert Smith coined the term "acid rain" to describe precipitation containing significant amounts of sulfuric acid. He attributed the acid to the burning of the coal that fueled the furnaces of the Industrial Revolution.

Rainwater naturally contains dissolved carbon dioxide from the atmosphere that forms weak carbonic acid, H_2CO_3. This acid ionizes, forming H^+ and HCO_3^- ions, resulting in rainfall that is normally slightly acid with a PH of about 5.6. The increase of carbon dioxide due to the burning of fossil fuels has some significant effects on atmospheric warning, but lowers pH only slightly [29] [30] .

However, the introduction of large amounts of sulfur oxides and nitrogen oxides into the atmosphere by the combustion of fossil fuels has a significant impact on the acidity of rainwater. Sulfur oxides, which react with water vapor to form sulfuric acid (H_2SO_4), are created when sulfur bearing-minerals in coal are burned. Nitrogen oxides, which react with water vapor to form nitric acid (HNO_3), are created as an unintentional by-product of high temperature combustion in an atmosphere containing 78 percent nitrogen. Small amounts of these two strong acids in rainfall can sharply lower the pH of rainwater [28] .

Acid Mine Drainage

This is refers to the outflow of acidic water from abandoned metal mine or coal mine.

Acid mine drainage is also the formation and movement of highly acidic water rich in heavy metals, see Figure 5. This acidic water forms through the chemical reaction of surface water and shallow subsurface water with rocks that contain sulfur-bearing minerals, resulting in sulfuric acid. Heavy metals can be leached from rocks that come in contact with the acid, a process that may be substantially enhanced by bacterial action. The resulting fluids may be highly toxic and, when mixed with groundwater, surface water and soil, may have harmful effects on humans, animals and plants [19] .

Impacts Mitigation

Mining is a relatively short-term activity, and much can be done to both limit environmental damage during mining and to restore the land when mining operations are complete:

All mining operations generate waste rock, often in very large amounts; the waste can be used in reclamation and backfilling when treated [1] .

Figure 5: Acid mine drainage. Source: www.worldcoal.org/coal/coal-mining/.

Meanwhile, planning the mining layout so as to have the least requirement of the forest land and take necessary steps for reclamation of the mined out land so that the forest land taken for mining purposes can be brought back to forest.

The noise and vibration producing activities in the mines and associated activities be planned to have the minimum possible intensity and impact on the living organisms in the surrounding area.

Use of dust extractors with the drill machines can be expected to minimize air pollution due to drilling. However, optimizing the blast design and proper maintenance of the haul roads can minimize the generation of air borne dust [16] .

Countries, both developing and developed, should aim at using cleaner, less polluting technologies in earth resources exploitation. Use of these technologies aids mitigation and could result in substantial reductions in CO_2 emissions. Also, use of scrubbers to reduce the escape of obnoxious gases into the atmosphere during burning of fossil fuels [30] .

CASE HISTORY OF ENVIRONMENTAL ASPECTS OF AGGREGATE EXPLOITATION IN IMO STATE, NIGERIA

Several organization and individuals are involved in aggregate production in Imo State. Most of the organizations obtain quarry leases from Imo State Government to operate quarries where construction materials are mined and processed [31] [32] . Some of the organizations involved in aggregate production in Imo State are shown as Table 1.

According to Okeke, the locations of the quarries are related to the geology of the area [31] . Two types of natural aggregates are produced in Imo State, and they include: sand/gravel aggregate from Benin Formation in relatively flat low-lying southern part of the state (Owerri, Isu and Oru L.G.As) and crushed sandstone in the undulating northern part of the state (Okigwe and Ihite Uboma L.G.As). Sands and gravels are excavated by hand tools of shovels and diggers on land and by panning and dredging from rivers (particularly rivers Otamiri and Njaba). Sand and gravels are usually produced together but later screened to separate them into different fractions. The quarrying of sandstone in Okigwe and Ihite Uboma areas involve three processes of removal of top soil, excavation of sandstone boulders, and crushing to suitable commercial sizes of usually 25 mm and 50 mm.

Coal Mining in Enugu Nigeria

In the Enugu coal mines, both the Onyeama mine and Okpara mine were flooded after the Nigerian crises (of 1967-1970). Coal was first discovered in Enugu in 1907. The Enugu coal is associated with the highly fractured

Table 1: Organizations involved in aggregate production in Imo state

Organization	Type of aggregate	Quarry location
Owerri Sand/Gravel Association	River sand/gravel	Nekede and Ihiagwa (Owerri West LGA)
Bathoway Dredging	River sand	Nekede
S. O Amadi and Co	Sand/gravel	Nekede
Electroshed Dredging	Sand/gravel	Nekede
Awomamma Sand/Gravel association	Sand/gravel	Awomamma (Oru East LGA)
Okwudor Sand/Gravel Association	Sand/gravel	Okwudor (Njaba LGA)
Ikperejere Crushed Rock Association	Sandstone	Ikperejere (Ihite Uboma LGA)

(Modified from Okeke, 2001).

Mamu Formation (Lower coal measures) with intercalations of shale and sandstone [33] . The Mamu Formation itself was underlain by Enugu Shale and overlain by frible Ajali Formation [34] . The Onyeama mine was opened in 1958 and mechanized in 1977. Unfortunately the longwall mining method at Onyeama mine intercepted the major fault systems that contributed to the flooding of the mines. The Okpara mine was also flooded by surface drainage through the permeable Ajali Formation. Only Okpara mine is in operation at the moment at Enugu. This is because its flooding was not as bad as that of Onyeama mine and its always controlled [35] [36] .

SUMMARY AND CONCLUSION

Earth resources are materials provided by the Earth and useable to humans to make more complex (human- made) products. In a more simplified explanation, earth resources also referred to as natural resources which are useful raw materials gotten from the Earth. Earth resources are basically divided into renewable, flow and non- renewable resources. Sustainable development has to do with meeting the basic needs of people today without ruining the chances of future generations to do the same. Exploitation of earth resources produces so many benefits to mankind as well as associated problems ranging from environmental impacts to their depletion. The issue of depletion plays an important role in the use of non renewable and renewable earth resources. For renewable resources, depletion occurs when extraction exceeds renewal rate. Flow and environmental resources are not depleted and always exist. Major consequence of our exploiting earth resources is that we are interfering with the balance of some natural geochemical cycles (for example, carbon/sulphur cycles). Geological, engineering, environmental and economic factors control mineral availability. The adequacy of world mineral reserves and resources is strongly affected by consumption, stockpiles and recycling. However, mineral recycling and substitution alongside transfer of technology from industrially developed to developing nations will contribute significantly to sustainable development of these resources [37] . Geoscientists can and must contribute to solving the basic problems of mankind in the next decade, even more than today.

ACKNOWLEDGEMENTS

We happily acknowledge the effort of Engr. Dr. O. C. Okeke whose guidance made this work a huge success.

REFERENCES

1. Craig, J.R., Vanghorn, D.J. and Skeinner, B.J. (2011) Earth Resources and the Environment. 72-88.

2. WCED (1987) Our Common Future. World Commission on Environment and Development United Nations.

3. Robinson, J.G., Francis, G., Legge, R. and Lemer, S. (1990) Defining a Sustainable Society. Alternatives, 2, 36-46.

4. Marconick, R. (1990) Environment and Concept of Sustainable Development. Proceedings, Workshop on the Environment and Sustainable Development in Nigeria, 25-26 April 1989, Abuja, 44-52.

5. Lawrence, D.P. (1997) Integrating Sustainability and Environmental Impact Assessment. Environment Management, 21, 23-42.

6. http://dx.doi.org/10.1007/s002679900003

7. McNicoll, G. (2007) Population and Sustainability. Handbook of Sustainable Development. Edward Elgar Publishing, 125-139.

8. http://dx.doi.org/10.4337/9781847205223.00016

9. Milbraith, L.W. (1989) Envisioning a Sustainable Society. State University of New York Press, Albany, 400 p.

10. Nju, W.Y., Lu, J.J. and Khan, A.A. (1993) Spatial Systems Approach to Sustainable Development: A Conceptual Frame Work. Environmental Management, 17, 179-186.

11. Pedro, A.M.A. (2004) Mainstreaming Mineral Wealth in Growth and Poverty Reduction Strategies. Economic Commission for Africa, Addis Ababa, 5-6.

12. Pegg, S. (2006) Mining and Poverty Reduction: Transforming Rhetoric

into Reality. Journal of Cleaner Production, 14, 376-387.

13. Planas, F. (2012) The Exploitation of Environmental Resources. Un An Pour La Planete.

14. WCED (1992) Our Common Future. World Commission on Environment and Development United Nations.

15. Environmental Protection Agency (2010) Assessing the Multiple States, Chapter 5.

16. Weber-Fahr, M., Strongman, J., Kunanayagam, R., McMahon, G. and Sheldon, C. (2001) Mining and Poverty Reduction. Noor Internationaal WB PRSP Sourcebook, 4-6.

17. Bray, J. (2003) Attracting Reputable Companies to Risky Environments: Petroleum and Mining Companies. Environmental Resources and Conflict: Options and Actions. World Bank Publications, 287-347.

18. Carla, W.M. (2011) Environmental Geology. McGraw-Hill Publishing, 511 p.

19. Adekoya, J.A. (2003) Environmental Effect of Solid Minerals Mining. J. Phys. Sci. Kenya, 625-640.

20. Craig, F. and Jonathan, S. (2001) How Oil Drilling Works. Newsletter of Science and Environmental Science.

21. www.worldcoal.org/coal/coal-mining/

22. Day, J. and Tylecote, R.F. (1991) The Industrial Revolution in Metals. Institute of Metals, London.

23. John, P.H., Barry, J.H., John, F.P., William, P.I. and Ramachandran, V. (1994) Extraction and Processing for the Treatment and Minimization of Wastes. The Mineral, Metals and Materials Society, Pennsylvania.

24. www.nigeriaalternativeenergy.org

25. Ibidapo-Obe, O. and Ajibola, O.E. (2011) Towards a Renewable Energy Development for Rural Power Sufficiency.

26. Temilade, S. (2008) Status of Renewable Policy and Implementation Nigeria. Institute for Science and Society, University of Nottingham, Nottingham.

27. Gleick, P.H. (1993) Water in Crisis. In: Gleick, P.H., Ed., A Guide to the Worlds Freshwater Resources. Pacific Institute for Studies in Development, Environment and Stockholm Environmental Institute, Oxford University Press, London.

28. Balkau, F. (2000) Cyanide Management in Gold Mining: An International Code. International Council on Metals and the Environment Newsletter,

8, 4.

29.   Brereton, D. and Forbes, P. (2004) Monitoring the Impact of Mining on Local Communities: A Hunter Valley Case Study. CSRM, 12-13.

30.   Aigbedion, I.N. (2005) Environmental Pollution in the Niger-Delta, Nigeria. Inter-Discplinary Journal of Enugu-Nigeria, 3, 205-210.

31.   Federal Environment Protection Agency (FEPA) (1989) National Policy on the Environment. Federal Environment Protection Agency Abuja, Nigeria.

32.   Federal Environment Protection Agency (FEPA) (1991) National Interior Guidelines and Standards for Industrial Effluents, Gaseous Emission and Hazardous Waste Management in Nigeria. Federal Environment Protection Agency, Abuja.

33.   Okeke, O.C. (2001) Assessing the Sustainability of Aggregate Production in Imo State, Nigeria. Journal of Management Technology, 3, 183-189.

34.   Okeke, O.C. and Uzoh, O.F. (2009) Towards Achieving Sustainable Water Resources Management in Nigeria. Global Journal of Geological Sciences, 7, 85-92.

35.   http://dx.doi.org/10.4314/gjgs.v7i1.45162

36.   Simposn, A. (1954) The Nigerian Coalfield. Geological Survey of Nigeria Bulletin, Abuja.

37.   Reyment, R.A. (1965) Aspects of Geology of Nigeria. Ibadan University Press, Ibadan.

38.   Okeke, O.C. (2005) Grouting and Dewatering in Engineering Practice. ADSON Educational Publishers, Onitsha.

39.   Okeke, O.C. (1991) Role of Mineral Raw Materials in the Development of Indigenous Building Materials Industry in Nigeria. Journal of Mining Geology, 27.

40.   (2011) International Conference on Innovations in Engineering and Technology (IET), 8-10 August 2011.

Chapter 6

APPLICATION OF GEOPHYSICAL METHODS FOR GEOTECHNICAL PARAMETERS DETERMINATION AT NEW BORG EL-ARAB INDUSTRIAL CITY, EGYPT

Alhussein A. Basheer, Abdelnasser M. Abdelmotaal, Hany S. Mesbah, Khamis K. Mansour

Geomagnetic and Geoelectric Department, National Research Institute of Astronomy and Geophysics, Helwan, Cairo, Egypt

ABSTRACT

Due to a rapid increase in the population during the last few decades, the banks of the Nile River and its delta have reached maximum capacity. As a consequence of this increase, the Egyptian Government has constructed a number of new urban areas and industrial cities outside the Nile Delta. New Borg El-Arab City is one of these new industrial cities. This city is located around 60 km southwest of Alexandria City. This industrial city is proposed to include an airport, a number of factories, worker settlements and heavy truck roads. Therefore, a detailed study of site characterization should be performed before construction being in order to. The main target of this study is to determine the dynamic characteristics and geotechnical parameters at the proposed site using seismic refraction and electrical resistively techniques. Analysis and interpretation of the obtained results reveal that the subsurface consists of three layers with a gentle general slope toward the Mediterranean Sea. The classification of rock material for engineering purposes reveals that the study area is divided into three zones.

INTRODUCTION

New Borg El-Arab industrial zone is one of the new industrial cities in Egypt. It lies between latitudes 30.75043221 and 31.04684093 N and longitudes 29.44052705 and 29.6723497 E, covering an area of about 90 km² (Figure 1). This new industrial city is an urbanization project that is relatively large-scale regional development project and it will be good if the determination the dynamic characteristics and geotechnical parameters being incorporates in the planning stage. So, the main target of the present study is investigating the shallow subsurface structure conditions, the dynamic characteristics and geotechnical parameters of subsurface rocks using shallow seismic refraction profiling and electrical resistivity tomography surveys. The output of this study is very important for solving problems, which associated with the construction of various civil engineering purposes, land use-planning and earthquakes resistant structure design. The integrated interpretation of these techniques classifies the subsurface succession into three layers.

GEOLOGICAL SETTING

According to (El Shaazly, 1964), the geological setting of the study area is similar environment consisting of Quaternary deposits (scattered), Oligocene (Sand), Beniolitic Clay, and Middle Miocene as Limestone (from top to bottom), Figure 2. The geomorphological units west of Alexandria showed by (Abd El Mawla, 2010) as in Figure 3.

METHODOLOGY

In the present study two from the advanced technuqes have been applied; electrical resistivity tomography and shallow seismic refraction. Electrical resistivity tomography is a geophysical technique for imaging and mapping the vertical and horizontal sub-surface structures from electrical resistivity measurements made at the surface. The shallow seismic refraction technique is considered as one of the most effective geophysical method, which can be applied in the field of engineering seismology i.e. tunnels, dam sites, landslides, quarries, roads, reclemaited lands, caves and cavities.

Electrical Resistivity Tomography

In the ERT method the distribution of the electrical resistivity of the subsoil is obtained by injecting electrical current (by the current electrodes) into the ground and measuring the potential difference (by the potential electrodes) at two determined points of the surface. The method is based on the application of Ohm's law:

$$\rho_a = k\left(\Delta V / I\right)$$

(1)

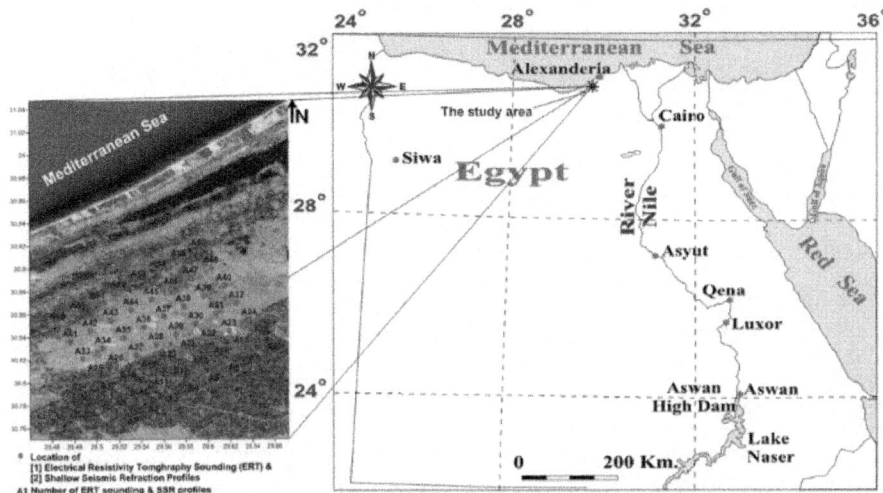

Figure 1: Location map of the study area and the site of ERT and SSRT stations.

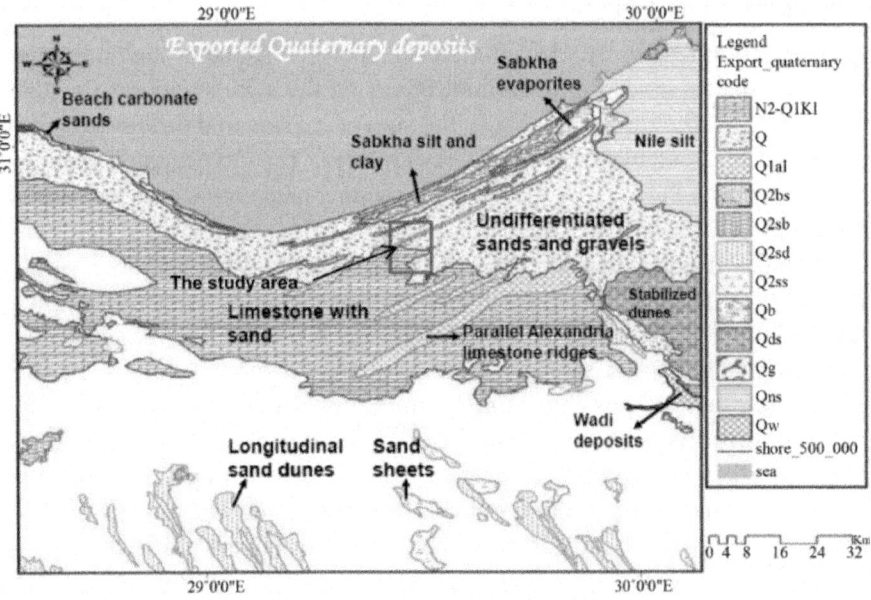

Figure 2: Geological map of the study area, west of Bourg El-Arab (after GNBCC, 2012).

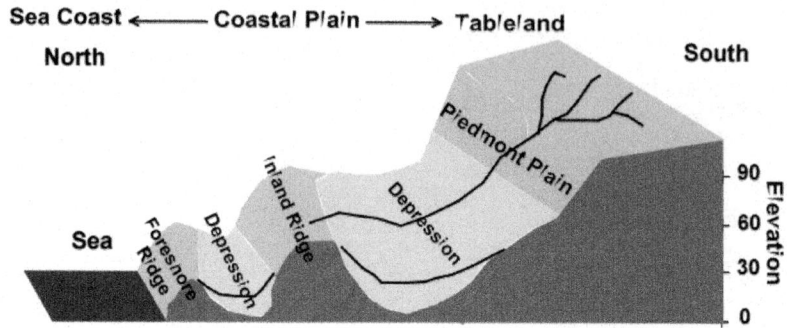

Figure 3: Geomorphological units west of Alexandria (after Abd Elmawla, 2010).

where ρ_a is the apparent resistivity (Parasnis, 1997), k is a geometric constant that depends only on the reciprocal positions of the current and potential electrodes; ΔV is the measured potential difference, and I is the intensity of the injected current. The apparent resistivity values depend on the true resistivity distribution.

The true resistivity distribution in the investigated medium can be estimated by an inversion procedure based on the minimization of a suitable function. This function is generally the sum of the squared difference between measured and calculated apparent resistivities. The investigated medium is discredited in a 2D (or 3D) grid of cells, where each cell is assigned an initial resistivity value. A finite-difference (Dey & Morrison, 1979a, b) or finite-element (Silvester & Ferrari, 1990)procedure computes the predicted apparent resistivity at the surface. The solution to the problem, as is well known, is not unique. For the same measured data set, there is wide range of models that can give rise to the same calculated apparent resistivity values. To narrow down the range of possible models, normally some assumptions are made concerning the nature of the subsurface (i.e. geology of the subsurface, whether the subsurface bodies are expected to have gradational or sharp boundaries) that can be incorporated into the inversion subroutine. The current method of solution minimizes the difference between measured and calculated apparent resistivities using the smoothness-constrained inversion formulation, which constrains the change in the model resistivity values to become smooth (Loke & Barker, 1996; Loke, 2001). The "smoothness-constrained robust inversion" method (Loke, 2001) has proved to be much more useful when the subsurface bodies have sharp boundaries (Loke, 2001).

ERT Data Acquisition

The ERT measurements are carried out along 56 profiles oriented approximately

N-S (Figure 1). For this type of data set, a series of 2D inversions is usually first carried out. The 3D inversion is then used on a combined data set with the data from all the survey lines. Apparent resistivity data were collected using the Syscal R2D instruments in multi-electrode configuration. The distance between profiles was 2 km and a total of 24 electrodes spaced at 5-m intervals are employed. Data are collected using the pole-pole array. The pole-pole array was chosen because, as is well known, it is very sensitive to horizontal changes in resistivity, and is therefore suitable for mapping vertical and horizontal structures (Loke, 2001). Figure 4 shows the principle of ERT technique.

ERT Data Processing and Interpretation

The least-square inversion by Jacobian matrix calculation was used in order to obtain an image of the true resistivity distribution as a function of depth. The least-square inversion method (Loke, 2001) minimizes the absolute changes (11-norm) in the model resistivity values (Loke, 2001). All pole-pole inverted resistivity sections produced low RMS errors (<13%). The Res2dinv software(Loke, 2001) is used to invert all 2D and 3D data sets. Figure 5 shows an example for inverted resistivity section with tomography data along profiles "1, 17, and 56".

As the measured sections are straightforward to combine all 2D images and produce resistivity slices for different depths. The Res3dinv resistivity inversion software (Loke, 2001) is used to automatically invert the apparent resistivity acquired data and to yield a three-dimensional resistivity model. The slicer-cubic software (Samuil, 2008) is used to automatically draw output interpreted resistivity data from Res3dinv software to acquiesce a three-dimensional resistivity model (Figure 6). From this figure it can be concluded that, the topmost layer presents a larger area with high resistivity variations over short distances (about 2100 Ωm to 1500 Ωm). In comparison, second layer, lying at a depth ranging from 13.35 to 16.75 m, shows more gradual lateral variations in the model resistivity values, this layer presents moderately resistivity values (ranging from 1500 Ωm to 600 Ωm). Third layer lying at a depth varies between 22.25 and 24.60 m. The resistivity values of it are clearly visible, ranging from about 600 Ωm to about 200 Ωm.

Shallow Seismic Refraction Tomography (SSRT)

Many researchers have used seismic refraction technique to determine the characteristics of the sites, and the necessary parameters for constructions (e.g. (Dutta, 1984), (Marzouk, 1995),(Mohamed, 1993), (El-Behiry, 1994), (Hatherly, 1986), (Sjogren & Sandberg, 1979) and(Abdelmotaal, 2010)).

Seismic exploration involves generation of seismic waves and recording the arrival times of these waves from the source to the series of geophones (Figure 7).

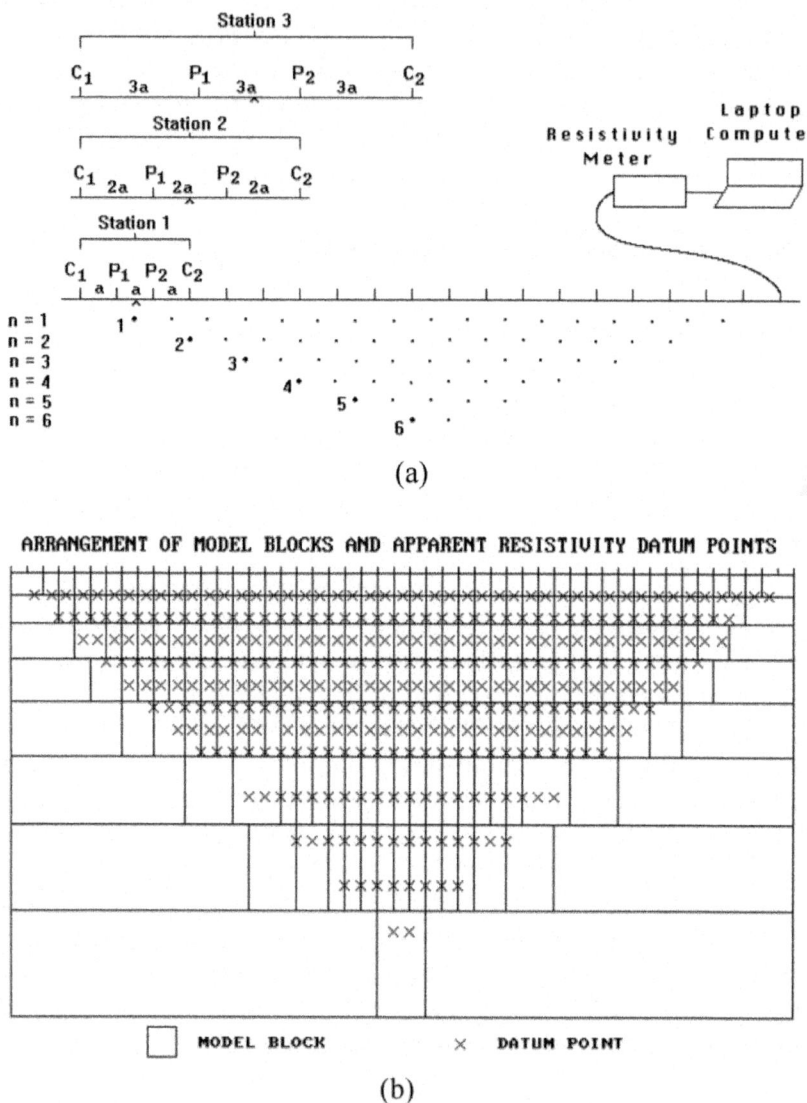

(a)

(b)

Figure 4: (a) Sequence of measurements to build up a pseudo section using a computer controlled multi-electrode survey setup; (b) Arrangement of blocks used in a model together with data points in pseudo section.

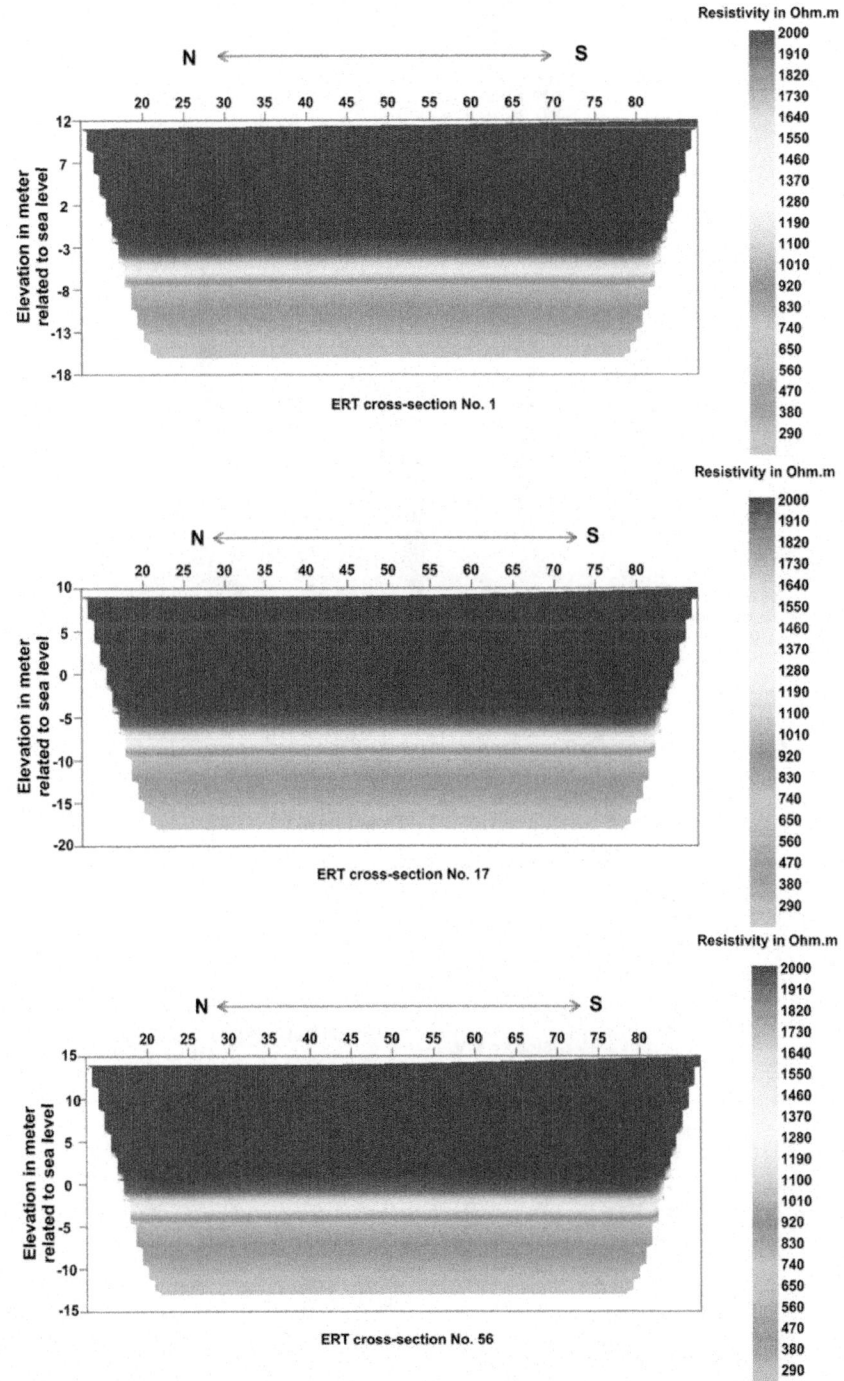

Figure 5: Two-dimensional resistivity topographical profiles No. 1, 17, and 56.

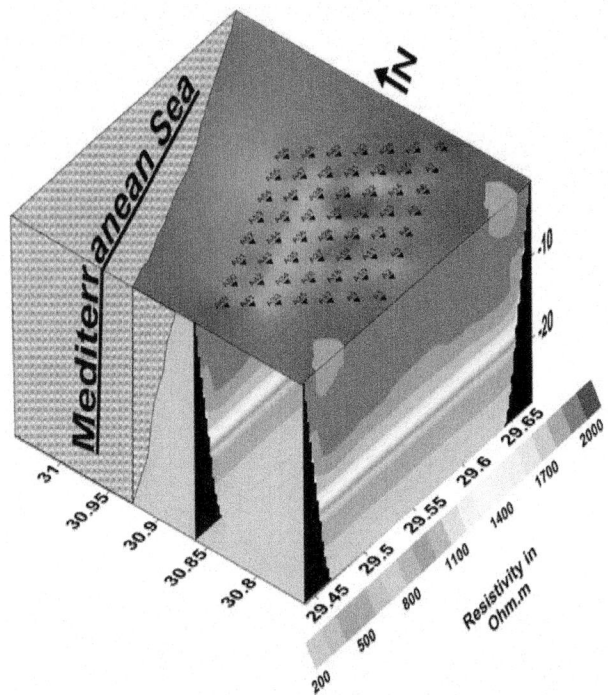

Figure 6. Three-dimensional resistivity model of the study area.

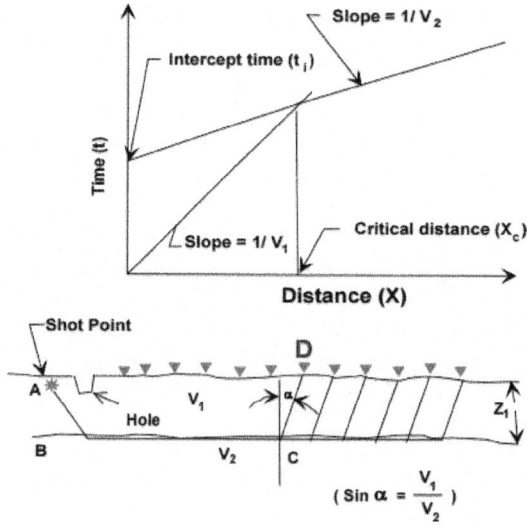

Figure 7: Velocity estimation using the slope method calculation with shot point and geophones array in field.

Seismic refraction is used to evaluate the necessary parameters for constructions, or to solve the problems related to the geological nature of sub-surfaces, mining works, and the environmental conditions overcame in the site (Stumpel et al., 1984).

In the shallow seismic refraction method, the seismic waves, created by artificial sources such as an auto-hammer viberator, propagate through the medium and are refracted at interfaces, where the seismic velocity or density changes. Geophones laid on a single line record the waves returning to the surface after travelling different distances through the ground. By measuring the travel time between the break and the recording of a seismic signal, the seismic velocity in the subsurface and the depth of the interfaces may be inferred. Figure 7 schematic diagram shows the principle of the shallow seismic refraction.

Conventional analysis of seismic refraction data sets makes simplified assumptions about the velocity structure that conflict with observed heterogeneity, lateral discontinuities, and gradients(Parasnis, 1997). Refraction tomography is designed to resolve velocity gradients and lateral velocity changes, enabling it to be applied in settings where traditional techniques fail.

The method used in this paper utilizes non-linear travel time tomography consisting of ray tracing for forward modeling and simultaneous iterative reconstruction technique (SIRT) for inversion. In this method, the velocity model is represented by quadrangle cells (Figure 8). The width of each cell is chosen as the receiver interval. First-arrival travel times and ray paths are calculated by the ray tracing method based on Huygen's principle (Parasnis, 1997). A ray is expressed as a line connecting the nodes arranged on the cell and the travel time between a source and a receiver is defined as the fasted travel time of all ray paths (Figure 8). A model is updated by the SIRT (for more information see (Morey & Schuster, 1999), (Nemeth et al., 1997).

SSRT Data Acquistion

The target to be executed was the determination of dynamic properties of soil and foundation rocks by using seismic refraction method and this implied firstly to throw light upon the end product of seismic refraction technique that was used in this study. Shallow seismic refraction technique was applied in the present study to determine the subsurface layering and its dynamic characteristics at the study area. The seismic refraction data have been acquired along 56 shallow seismic refraction profiles spread over the same area proposed for the R2D resistivity imaging survey (Figure 1). The survey has been accomplished using 24 channels signal enhancement seismograph "GEOMETRICS SMARTSEIS" along 120 m length profiles. Three shots are selected on the seismic line for measuring perpendicular and surface waves: two near shots at the edges of the line, and one shot at the middle of the line. In the surface P and SH wave velocities have been measured. The hammer connects to sensor to provide the time break to the seismograph. The power of the hammer helps to avoid loosing of waves strength that may be caused by Blind layer in some condition. The low pass filter in the recording system has 7 - 10 MHz as frequency response, which is suitable for the recording condition, positioned before the analog-digital conversion circuit. Data were stacked at least three times for each source. Figure 9 shows the measured raw data (typical recorded seismic traces) of profiles number 1, 17, and 56 as examples.

SSRT Data Processing and Interpretation

For interpretation of refraction data an initial depth model is obtained using Gardner's method(Gardner, 1939). To perform the processing and interpretation of the seismic refraction tomography data in the current study, (SIPEEDIT (V. 3) 2002) software developed by OYOO Company is used. Therefore, an initial velocity model is estimated. In this case, the initial velocity model is represented by the results obtained from the simple interpretation of refraction data (such as the model shown in Figure 9). The model is represented by 1-1 m cells. Refraction rays are traced through this model to give calculated travel times. A misfit function, consisting of the squared difference between the observed and computed travel times, is calculated. The model is adjusted until the misfit is minimized. The iterations are stopped when the RMS travel time residual (difference between the calculated travel times for the initial model and the observed ones) is less than the average travel time pick error.

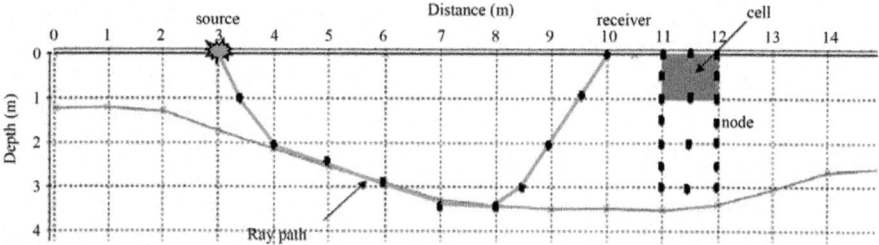

Figure 8: Principle of the ray tracing.

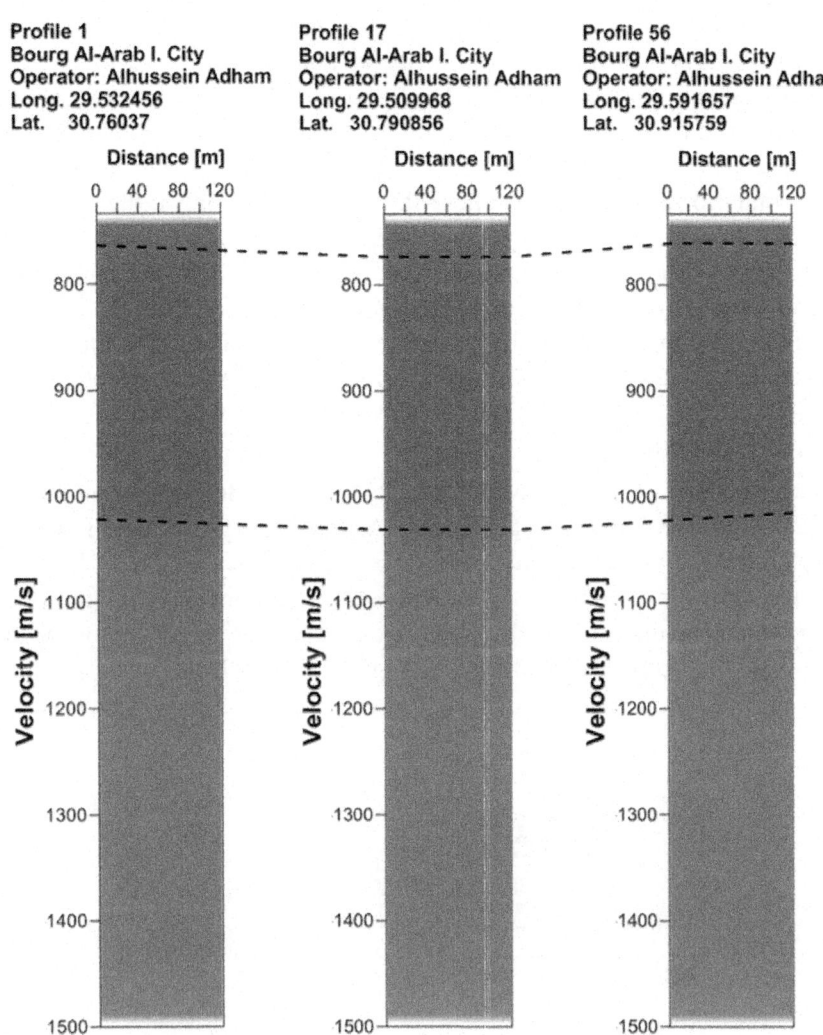

Figure 9: Typical recorded seismic traces for lines 1, 17, and 56.

SSRT Velocity

The model generated by seismic refraction tomography shows the variation of the Vp and Vs in the subsoil. The seismic refraction tomography survey points out that the shallow subsoil may be divided into three main layers. The surface layer exhibits a P-wave velocity of about 734 to 772.2 m/s, and a Vs velocity of about 389 to 408 m/s, these velocities correspond to surface sandy weathered zone with thickness ranges from 11 to 13.4 m. This layer corresponds to the area in which the ERT profile shows high resistivity values (Figures 5 and 6).

These velocity values are typical of more weathered subsoil (Leucci, 2004). The second layer shows a P-wave velocity ranges from 1000 to 1120 m/s, and a Vs velocity ranges from 521 to 606 m/s, these velocities correspond to calcareous sands weakly cemented with thickness varies between 15 to 16 m. This layer is characterized by the mainly high seismic velocities (Vp and Vs), corresponds to the area about where the ERT profile shows the moderately low resistivity values. The third layer shows a P-wave velocity between 1400 and 1550 m/s, and a Vs velocity between 669 to 781 m/s, those correspond to sandy marl layer strongly cemented with depth vary from 25 and 26 m where the ERT profile shows the moderate resistivity values. The third layer is characterized by the highest seismic velocities (Vp and Vs), corresponds to the area about where the ERT profile shows the lowest resistivity values. The grey zone presents the no-ray area.

The increase of seismic wave velocity could be due to a more compact soil (calcareous sands weakly cemented). According to the variation in Vp and Vs velocities are detected, four cross-sections, over the study area, have been drawn to show the distribution of these layers in the four sides (Figures 10-13). Three dimension maps of the distribution of Vp and Vs velocities for the three layers over the study area are shown in Figures 14 and 15.

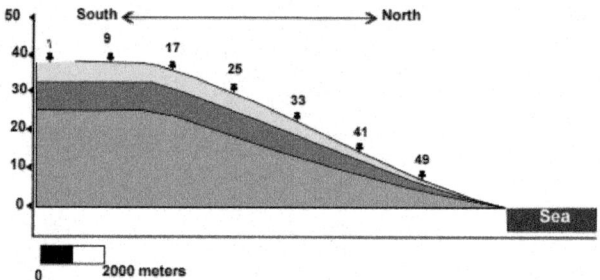

Figure 10: Depth cross-section along profiles (1, 9, 17, 25, 33, 41, and 49).

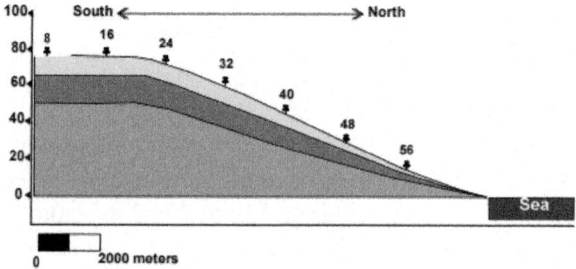

Figure 11: Depth cross-section along profiles (8, 16, 24, 32, 40, 48, and 56).

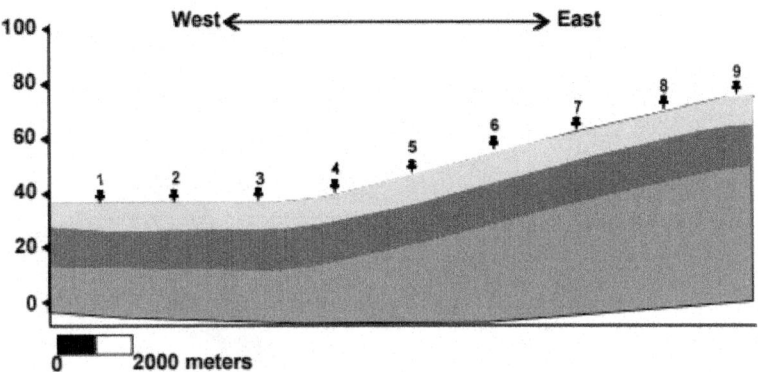

Figure 12: Depth cross-section along profiles (1, 2, 3, 4, 5, 6, 7, 8, and 9).

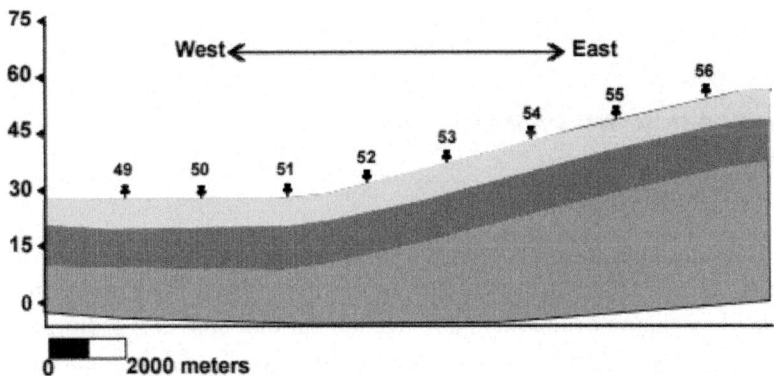

Figure 13: Depth cross-section along profiles (49, 50, 51, 52, 53, 54, 55, and 56).

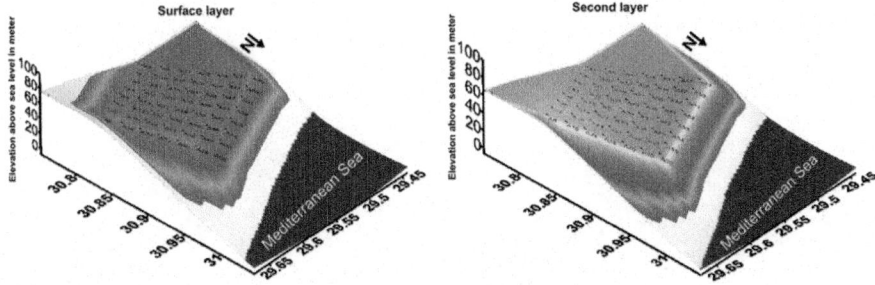

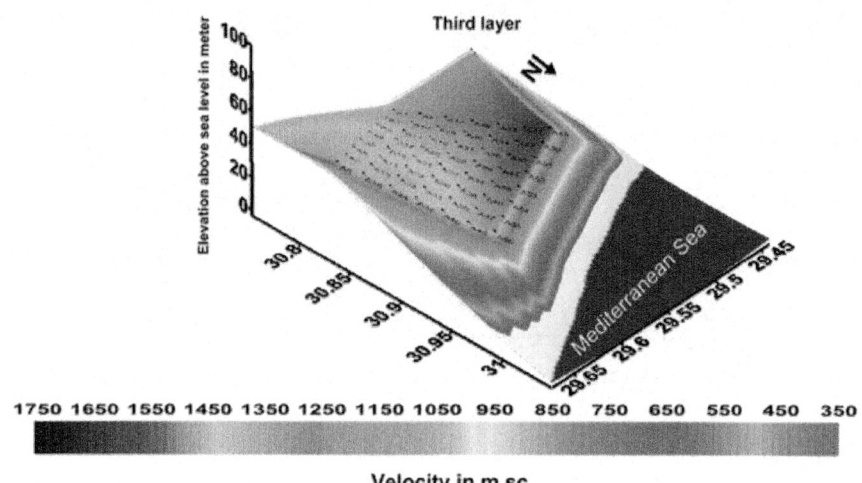

Velocity in m.sc

Figure 14: Three-dimension velocity map for seismic perpendicular waves.

GEOTECHNICAL PARAMTERS

Density Gradient (Di)

There is a direct relation between the material competence and the value of the density gradient and the compression velocity (Vp). "The compaction" can be defined as "name of the physical mechanism that converts sediments from their initial state to a progressively denser state under the influence of their own weight or of tectonic movements" (Cordier, 1985). The density gradient (Di) relates to how much consolidation or settlement will take place. Many relations can express it "according to Stumpel et al. (Stumpel et al., 1984)"

$$Di = \left(Vp2 - 4/3\,Vs2\right) - 1 \qquad (2)$$

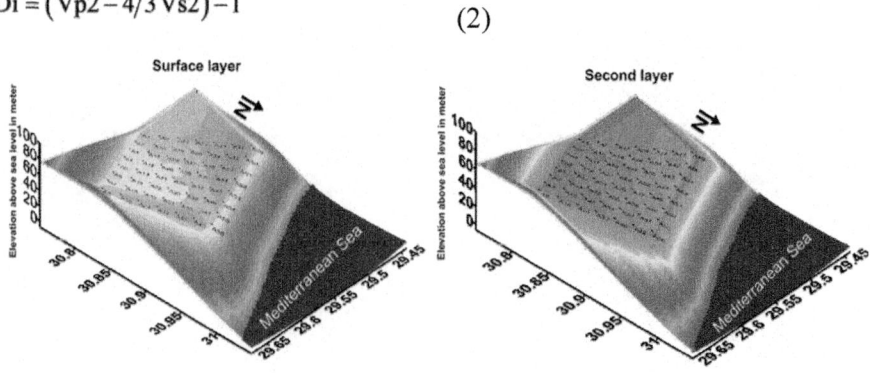

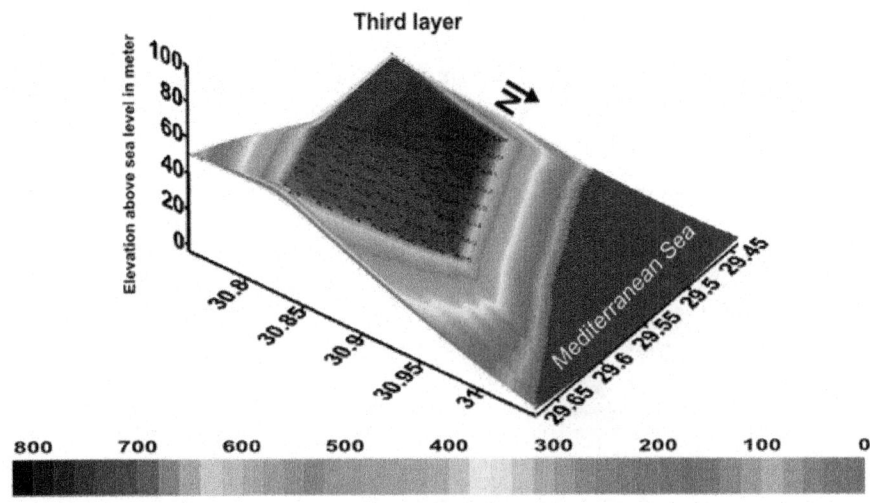

Figure 15: Three-dimension velocity map for seismic surface waves (Vs).

The distribution of the density gradient in the subsoil layers has been shown in Figure 16. Generally over the three layers, the relatively low values are observed in the southeastern part of the study area, which indicates the relatively moderate competent materials. The middle values are occupied in the middle part that may be due to difference in sorted materials. While the high values are observed in the northwestern part of the study area, they reflect the fairly moderate competent materials in this part of the study area. Table 1 shows the distribution of the density gradient.

Standard Penetration Test (SPT) [N-Value]

The Standard Penetration Test (SPT) is geotechnically known as "the resistance to penetration by normalized cylindrical bars under standard load". N-value is geophysically evaluated using Imai's(Imai, 1976; Stumpel et al., 1984), which is given by the formula:

$$Vs = 89.9N0.341 \qquad (3)$$

Table 2 shows the distribution of N-Value. In general for the three layers of the study area, Figure 17 shows the comparatively low values that are experimented in the southeastern part of the study area, which point toward the relatively moderate competent materials. The middle values are inhabited in the middle part that may be suitable to difference in sorted materials. Whereas the high values are observed in the northwestern part of the study area, they reflect the fairly moderate competent materials in this part of the study area.

Kinetic Bulk Modulus (K)

"Sheriff" defined the Kinetic Bulk modulus as "the stress-strain ratio under simple hydrostatic pressure" (Sheriff, 1991). The law can calculate the different values of Bulk modulus to each layer:

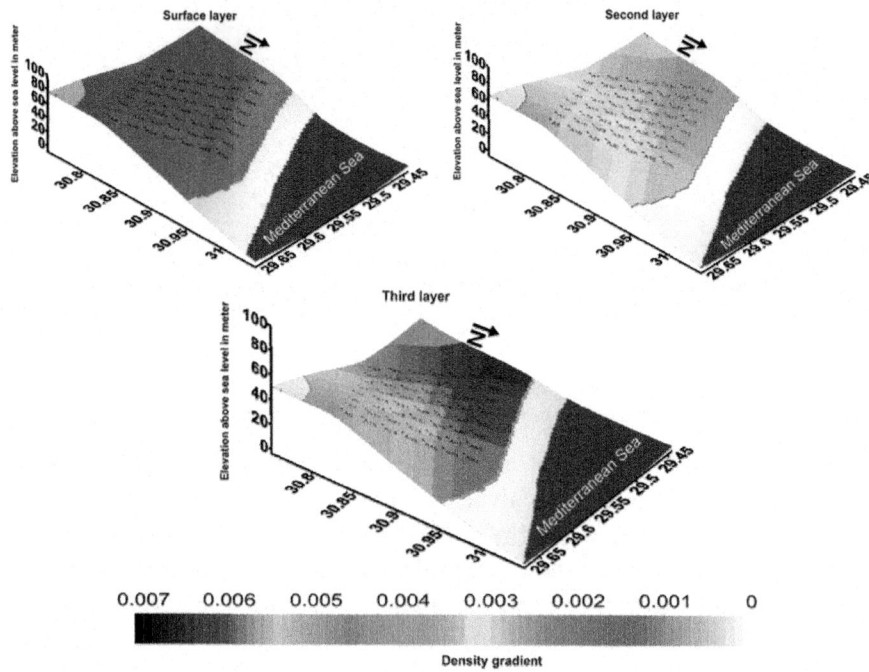

Figure 16: Three-dimension map of density over the study area.

Table 1: The distribution of the density gradient of the study area

Subsoil layers	Density gradient
Surface layer	2.67×10^{-6} to 2.97×10^{-6}
Second layer	1.36×10^{-6} to 1.65×10^{-6}
Third layer	6.72×10^{-7} to 7.67×10^{-7}

Table 2: The distribution of (N-Value) over the study area

Subsoil layers	Standard penetration test (N-Value)
Surface layer	78 to 89
Second layer	172 to 269
Third layer	404 to 567

$$K = \rho \cdot \left[Vp2 - (3/4) Vs2 \right] \tag{4}$$

where ρ is the density.

Table 3 shows the division of Bulk modulus. Over the three subsoil layers, the minimums values of Bulk modulus are observed in the southeastern part of the study area. These values increase to be moderate ones in the center part, while the maximum values are observed in the western and the northwestern corners of the study area (Figure 18).

Ultimate Bearing Capacity (Qult)

The ultimate bearing capacity (Qult) can be defined as "the maximum load required for shear failure or sand li- quefaction". This capacity is controlled by shear strength factor.

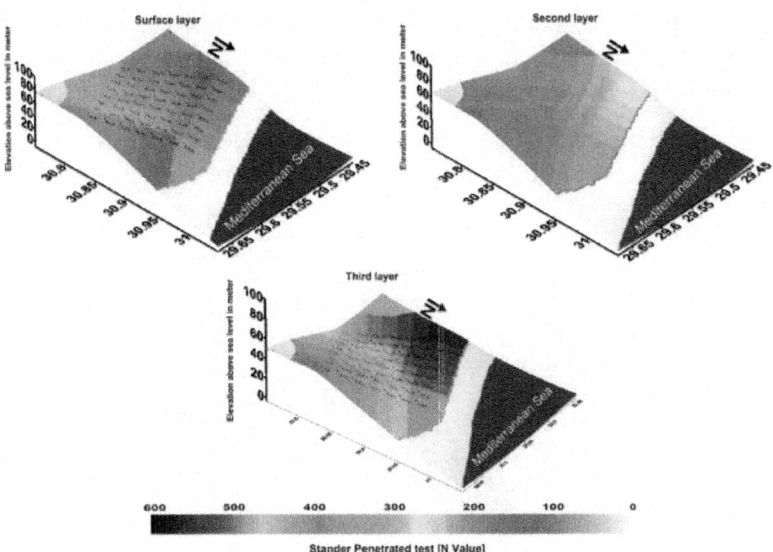

Figure 17: Three-dimension map of Standard penetration Test [N-Value] over the study area.

Table 3: The distribution of Kinetic Bulk modulus (K) over the study area

Subsoil layers	Kinetic Bulk modulus
Surface layer	528 to 648 Dyn/Cm
Second layer	1803 to 3387 Dyn/Cm
Third layer	6016 to 9681 Dyn/Cm

The ultimate bearing capacity for cohessionless soils can be calculated by using the standard penetration test (SPT) from Parry's formula (Parry, 1977) as:

$$Qult = 30N \qquad (5)$$

where "N value" is the resistance to penetration by normalized cylindrical bars under standard load.

Figure 19 shows the overview of the three layers in the study area. The minimums values of ultimate bearing capacity are observed in the southeastern part of the study area. These values increase to be moderate ones in the center part, while the maximum values are observed in the north and the northwestern corners of the study area. Table 4 reveals the distribution of the ultimate bearing capacity (Qult) in the subsoil layers of the study area.

Allowable Bearing Capacity (Qa)

The allowable bearing capacity is the maximum load to be considerable to avoid shear failure or sand liquefaction. It can be termed as allowable bearing pressure too, the foundation materials are affected by both strength and deformation characteristics.

The allowable bearing capacity can be calculated by dividing the ultimate bearing capacity value (Qult) by suitable factor of safety (Abd Elrahman, 1989) as:

$$Qa = Qult/F \qquad (6)$$

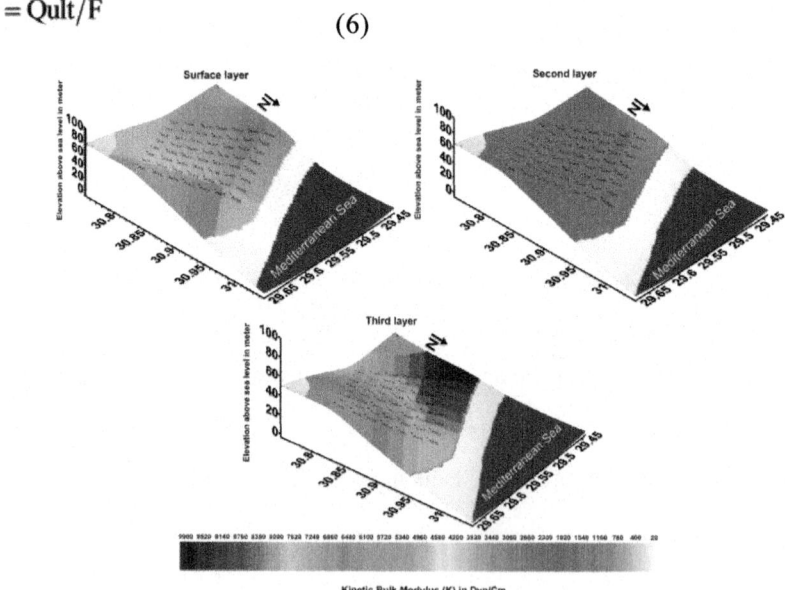

Figure 18: The allotment of Kinetic Bulk Modulus (K) in the study area.

The safety factor (F) equals 2 when the soil is cohessionless and equals 3 when the soil is a cohesive material. Table 5 reveals the distribution of the allowable bearing capacity (Qa) in the subsoil layers of the study area. The minimum values of allowable bearing capacity are observed in the southeastern part of the study area. These values increase to be moderate ones in the center part, while the maximum values are observed in the north and the northwestern corners of the study area (Figure 20).

Rock Material Calssification

This study suggests the classification of foundation rock materials in the study area for engineering purposes based on all the calculated moduli and parameters (Figure 21). This study is divided (the selected study area only) into:

1-Zone (A): (competent material zone):

This zone covers the northwestern corner and the small strip in the middle quarter of the study area. Good competent materials characterize this zone suggesting suitability for any engineering purposes.

2-Zone (B): (Moderated competent material zone):

This zone covers the middle quarter with small strip in the southwestern part of the study area. This zone suggests suitability for moderately and low-load engineering purposes.

3-Zone (C): (non-competent material zone):

This zone covers the eastern part of the study area with some extent in the north. This zone consists of less competent materials not suitable for limited construction purposes. This zone is suggesting non-suitability for any engineering purposes that must be kept away from any constructions or loadness and watery activities.

DISCUSSIONS AND CONCLUSIONS

In order to map the Bourg Al-Arab industrial city (Alexandria, Egypt), electrical resistivity tomography (ERT) and shallow seismic-refraction tomography (SSRT) methods are used.

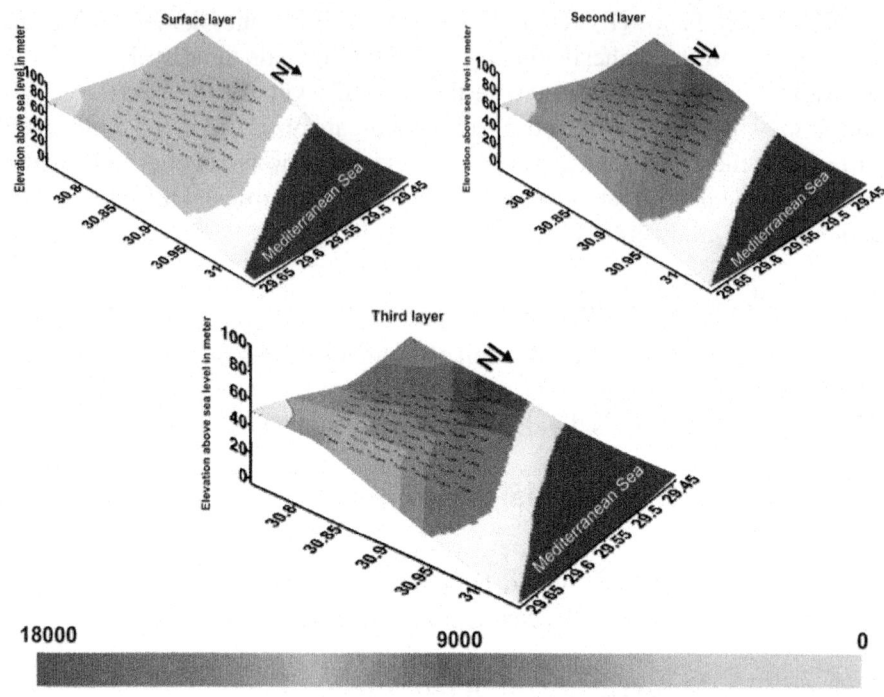

Figure 19: Three-dimension map of the ultimate bearing capacity over the study area.

Table 4: The ultimate bearing capacity (Qult) over the study area

Subsoil layers	The ultimate bearing capacity (Qult)
Surface layer	2211 to 2950 K.Pa
Second layer	5827 to 7793 K.Pa
Third layer	12131 to 1707 K.Pa

Table 5: The allowable bearing capacity (Qa) over the study area

Subsoil layers	The allowable bearing capacity (Qa)
Surface layer	1105 to 1475 K.Pa
Second layer	2594 to 3915 K.Pa
Third layer	6065 to 8503 K.Pa

The results highlight the reliability of the integrated data interpretation based on two physical parameters with different resolution and sensibility.

Electrical survey, effectively visualized as iso-resistivity surfaces, classified the area into three layers. Shallow seismic refraction tomography correlates well with ERT. The integrated interpretation of seismic refraction and resistivity tomography makes it possible to reduce ambiguity. The proposed method of 3D visualization (iso-resistivity and iso-velocity) confirms the interpretation of the 2D standard horizontal sections.

Measuring Vp and Vs velocities is able to calculate the geotechnical parameters of the study area; these parameters classify the area, according to the compaction, hydrostatic pressure, the resistance to load penetration, and load of shear failure or sand liquefaction, into three zones. The geotechnical parameters categorize the area into three zones of foundation rock material related to the engineering purposes.

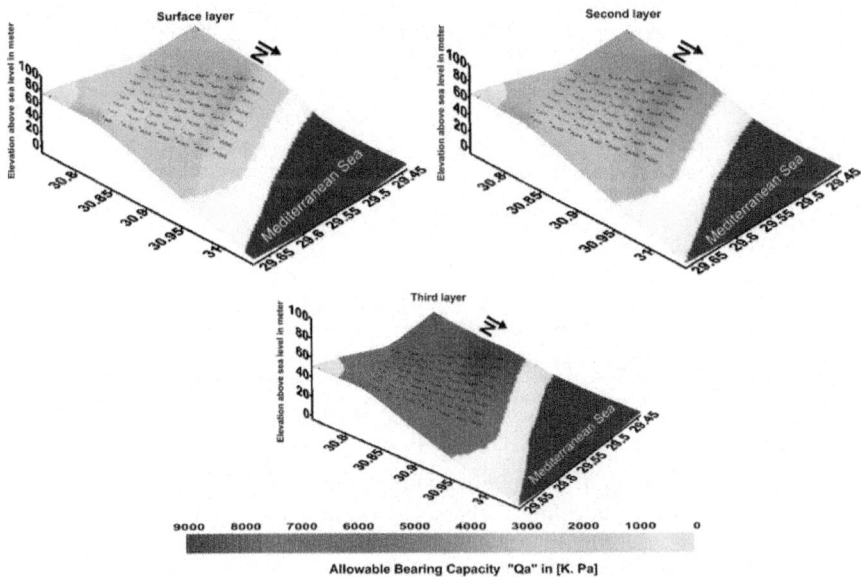

Figure 20: Three-dimension map of the allowable bearing capacity over the study area.

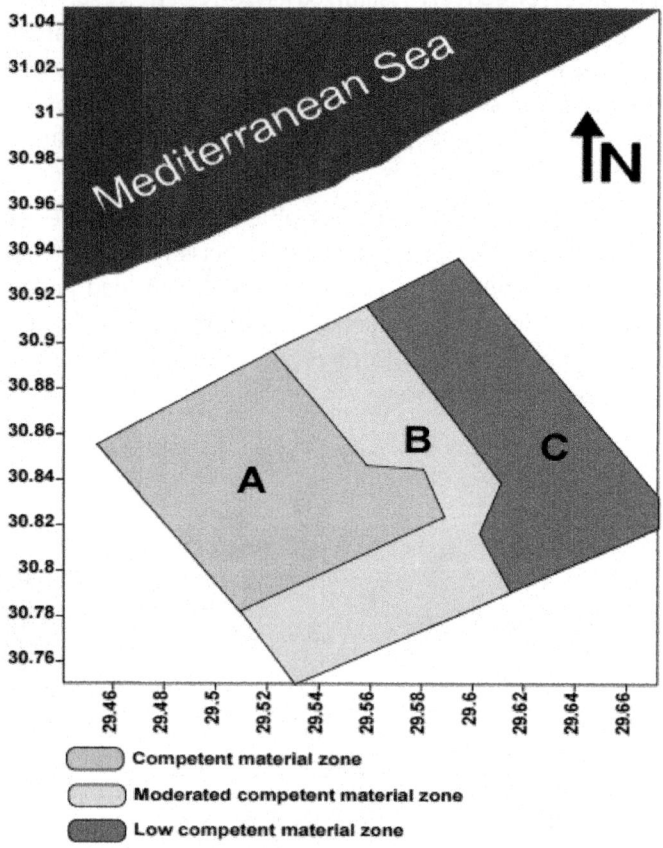

Figure 21: Classification of the foundation rock material quality for engineering purposes according to the geotechnical characteristics in the study area.

REFERENCES

1.   Abd Elmawla, S. H. (2010). National Workshop. Alexandria.

2.   Abdelmotaal, A. M. (2010). Engineering Seismology Studies for Land-Use Planning at the Proposed Tushka New City Site, South Egypt. Ph.D. Thesis, Qena: South Valley University.

3.   Abd Elrahman, M. M. (1989). Evaluation of the Kinetic Moduli of the Surface Materials and Application to Engineering Geologic Maps at Ma'Barrisabah area (Dhamar Province), Northern Yemen. Egyptian Journal of Geology, 33, 228-252.

4.   Cordier, J. P. (1985). Velocities in Refraction Seismology (276 p).

Holland: Radial Publisher Company. http://dx.doi.org/10.1007/978-94-017-3641-1

5.   Dey, A., & Morrison, H. F. (1979a). Resistivity Modelling for Arbitrary Shaped Two-Dimensional Structures. Geophysical Prospecting, 27, 1020-1036.http://dx.doi.org/10.1111/j.1365-2478.1979.tb00961.x

6.   Dey, A., & Morrison, H. F. (1979b). Resistivity Modelling for Arbitrarily Shaped Three-Dimensional Shaped Structures. Geophysics, 44, 753-780. http://dx.doi.org/10.1190/1.1440975

7.   Dutta, N. P. (1984). Seismic Refraction Method to Study the Foundation Rock of a Dam. Journal of Geophysical Prospecing, 32, 1103-1110. http://dx.doi.org/10.1111/j.1365-2478.1984.tb00757.x

8.   El-Behiry, M. G., Hosney, H., Abdelhady, Y., & Mehanee, S. (1994). Seismic Refraction Method to Characterize Engineering Sites. EGS/SEG Proceedings of the 12th Annual Meeting, 85-94.

9.   El Shaazly, M. M. (1964). Geology and Hydrology of Mersa Matrouh Area, Western Mediterranean, Littoral U. A. R. Ph.D. Thesis, Cairo: Cairo University.

10. Gardner, L. W. (1939). An Areal Plan of Mapping Subsurface Structure by Refraction Shooting. Geophysics, 4, 247-259. http://dx.doi.org/10.1190/1.1440501

11. GNBCC (2012). Geology of the Nile Basin Countries Conference. Alexandria.

12. Hatherly, P. J. (1986). Attenuation Measurements on Shallow Seismic Refraction Data. Geophysics, 51, 250-254. http://dx.doi.org/10.1190/1.1442084

13. Imai, L. (1976). The Functions of Seismic Wave in Ground Material and Its Interpretations. Geophysics, 41, 745-797.

14. Leucci, G. (2004). I metodi elettromagnetico impulsivo, elettrico e sismico tomografico a rifrazione per la risoluzione di problematiche ambientali: sviluppi metodologici e applicazioni. Ph.D. Thesis.

15. Loke, M. H., & Barker, R.D. (1996). Rapid Least-Squares Inversion of Apparent Resistivity Pseudosections Using a Quasi-Newton Method. Geophysical Prospecting, 44, 131-152.http://dx.doi.org/10.1111/j.1365-2478.1996.tb00142.x

16. Loke, M. H. (2001). Electrical Imaging Surveys for Environmental and Engineering Studies. A Practical Guide to 2-D and 3-D Surveys. RES2DINV Manual, IRIS Instruments.www.iris-instruments.com

17. Marzouk, I. A. (1995). Engineering Seismological Studies for Foundation

Rock for El-Giza Province, Bull. of National Research Institute of Astronomy and Geophysics (NRIAG). B. Geophysics, 11, 265-295.

18. Mohamed, A. A. (1993). Seismic Microzoning Study and Its Applications in Egypt. Ph.D. Thesis, Cairo: Ain Shams University.

19. Morey, D., & Schuster, G. T. (1999). Paleoseismicity of the Oquirrh fault, Utah from Shallow Seismic Tomography. Geophysical Journal International, 138, 25-35.http://dx.doi.org/10.1046/j.1365-246x.1999.00814.x

20. Nemeth, T., Normark, E., & Qin, F. (1997). Dynamic Smoothing in Cross-Well Traveltime Tomography. Geophysics, 62, 168-176. http://dx.doi.org/10.1190/1.1444115

21. Parasnis, D. S. (1997). Principles of Applied Geophysics (5th ed.). London: Chapman and Hall.

22. Parry, R. H. C. (1977). Estimating Bearing Capacity of Sand from SPT Values. JGED, ASCE, 103, 1013-1045.

23. Samuil, Q. J. (2008). Slicer Cubic 5.0 Manual, Slicer-Cubic Software. Landsub 5, Germany.

24. SEIPEEDIT (2002). Seismic Interpretation Program Software. New York: OHOO Company.www.Ohoo.com

25. Sheriff, R. E. (1991). Encyclopedic Dictionary of Exploration Geophysics (3rd ed.). Society of Exploration Geophysicists.

26. Silvester, P. P. and Ferrari, R. L. (1990). Finite Elements for Electrical Engineers (2nd ed.). Cambridge: Cambridge University Press.

27. Sjogren, B. O., and Sandberg, J. (1979). Seismic Classification of Rock Mass Qualities. Geophysical Prospecting, 27, 409- 442. http://dx.doi.org/10.1111/j.1365-2478.1979.tb00977.x

Chapter 7

GEOTECHNICAL DISTINCTION OF LANDSLIDES INDUCED BY NEAR-FIELD EARTHQUAKES IN NIIGATA, JAPAN

Hirofumi Toyota and Susumu Takada

Department of Civil and Environmental Engineering, Nagaoka University of Technology, Nagaoka, Niigata 940-2118, Japan

ABSTRACT

Landslides triggered by near-field earthquakes with epicentres directly beneath towns have attracted intense attention since the 2004 Mid-Niigata (Niigata-ken Chuetsu) Earthquake. Hilly and mountainous areas sustained heavy damage. Social problems developed when many towns became isolated because landslides cut off traffic and public service lifelines. Soil from landslides closed river channels and formed natural dams. The natural dams submerged some towns. Emergency measures were undertaken promptly to prevent debris flows caused by natural dam breaks. Subsequently, the 2007 Mid-Niigata Offshore (Niigata-ken Chuetsu-oki) Earthquake and the 2011 Northern Nagano Earthquake struck the Niigata region. Landslides triggered by those earthquakes differed in terms of their number, scale, and location. Therefore, characteristics of the landslide sites of the respective earthquakes were examined to ascertain their topographical and geological features. Furthermore, differences in groundwater level and damage related to compound disasters were explained for discussion of the stability progress of damaged slopes.

INTRODUCTION

Niigata prefecture is located in western and central Japan (Figure 1). In the midlands of Niigata, the Chuetsu region, various natural disasters have frequently struck in recent years. At 17:56 on 23 October, 2004, the Mid-Niigata (Niigata-ken Chuetsu) Earthquake, the main tremor of which had a magnitude of 6.8, struck central Niigata-ken (Chuetsu area) and severely damaged the infrastructure of hilly and mountainous areas including Kawaguchi town, Ojiya city, Nagaoka city, and their environs. Numerous landslides occurred especially in the "Yamakoshi area" (Figure 1). That village was isolated by the cutting of all roads and infrastructural lifelines. The epicentral thrust-fault earthquake had a hypocentre of about 13 km depth, with characteristic frequent strong aftershocks that engendered further damage. In addition, rainfall of more than 100 mm was recorded from typhoon number 23, which had passed through the Chuetsu region three days priorly. Data from the Japan Meteorological Agency (JMA) indicated that the daily rainfall of 21 October 2004 at Nagaoka city reached 115 mm. Under those circumstances, more than 3,000 landslides occurred in the hilly area close to the seismic centre during the earthquake. The Japan Society of Civil Engineers [1] and Toyota et al. [2] reported specific damage caused by the disaster. Moreover, Tsukamoto et al. [3] and Rathje et al. [4] estimated the ground movement caused by landslides during the earthquake using investigation or remote sensing data. Geological and geomorphological features and special distributions of landslides in this area were reported by Chigira and Yagi [5] and Wang et al. [6]. Analyses of landslides including discontinuous dip layer were conducted by Onoue et al. [7] and Deng et al. [8]. They assumed that the landslides had been triggered by increase of pore water pressure in seam layers [9] and analysed the displacement using cyclic loading test results or Newmark's method [10].

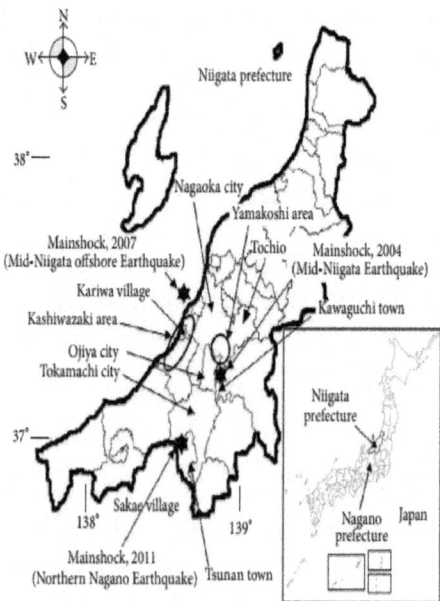

Figure 1: Map of Niigata prefecture.

Heavy snow fell during the two winters following the Mid-Niigata Earthquake. Yearly snowfall is usually about 400 cm in Nagaoka, but it reached nearly 700 cm per year in each of the two years following the earthquake. The snow in one year was the heaviest in 19 years. The collapsed slopes exhibited interesting behaviour during snow and snow-melting seasons. Actually, this region has been notorious as a tertiary type landslide area. Landslides have frequently occurred in April and May: the snow-melting season.

Just three years after the Mid-Niigata Earthquake, at 10:30 on 16 July 2007, the Mid-Niigata Offshore (Niigata-ken Chuetsu-oki) Earthquake, the main tremor of which had a magnitude of 6.8, struck offshore about 30 km northwest of the epicentre of the Mid-Niigata Earthquake. The earthquake was caused by a thrust fault with a hypocentre of about 17 km depth. The JMA Seismic Intensity was recorded as "upper 6" in Kashiwazaki city, Nagaoka city, and Kariwa village. Major landslides were concentrated almost entirely along the "Kashiwazaki area" coastline (Figure 1). Earthquake damage was reported by the JGS [11], Onoue and Toyota [12], and Toyota and Onoue [13].

At 3:59 on 12 March 2011, immediately after the 2011 off the Pacific coast of Tohoku Earthquake (14:46 on 11 March 2011) [14], another earthquake (designated as the Northern Nagano Earthquake) occurred at the border of Nagano and Niigata prefectures [15]. Its main tremor had a magnitude of 6.7. The earthquake, caused by a thrust fault, had a hypocentre of about 8 km depth. The JMA Seismic Intensity was recorded as "upper 6" in Sakae village (Figure 1). Some landslides occurred in Sakae village, Tokamachi town, and Tsunan town under the remaining snow conditions.

For those successive earthquakes which occurred in close proximity, the landslide types are investigated for each earthquake and are compared considering the importance of groundwater levels on landslides. Moreover, the actual conditions of compound natural disasters were examined in the Chuetsu area to assess the compounded damage of these successive natural disasters.

Geological and Geomorphological Features

In the Chuetsu region, thick alluvium covers plains; hilly areas are composed mainly of soft mudstone of quaternary and tertiary deposits (Figure 2). Quaternary deposits are new strata formed from 2 million years ago. Tertiary deposits are geological structures formed between 24 million and 2 million years ago. The geomorphology formed by folding presents a prominent landslide area in this region.

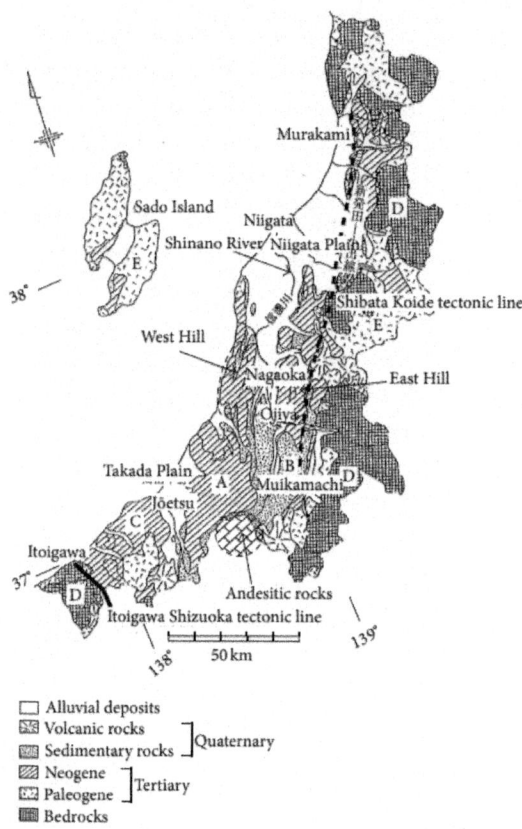

Figure 2: Geological map of Niigata [16].

The Chuetsu region was originally notorious as a tertiary type landslide area [16]. This area has been struck by many natural disasters in recent years. In the Chuetsu area, including Nagaoka and Ojiya (Figure 1), thick alluvium covers the Niigata plain created by the Shinano River. Hilly areas are composed mainly of soft mudstone of quaternary and tertiary deposits. This region has been compressed along the northwest-southeast axis because of crustal movements. Its folded mountains present prominent landslide configurations such as the cuesta landform. Earth-related disasters are generally concentrated in the "West Hills" and "East Hills" (Figure 2). The latter region includes a catchment area dotted with many ponds and rice terraces.

Tertiary mudstone is distributed in the Tokamachi and Tsunan areas, which are near the Northern Nagano Earthquake epicentre. This geology resembles that of the East Hills and is also notorious as a tertiary type landslide area. Nevertheless, it is interesting that Andesitic rocks mainly compose mountainous areas in the northern part of Nagano when crossing the prefectural border separating Niigata and Nagano (Figure 2).

CHARACTERISTICS OF LANDSLIDES

The Mid-Niigata Earthquake

Old Landslides

"Tanesuhara," "Mushigame," and "Asahikawa" in Yamakoshi area have been described as representative designated landslide areas. Figure 3 presents a landslide map of those areas during the Mid-Niigata Earthquake provided by the Geospatial Information Authority of Japan (GSI). River-clogging landslides along the Imo River during the earthquake are the "Terano," "Nampei," "Naranoki," "Higashi-takezawa," and "Junidaira" landslides from upstream (Figure 3). Among those landslides, Terano and Higashi-takezawa were large-scale landslides, which required urgent countermeasures against river clogging.

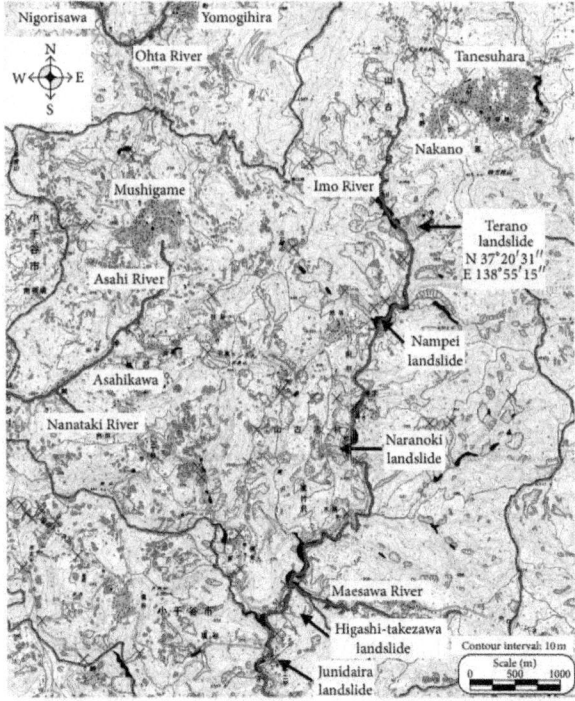

Figure 3: Disaster map around Yamakoshi area during Mid-Niigata Earthquake (provided by GSI).

Table 1 presents the number of landslides occurring during 1949–2002 around Yamakoshi area [16]. Many landslides occurred in the Asahi River basin, extending from Mushigame, and in the Ohta River, which has Yomogihira and Nigorisawa as designated landslide areas. Although 17 landslides occurred in the Imo River basin, almost all occurred in the Tanesuhara or Nakano landslide area located upstream. Landslides were reportedly only a few in Nampei, Higashi-takezawa, or Junidaira, where natural dams were created as a result of the earthquake. When classified by region (Table 1(b)) from most to least numerous, they are Asahikawa, Tanesuhara, and Mushigame

Table 1: Number of landslides during 1949–2002

(a) Classified by basin

	Number of landslides
Asahi River	29
Ohta River	19
Nanataki River	19
Imo River	17

(b) Classified by region

	Number of landslides
Asahikawa	32
Tanesuhara	15
Mushigame	7

The history of the former Yamakoshi village is well recorded [17]. Its history from 1700 indicates Tanesuhara as the site of the greatest recorded earth-flow disaster. Table 2 shows records of landslides in Tanesuhara. The landslides are only recorded in Tanesuhara because no such large villages existed in Yamakoshi area. Although landslides in prehistoric times are unknown, the first landslide recorded in Tanesuhara occurred in 1824. An exogenous factor causing landslides is snow-melt water in early spring. At that time, slopes were destroyed and two large clogging ponds were created in the Imo River. About a century after that event, a landslide occurred during the snow-melting season at almost the same place, thereby forming a natural dam. During the Mid-Niigata Earthquake, which occurred about 80 years after the previous event, the river-clogging landslide was broken at Terano near Nakano. The decisive difference is that the expected exogenous factor was not snow-melt runoff, but an earthquake. As described above, river clogging caused by landslides is a common phenomenon in the Imo River of Tanesuhara because large landslides that clog the river have occurred every century.

Table 2: Old landslides in Yamakoshi area

Number	Year	Disaster	Place	Conditions
1	1824 April: thawing season	Landslide length 1.4 km, width 700 m	Nakano	Disaster throughout the village. 50% of rice fields could not be cultivated. Flood induced by river clogging. Two ponds of 100 m length and 10 m depth were created.
2	1926 May	Landslide length 1.1 km, width 180 m	Nakano	Fields and mountains of more than 2 km² were damaged. Floods induced by river clogging.
3	1929 April	Landslide width 50 m	Terano	Prefectural route (Tochio-Ojiya) was severed; bridge collapsed. River clogging is unclear.
4	1932	Landslide	Between Nakano and Terano	A large landslide occurred during construction of landslide measures. Details are unclear.

Geological Features

Related to unrecorded landslides of prehistoric times, the landslide history of Yamakoshi area was investigated using a topographical map that was compiled using information from an aerial photograph.

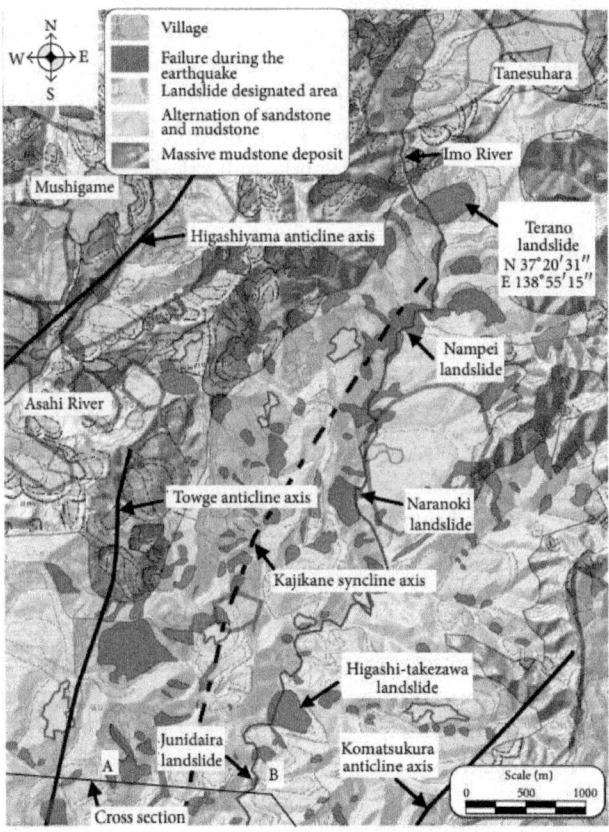

Figure 4: Landslide distribution and geological map around Yamakoshi area during Mid-Niigata Earthquake (provided by NIED and JST).

Figure 4 portrays landslide locations during the Mid-Niigata Earthquake presented on the map of old landslide topography (provided by the National Research Institute for Earth Science and Disaster Prevention (NIED) and the Japan Science and Technology Agency (JST) [18]) in Yamakoshi area. Moreover, the figure was overlaid on a simplified geological map provided by Takeuchi et al. [19] and on landslide-designated areas obtained from a conservation map of Niigata prefecture [20]. Many designated landslide areas are located in the East Hill area. Large-scale landslides that occurred during the earthquake such as those of Terano or Higashi-takezawa might be of a reactivated type of landslide because they coincide completely with the old landslide topography [18].

The west side of the map, which is classified in the Asahi River basin (Figure 4), is geologically an Araya deposit: massive dark grey mudstone. At the eastern side of the map, the Imo River basin (Figure 4), the deposit changes to alternating sandstone and massive mudstone, designated as Kawaguchi and Wanatsu deposits. Alternation of sandstone and mudstone is distributed mainly along the Imo River, except in its upper course. As presented in Figure 4, more numerous landslides occurred during earthquakes in areas with alternated layers of sandstone and mudstone than in areas with massive mudstone deposits. This finding implies that sandy natural slopes are more fragile than clayey natural slopes during earthquakes. However, landslide-designated areas, in which landslides are induced by snowmelt waters, are distributed mainly in the massive mudstone deposits. Moreover, the notable geological features of this region are syncline and anticline structures. They form a complex topography in which synclinal axes and anticlinal axes are arranged with a short interval (Figure 4). For that reason, peculiarly cuesta topography is apparent in this region. Fragile and weak slopes are therefore formed easily. In addition, the river scours the riverbed and the slope toe. Thereby, the slope becomes unstable. For those reasons, it is considered that numerous landslides occurred during the earthquake. Figure 5 depicts a geological cross section of the area south of Yamakoshi area (Junidaira) published by the National Institute of Advanced Industrial Science and Technology (AIST). The location of this sampled geological profile is presented in Figure 4. Sandstone (W) and sandy mudstone (S, Ku_2) are distributed widely around the Imo River. This representative folded mountain area is composed of syncline and anticline. Geological cross sections clarify that a dip slope is apparent on the left bank of the Imo River. A reverse-dip slope is apparent on the right bank of the river. Peculiar slope failures occurred at the right and left banks of the Imo River. Surface failures occurred frequently at the right bank having reverse-dip slope strata, whereas problematical large landslides occurred occasionally at the left bank, which is a dip gentle slope.

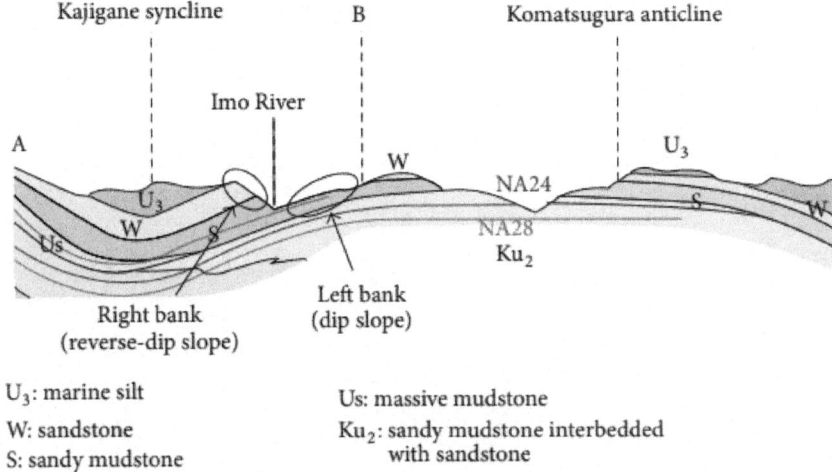

Figure 5: Geological cross section of the south of Yamakoshi area published by the National Institute of Advanced Industrial Science and Technology (AIST).

Mid-Niigata Offshore Earthquake

According to a report of the Ministry of Land, Infrastructure, Transport and Tourism (MLIT) given on 6 August 2007, 108 slope failures occurred during the Mid-Niigata Offshore Earthquake. The main slope failure sites are concentrated in the steep slope of coastal terrace from Shiiya to Hijirigahana, which is about 25 km distant, as shown in Figure 6. Although almost all slides were surface collapses, large mass movements rarely occurred in inland areas as described in Toyota and Onoue [13] because the earthquake occurred not under a mountainous area but under an offshore area. The fault extended from the epicentre to the south. However, the landslides are much fewer than the more than 3,000 which occurred during the Mid-Niigata Earthquake, despite the similarity of the earthquakes.

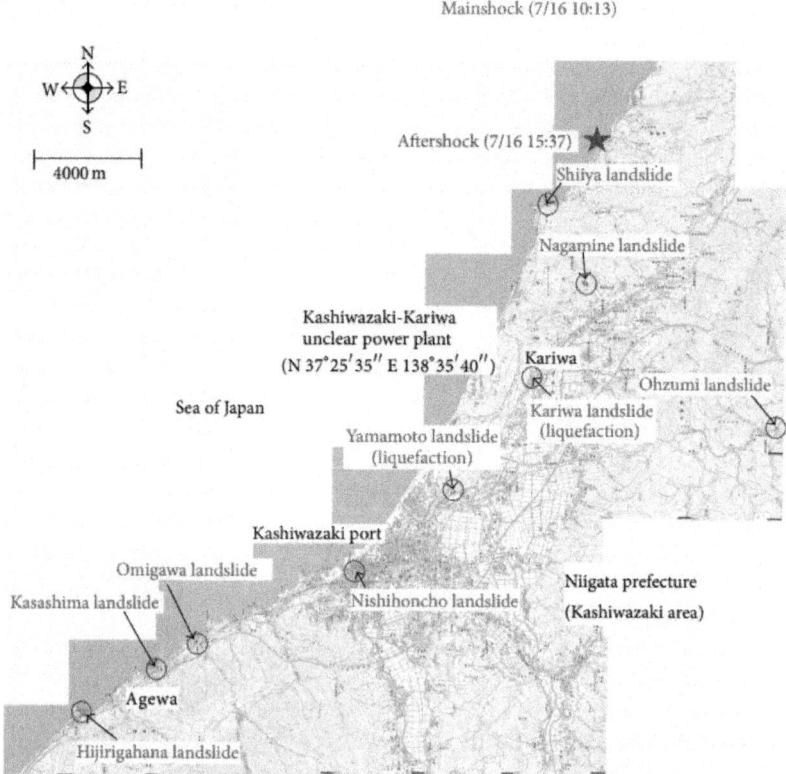

Figure 6: Landslide locations during the Mid-Niigata Offshore Earthquake (provided by GSI).

Northern Nagano Earthquake

The earth-slide disasters during the earthquake were not perfectly clear because the snow remained deeper than 2 m in April in mountainous areas. According to the Niigata prefectural office, 19 slope failures threatening residential life were reported on 20 March 2011. The slope failures were concentrated in Tokamachi city and Tsunan town, which have tertiary mudstone as their main geology (Figures 1 and 2). The Nagano prefectural office reported on 16 March 2011 that seven severe slope failures during the earthquake occurred in Sakae village, which has main geology of Andesitic rocks (Figures 1 and 2). Although the same type of earthquake occurred with the Mid-Niigata Earthquake that occurred directly underneath the mountainous area, it remains unclear why so few earth disasters were generated by the earthquake.

IMPORTANCE OF GROUNDWATER LEVEL

Water level and pore water pressure have been measured at the author's laboratory since 1995 in the old landslide area at the former Tochio city (Figure 1). The data were acquired using a data logger every hour on the hour. A detailed topographical map of this area is presented in Figure 7. The observation point is inferred to be the upper part of the landslide. The gentle slope of a river terrace, which was formed by the meandering flow of river, spreads in the lower part of the landslide. Many infiltration wells have been made in this area as landslide countermeasures. After their installation, no remarkable mass movement has been reported to date. Figure 8 shows the depth of ground-installed water pressure sensors and their geologic column. A borehole with a strainer, which is 27 m deep, was made in the slope. Then a water pressure transducer was set at 12 m depth in the borehole for estimation of water level. For pore water pressure measurement, a water-pressure transducer in the borehole was buried in sand. Their upper and lower sides were sealed with bentonite. The ground consists of soft silty soil up to 10.3 m deep from the surface and a sandy gravel layer of about 70 cm. Soft underlying rock comprises sand and silt in sections deeper than 11 m.

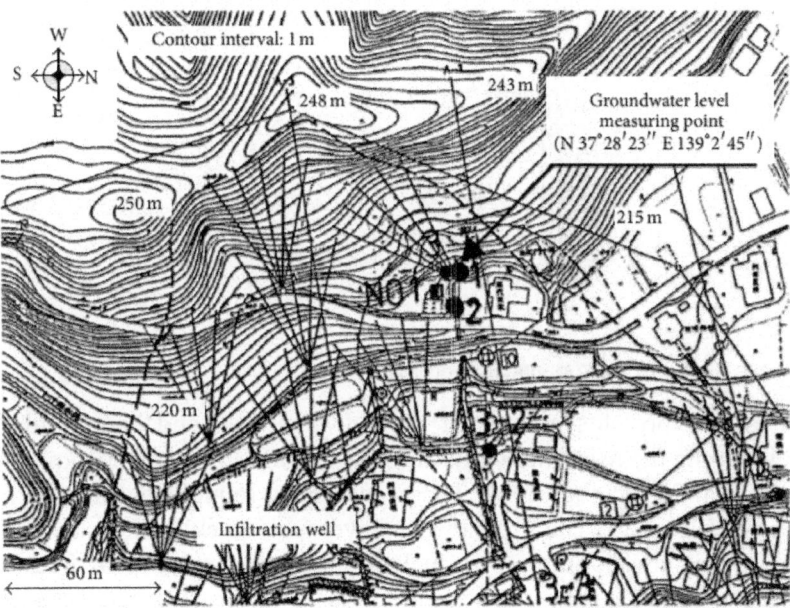

Figure 7: Topographical map (old landslide area at Tochio).

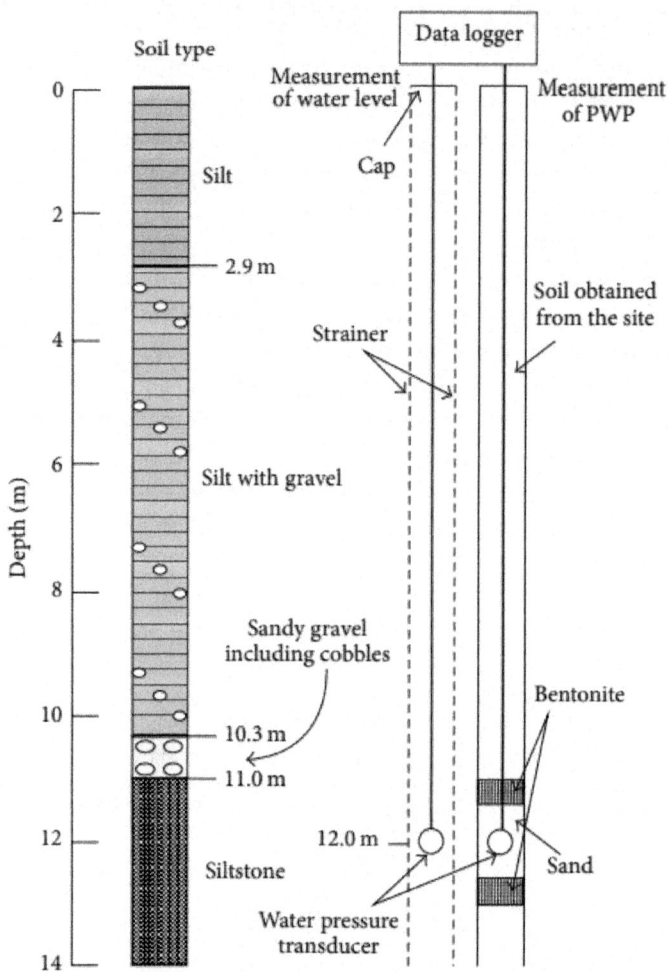

Figure 8: Depth of sensors and soil profiles.

Figure 9 presents records of groundwater level and hourly rainfall at Tochio during the Mid-Niigata Earthquake. Rainfall from typhoon number 23 occurred on 20 October 2004. The accumulated rainfall reached 100 mm in this area (Figure 9). The groundwater level rose quickly about 4 m with the rainfall. This area is inferred to include many catchment areas because the elevated groundwater levels descend gradually over a few days. The rise in the groundwater table induced by the typhoon had dropped by about 3 m in Tochio by the time of the Mid-Niigata Earthquake. It is apparent that this area still contained more water during the Mid-Niigata Earthquake than it did under ordinary conditions.

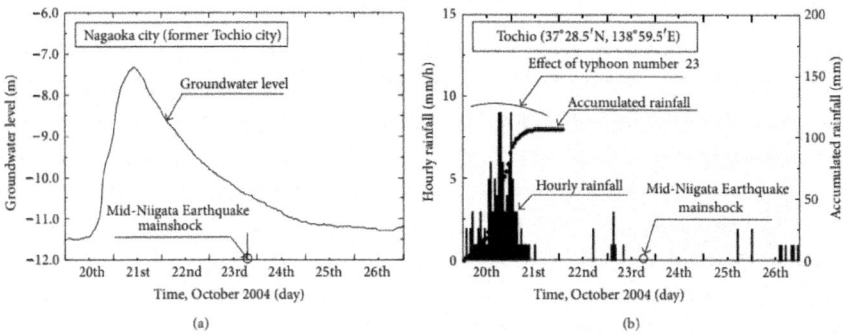

Figure 9: (a) Groundwater level and (b) rainfall during 20–26 October 2004.

Groundwater level fluctuation (October 2004–May 2005) is shown in Figure 10. The groundwater level at three days before the Mid-Niigata Earthquake was the highest in the displayed period of time. The snow season starts from mid-December. The total amount of snowfall reached 699 cm at Nagaoka during that winter. Snow melting accelerates from March with the coming of spring. The groundwater level increases during the snow-melting season because of soaking of snow water and suddenly drops after snow melting (in May), as shown in Figure 10. For that reason, tertiary type landslides have occurred frequently in this area in April and May: the snow-melting season. Therefore, it was feared that the slope damaged by the Mid-Niigata Earthquake starts moving again. Observations were continued during and after the snow season.

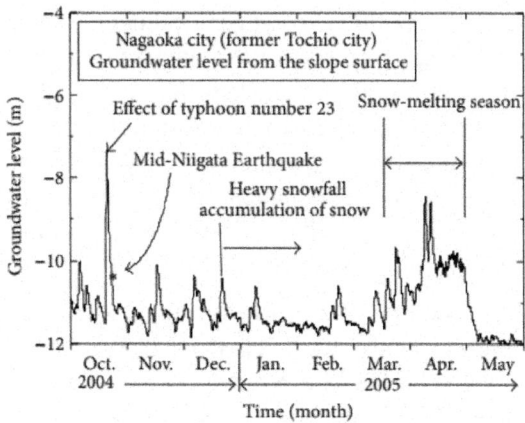

Figure 10: Groundwater level fluctuation.

FOLLOW-UP OF DAMAGED SLOPE

The Chuetsu region is a heavy snow area. The risk of snow avalanche has been increasing because the Mid-Niigata Earthquake damaged snow-protection facilities and natural vegetation. According to Niigata prefecture, 228 avalanche mitigation fences and 8 snow sheds were damaged during the earthquake. Figure11(a) shows an example of landslide in which soil mass slipped down with snow avalanche fences. Surface failures scoured the fence foundations. The fences were deformed because of falling soil. Soil accumulation behind fences and above snow sheds was also severe problem. As an emergency restoration, the accumulated soils were removed and large sandbags were emplaced to protect the infrastructure from the snow avalanche before the winter.

(a)   (b)

Figure 11: (a) Damage to snow avalanche fence; (b) snow-earth avalanche [21].

Snow avalanches occurred more frequently that year than in a normal winter. Some slopes were not covered with deep snow because a small snow slide occurred at all snowfalls. However, frequent patrols were conducted, preventing heavy damage. The risk of avalanche was reduced by the removal of unstable snow, construction of snow walls, and road closures during the winter. A new type of avalanche, where snow fell together with surface soil stripped from the damaged slope, was observed in Tokamachi City (Figure 1) as shown in Figure 11(b). An avalanche of this type, called a "snow-earth mixture avalanche," is regarded as a compound natural disaster, causing extensive damage to residences and roads [21].

Landslides have occurred frequently during the snow-melting season in this area because of high groundwater levels (Figure 10). Therefore, it was feared that the slope damaged by the Mid-Niigata Earthquake starts moving again as a secondary disaster. Therefore, after snow melting, follow-up investigations were conducted to observe the slope failure sites. However, Toyota [22] reported that no large mass movement occurred during the snow-melting season. These sliding masses induced by the Mid-Niigata Earthquake were therefore stable except for erosion from rain and snow melting. Situations of slope failures can be compared for those before winter and after the snow-melting season in Figure 12. There is no apparent difference between the photographs.

Figure 12: Slope failures: upper row, before winter, and lower row, after snow-melting season.

EARTHQUAKE DAMAGE COMPARISONS

Comparisons between the Mid-Niigata Earthquake, the Mid-Niigata Offshore Earthquake, and the Northern Nagano Earthquake are presented in Table 3. All earthquakes are mutually similar except for the frequency of the aftershocks: they were frequent after the Mid-Niigata Earthquake, but not after other earthquakes. Many evacuees, to escape from the fear of repeated strong aftershocks, were compelled to reside for extended periods in inconvenient evacuation areas. Some were afflicted with phlebothrombosis, often called "economy class syndrome."

Table 3: Comparisons of earthquakes

	Mid-Niigata	Mid-Niigata Offshore	Northern Nagano
Date	23 Oct. 2004	16 July 2007	12 March 2011
Epicentre	Chuetsu area	Chuetsu offshore	Northern Nagano
Depth	13 km depth	17 km depth	8 km depth
Cause	Thrust fault	Thrust fault	Thrust fault
Magnitude	M6.8	M6.8	M6.7
JMA intensity	7 (at Kawaguchi)	6 upper (at Kashiwazaki)	6 upper (Sakae)
Max. acceleration	1750.2 gal (at Tokamachi)	812.7 gal (at Kashiwazaki)	803.5 gal (at Tsunan)
Aftershock	Frequent	Rare	Rare
Landslides	Thousands	Medium	Medium

However, the types of damage related to the earth disaster were different among the earthquakes in spite of being only 60 km distant from the epicentre. Although the main landslide sites are hilly and mountainous areas and although more than 3,000 slope failures occurred during the Mid-Niigata Earthquake, 108 slope failures occurred mainly in the steep slope of the coastal terrace during the Mid-Niigata Offshore Earthquake according to a report of MLIT on 6 August 2007 because the earthquake occurred not under a mountainous area but under an offshore area. The Northern Nagano Earthquake occurred under mountainous areas between Nagano and Niigata. However, few landslides were reported, as described in Section 3.3. Two possible reasons can be considered. The first reason is that the main geology of Sakae Village in the northern part of Nagano is Andesitic rock, which is different from the tertiary mudstone of the Chuetsu area (Figure 2). Examples of slope failures are exhibited in Figure 13. Figure 13(a) shows a river-clogging landslide at Yamakoshi area during the Mid-Niigata Earthquake. The weak soft deposit moved down and closed the Imo River channel. Figure 13(b) presents river-clogging landslides at Sakae Village during the Northern Nagano Earthquake. Apparently, the weathered part of a steep cliff fell and clogged a narrow valley. Then the dammed lake was created by the moving mass. However, the tertiary mudstone is distributed in Tokamachi and Tsunan areas (Figures 1 and 2) where strong seismic motion was recorded during the earthquake. Therefore, the second considered reason is that groundwater level might not be high during the Northern Nagano Earthquake compared with that of the Mid-Niigata Earthquake, as shown in Figure 9(a).

(a)   (b)

Figure 13: Typical river-clogging landslides that occurred (a) during the Mid-Niigata Earthquake and (b) during the Northern Nagano Earthquake.

From Figure 10, it is inferred that the ground water level remained low immediately before snow melting during the Northern Nagano Earthquake. Although snow remained deeper than 2 m in mountainous areas, few snow avalanches were triggered by the earthquake. Accumulated snow is apparently quite stable against earthquakes during this season (March). Consequently, the relations between the epicentre, the geography including geology, and groundwater level are extremely important factors to discuss to assess the risk of landslide damage.

CONCLUSIONS

Several earthquakes struck the Chuetsu area of Niigata successively in recent years. Damage investigations have been reported individually and have been mutually compared. Information about damage related to geotechnical engineering has been assembled and discussed from the viewpoint of compound natural disasters. Although no severe damage has been certified as a secondary disaster, these disasters underscore the possibility of compound natural disasters or slight damage such as erosion occurring in the damaged slopes. A summary of the main findings from the study is presented below.(1) The importance of groundwater level during landslide disasters was assessed using field measurement results. Results show that groundwater during the Mid-Niigata Earthquake was higher than that under normal conditions.(2) Large-scale landslides during the Mid-Niigata Earthquake were regarded as reactivated type landslides because they occurred on old landslide traces.(3) Most landslides during the Mid-Niigata Earthquake occurred at alternating sandstone and mudstone strata, but landslides that moved gradually during the snow-melting season usually involved mudstone in large quantities from tertiary deposits.(4)Landslides during the Mid-Niigata Offshore Earthquake were fewer than those which occurred during the Mid-Niigata Earthquake, perhaps because the main tremor occurred not under a mountainous area but under an offshore area during the Mid-Niigata Offshore Earthquake. (5)Landslides that occurred during the Northern Nagano Earthquake were also fewer than those that occurred during the Mid-Niigata Earthquake. The reasons considered are that the main geology of the mountainous area in northern Nagano is not tertiary mudstone but Andesitic rock. Moreover, the groundwater level might not be high compared to that prevailing during the Mid-Niigata Earthquake. Only a few snow avalanches triggered by the earthquake were reported in spite of snow that remained on the ground.(6) The risk of snow avalanche increased after the earthquakes because snow-protection facilities and natural vegetation were damaged.(7)Some additional damage occurred such as surface erosion and surface slides during rainfall and snow-melting season in the slopes damaged and loosened by the earthquakes. Carefully conducted daily observations are important to prevent a secondary disaster because some indications (slight damage) will precede severe damage. Timely countermeasures taken for cases of slight damage must be taken to reduce the risk of severe damage.

CONFLICT OF INTERESTS

The authors declare that there is no conflict of interests regarding the publication of this paper.

REFERENCES

1. Japan Society of Civil Engineers, Ed., Report on the 2004 Niigata Chuetsu Earthquake, CD Rom, Japan Society of Civil Engineers, 2006, (Japanese).

2. H. Toyota, J. Wang, K. Nakamura, and N. Sakai, "Evaluation of natural slope failures induced by the 2004 Niigata-ken Chuetsu Earthquake," Soils and Foundations, vol. 46, no. 6, pp. 727–738, 2006. View at Publisher · View at Google Scholar · View at Scopus

3. Y. Tsukamoto, K. Ishihara, and Y. Kobari, "Evaluation of run-out distances of slope failures during 2004 Niigata-ken Chuetsu Earthquake," Soils and Foundations, vol. 46, no. 6, pp. 713–725, 2006. View at Publisher · View at Google Scholar · View at Scopus

4. E. Rathje, R. Kayen, and K.-S. Woo, "Remote sensing observations of landslides and ground deformation from the 2004 Niigata Ken Chuetsu earthquake," Soils and Foundations, vol. 46, no. 6, pp. 831–842,

2006. View at Publisher · View at Google Scholar · View at Scopus

5.   M. Chigira and H. Yagi, "Geological and geomorphological charactetristics of landslides triggered by the 2004 mid Niigata prefecture earthquake in Japan," Engineering Geology, vol. 82, no. 4, pp. 202–221, 2006. View at Publisher · View at Google Scholar · View at Scopus

6.   H. B. Wang, K. Sassa, and W. Y. Xu, "Analysis of a spatial distribution of landslides triggered by the 2004 Chuetsu earthquakes of Niigata Prefecture, Japan," Natural Hazards, vol. 41, no. 1, pp. 43–60, 2007. View at Publisher · View at Google Scholar · View at Scopus

7.   Onoue, A. Wakai, K. Ugai et al., "Slope failures at Yokowatashi and Nagaoka College of Technology due to the 2004 Niigata-ken Chuetsu Earthquake and their analytical considerations," Soils and Foundations, vol. 46, no. 6, pp. 751–764, 2006. View at Publisher · View at Google Scholar · View at Scopus

8.   J. Deng, H. Kameya, Y. Miyashita, J. Kuwano, R. Kuwano, and J. Koseki, "Study on dip slope failure at Higashi Takezawa induced by 2004 Niigata-Ken Chuetsu earthquake," Soils and Foundations, vol. 51, no. 5, pp. 929–943, 2011. View at Publisher · View at Google Scholar · View at Scopus

9.   K. Sassa, H. Fukuoka, G. Scarascia-Mugnozza, and S. Evans, "Earthquake-induced landslides: distribution, motion and mechanisms," Soils and Foundations, pp. 53–64, 1996. View at Google Scholar

10.   N. M. Newmark, "Effects of earthquakes on dams and embankments," Géotechnique, vol. 15, no. 2, pp. 139–160, 1965. View at Publisher · View at Google Scholar

11.   Japanese Geotechnical Society, Ed., Report on the 2007 Niigata-ken Chuetsu-oki Earthquake, Japanese Geotechnical Society, 2009, (Japanese).

12.   Onoue and H. Toyota, "Damage induced by the 2007 Niigataken Chuetsu-oki earthquake," inProceedings of the 14th World Conference on Earthquake Engineering, Paper ID: S26-15, pp. 1–8, Beijing, China, 2008.

13.   H. Toyota and A. Onoue, "Characterisation of slope failures during the 2004 Niigata-ken Chuetsu and the 2007 Niigata-ken Chuetsu-oki Earthquake," in Proceedings of the 14th World Conference on Earthquake Engineering, PaperID: S26-18, pp. 1–8, Beijing, China, 2007.

14.   Joint Editorial Committee for the Report on the Great East Japan Earthquake Disaster, Report on the Great East Japan Earthquake Disaster, Fundamental Aspects 3, Geohazards (Abstract), JGS, 2014, (Japanese).

15. Japan Metrological Agency, "Report on the Northern Nagano Earthquake (2nd issue)," 2011,http://www.jma.go.jp/jma/press/1103/12d/201103120800.html.

16. Niigata Branch of the Japan Landslide Society, Ed., Record of Landslide Disasters in Niigata, CD Rom, Niigata Branch of the Japan Landslide Society, 2003, (Japanese).

17. Editorial Committee of History of Yamakoshi Village, History of Yamakoshi Village, Editorial Committee of History of Yamakoshi Village, Yamakoshi, Japan, 1981, (Japanese).

18. National Research Institute for Earth Science and Disaster Prevention (NIED) and Japan Science and Technology Agency (JST), "Landslide map," 2004,http://lsweb1.ess.bosai.go.jp/jisuberi/jisuberi_mini/jisuberi_top.html.

19. K. Takeuchi, Y. Yanagisawa, J. Miyazaki, and M. Ozaki, "1:50,000 digital geological map of the Uonuma region, Niigata Prefecture (Ver. 1)," GSJ Open-File Report 412, 2004,http://www.gsj.jp/GDB/openfile/files/no0412/index.html. View at Google Scholar

20. Niigata Prefecture, Ed., Conservation Map, Niigata Prefecture, Niigata, Japan, 1982, (Japanese).

21. Kamiishi and M. Machida, "Snow avalanche by the Chuetsu Earthquake and heavy snow," inEarthquake Disaster and Prevention in Snowy Areas, pp. 45–67, Japan Society for Snow Engineering, Joshinetsu, Japan, 2008, (Japanese). View at Google Scholar

22. H. Toyota, "Stability progress of slopes damaged by natural disaster—case study in the Chuetsu area of Niigata, Japan," in Proceedings of the 17th Southeast Asian Geotechnical Conference, vol. 2, pp. 113–122, Taipei, Taiwan, 2010.

Chapter 8

GEOTECHNICAL CHARACTERIZATION OF MINED CLAY FROM APPALACHIAN OHIO: CHALLENGES AND IMPLICATIONS FOR THE CLAY MINING INDUSTRY

Anthony R. Moran[1] and Hiroshan Hettiarachchi[2]

[1]CH2M Hill, 7927 Nemco Way, Suite 120, Brighton, MI 48116, USA

[2]Department of Civil Engineering, Lawrence Technological University, 21000 West Ten Mile Road, Southfield, MI 48075, USA

ABSTRACT

Clayey soil found in coal mines in Appalachian Ohio is often sold to landfills for constructing Recompacted Soil Liners (RSL) in landfills. Since clayey soils possess low hydraulic conductivity, the suitability of mined clay for RSL in Ohio is first assessed by determining its clay content. When soil samples are tested in a laboratory, the same engineering properties are typically expected for the soils originated from the same source, provided that the testing techniques applied are standard, but mined clay from Appalachian Ohio has shown drastic differences in particle size distribution depending on the sampling and/or laboratory processing methods. Sometimes more than a 10 percent decrease in the clay content is observed in the samples collected at the stockpiles, compared to those collected through reverse circulation drilling. This discrepancy poses a challenge to geotechnical engineers who work on the prequalification process of RSL material as it can result in misleading estimates of the hydraulic conductivity of the samples. This paper describes a laboratory investigation conducted on mined clay from Appalachian Ohio to determine how and why the standard sampling and/or processing methods can affect the grain-size distributions. The variation in the clay content was determined to be due to heavy concentrations of shale fragments in the clayey soils. It was also concluded that, in order to obtain reliable grain size distributions from the samples collected at a stockpile of mined clay, the material needs to be processed using a soil grinder. Otherwise, the samples should be collected through drilling.

INTRODUCTION

Clayey soil found beneath a coal layer, which is also known as underclay, is an ideal construction material for Recompacted Soil Liners (RSLs) in landfills due to its low hydraulic conductivity and relatively low cost. Coal mines in Appalachian Ohio typically encounter large volumes of underclay due to the depositional nature of coal formations there. Clay mines make a profit and increase usable space at the mines by mining and selling the underclay to landfills that do not have sufficient clay borrow sources located within their property boundaries or permitted limits of waste. Landfills can save money utilizing the mined clay if the haul distance is short enough with respect to the expense of using additional geosynthetic materials.

The research described in this paper comes to light from experiences with underclay from Appalachian Ohio, which has shown some unique characteristics in its composition. The Ohio region, at the base of the Appalachian basin, is comprised of shale enriched clay deposits found beneath coal layers [1]. These clay deposits are classified with a layer number. The layering system is based on the cyclothemic depositions of the coal layers in this region. The deeper the clayey soil layer, the lower the number identification for the layer [1].

The solid waste division of the Ohio Environmental Protection Agency (OEPA) requires that RSL material be pre-qualified prior to use in construction [2]. This means that samples are analyzed for grain-size distribution, Atterberg limits, compaction characteristics and hydraulic conductivity prior to their acceptance for use. For hydraulic conductivity, remolded samples satisfy the OEPA requirement for the establishment of the RSL's hydraulic conductivity [2]. Clay mines typically pay for the cost of the drilling exploration and allow consultants onto their sites to monitor the drilling of potential RSL borrow sources. The method of collecting samples for laboratory testing through drilling as well as collecting samples from representative stockpiles are deemed acceptable.

Standard field sampling methods and laboratory sample preparation procedures, should not produce different grain-size distributions for the same material. An engineering mind would expect that a sample collected within the borrow location at a clay mine would classify the same and contain the same clay content as when the soils are excavated and hauled to a stockpile at a landfill and re-tested. But experience with the mined clay from Appalachian Ohio has revealed that this is not always the case, and the mined clay from Appalachian Ohio has shown drastic differences in particle size distribution depending on the sampling and/or laboratory processing methods. More particularly, the clay content of the material, *i.e.* particles finer than 0.002 mm as per AASHTO [3], has been shown to vary substantially. The research described in this paper was conducted to investigate how the sampling and/ or laboratory processing methods could affect the grain-size distribution of mined clay from Appalachian Ohio.

MATERIALS AND METHODS

Underclay samples were collected from a location in Appalachian Ohio that was previously used for coal mining. This underlay layer was specifically identified as Number 5 clay [1]. The clay source was originally under approximately 60 feet of overburden.

Sample Collection

About 100 soil samples were collected at different depths in the clayey soil deposit using a hollow stem auger with reverse circulation drilling. Before transporting to the laboratory, soil samples were first visually classified in the field and then collected in one gallon ziplock bags and labeled in numerical order by collection. Samples retrieved by this method were observed to have flour-like consistency as can be seen in Figure 1(a). All soil samples collected by drilling visually appeared nearly identical. The drilling efforts spanned approximately three days at the proposed borrow location at the clay mine. The samples collected through reverse circulation drilling were designated as D-samples during this research.

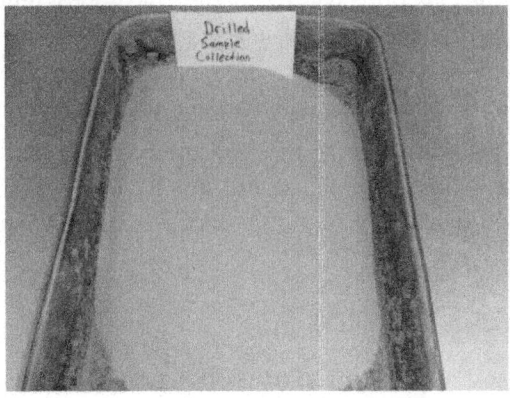

(a)

(b)

Figure 1: (a) Samples collected through drilling (D-samples); (b) Samples collected at the stockpiles (S-samples).

Once the clayey soil was mined and stockpiled, a second set of samples were collected from the stockpiles. They were designated as S-samples to indicate the method of collection. During stockpiling, accurate records were kept so that the original location and the depth of the soil could be easily identified to compare S-samples with previously collected samples D-samples. Depth and location identification was conducted by utilizing a hand held GPS unit. When the area of the clay mine was excavated, the approximate location of this material was flagged in the stockpiles. By doing this, it was ensured that the S-soil samples were also from the same locations where the D-samples were collected. S-samples were collected and transported in five gallon buckets.

About 65 S-samples were finally collected to compare with D-samples. Samples retrieved from this method did not have a fine consistency as the D-samples, and instead the S-samples contained many clay clods, as seen in Figure 1(b).

Sample Processing

For the initial soil characterization, soil samples were processed either completely manually or using a soil processor. During manual processing, soil samples were first allowed to air dry at room temperature then placed over a #4 screen and pushed through by hand. This manual preparation was designated as M-processing. During sample preparation using the soil processor, air dried samples were first placed in the soil processor. Then the weighted lid of this machine applied a normal dead load to the sample as an air-operated arm moves the #4 screen perpendicular to the sample. The results produced by this method were more or less identical to the soil samples produced by the above process and hence this preparation method was also designated as M-processing.

D-samples did not need much of a preparation other than air drying as they were already fine in consistency. However, to make the research process consistent, D-samples were also nominally processed manually by sieving through the #4 screen. Soil clods in S-samples needed to be processed before testing. An example of a manually processed S-sample is shown inFigure 2(a). The #4 screen used in M-processing can be seen in Figure 2(b).

(a)

(b)

Figure 2: (a) A stockpile sample processed manually using #4 screen (SM) (b) #4 screen used in manual processing.

Geotechnical Characterization

ASTM test protocols were followed during geotechnical testing. To characterize the material, particle size analysis (ASTM D421 and D422), Atterberg limits (ASTM D4318), classification (ASTM D2487), modified Proctor (ASTM D1557) and hydraulic conductivity (ASTM D5084) tests were carried out as per the ASTM test protocols [4]. Hydraulic conductivity tests were slightly modified to suit the requirements of this research. Before testing for hydraulic conductivity, the samples were placed in a sealed container to temper for at least 16 hours. Then samples were remolded (mold dimensions: 2.79" diameter × 3" height) and compacted to 90 percent of the maximum dry unit weight, as determined by the modified Proctor test. Remolded specimen was then placed within a flexible wall permeameter cell.

RESULTS, ANALYSIS AND DISCUSSION

All D-samples and S-samples were first processed manually (hence designated as DM and SM) and subjected to mechanical analysis to obtain their gradation.

Size distributions of all theses samples are given in Moran [5]. Based on the results of mechanical analysis, 13 DM-samples with similar clay content (approximately 28%) were selected for further analysis. Size distributions of these 13 DM-samples are given in Table 1. SM-samples, those were originated from the same locations as in DM-samples were also selected for further processing. Size distributions of the 13 SM-samples are given in Table 2. As is evident from Tables 1 and 2, the grain size distributions of DM-samples do not match with the grain size distributions for SM-samples, though they originated from the same locations. The difference in clay content was nearly 10 percent. To illustrate this difference in clay content, grain-size distributions for DM-1 and SM-1 are compared in Figure 3.

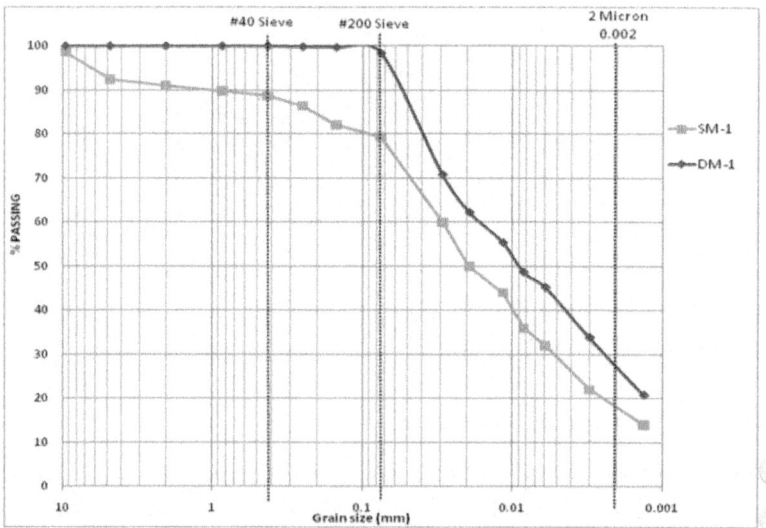

Figure 3: Comparison of grain size distributions: manually processed drilled sample (DM-1) and manually processed sample collected at the stockpile (SM-1).

Table 1: Geotechnical characteristics of DM-samples selected for further processing

Sample Number	Grain Size Distribution						Atterberg Limits[1]			Uscs Soil Type	Compaction Characteristics Based On Modified Proctor Test[2]		Hydraulic Conductivity (cm/sec)
	% Finer	% Finer	% Finer	% Finer	% Finer	% Finer	LL	PL	PI		γ_{dmax} (kN/m³)	OMC (%)	
	12.5 mm	9.5 mm	#4	#40	#200	.002 mm							
DM-1	100.0	100.0	100.0	100.0	98.3	27.5	30	16	14	CL	–	–	–
DM-2	100.0	100.0	100.0	82.1	69.4	27.7	29	16	13	CL	21.2	6.6	7.6×10^{-8}
DM-3	100.0	100.0	100.0	99.5	93.2	27.5	27	17	10	CL	22.0	7.1	9.0×10^{-8}
DM-4	100.0	100.0	100.0	94.9	83.2	28.3	28	15	13	CL	–	–	–
DM-5	100.0	100.0	100.0	99.7	96.1	28.1	24	14	10	CL	–	–	–
DM-6	100.0	100.0	100.0	98.9	92.6	28.8	23	15	8	CL	–	–	–
DM-7	100.0	100.0	100.0	83.3	73.3	29.8	26	16	10	CL	–	–	–
DM-8	100.0	100.0	100.0	96.5	89.1	27.7	28	16	12	CL	–	–	–
DM-9	100.0	100.0	100.0	98.9	94.6	27.2	28	18	10	CL	–	–	–
DM-10	100.0	100.0	100.0	80.7	74.4	27.7	27	17	10	CL	–	–	–
DM-11	100.0	100.0	100.0	98.9	95.9	27.6	33	19	14	CL	–	–	–
DM-12	100.0	100.0	100.0	99.8	95.4	28.2	25	17	8	CL	–	–	–
DM-13	100.0	100.0	100.0	99.8	93.8	27.6	26	17	9	CL	–	–	–

[1] LL = liquid limit, PL. = plastic limit, PI = plasticity index;
[2] γ_{dmax} = maximum dry unit weight, OMC = optimum moisture content.

Table 2: Geotechnical characteristics of SM-samples selected for further processing

Sample Number	Grain Size Distribution						Atterberg Limits[1]			Uscs Soil Type	Compaction Characteristics Based On Modified Proctor Test[2]		Hydraulic Conductivity (cm/sec)
	% Finer	% Finer	% Finer	% Finer	% Finer	% Finer	LL	PL	PI		γ_{dmax} (kN/m³)	OMC (%)	
	12.5 mm	9.5 mm	#4	#40	#200	.002 mm							
SM-1	100.0	98.5	92.4	88.7	79.3	17.3	30	17	13	CL	21.9	7.3	8.5×10^{-8}
SM-2	100.0	100.0	94.6	87.8	75.7	17.4	28	18	10	CL	21.3	5.9	7.6×10^{-8}
SM-3	100.0	100.0	95.5	87.1	76.9	17.2	28	18	10	CL	21.9	8.0	9.1×10^{-8}
SM-4	100.0	100.0	92.5	91.8	83.1	18.2	29	17	12	CL	21.7	7.0	6.5×10^{-8}
SM-5	100.0	100.0	98.7	94.6	85.6	17.3	29	18	11	CL	–	–	–
SM-6	100.0	99.4	96.0	89.0	78.1	18.5	24	16	8	CL	–	–	–
SM-7	100.0	100.0	99.5	93.9	83.7	20.6	27	17	10	CL	–	–	–
SM-8	100.0	99.2	94.1	89.2	79.0	18.8	28	16	12	CL	–	–	–
SM-9	100.0	100.0	95.6	93.2	80.2	19.5	28	19	9	CL	–	–	–
SM-10	100.0	99.2	97.5	90.7	78.6	19.3	28	17	11	CL	–	–	–
SM-11	100.0	100.0	100.0	92.8	82.4	19.1	31	19	12	CL	–	–	–
SM-12	100.0	100.0	98.4	88.9	78.1	17.5	25	17	8	CL	–	–	–
SM-13	100.0	99.4	94.5	88.1	78.0	17.6	27	17	10	CL	–	–	–

[1] LL = liquid limit, PL = plastic limit, PI = plasticity index;
[2] γ_{dmax} = maximum dry unit weight, OMC = optimum moisture content.

Both DM and SM soil samples were then subjected to further geotechnical characterization through Atterberg limits, classification per USCS, modified Proctor and hydraulic conductivity tests. According to the results of these tests summarized in Table 1 and Table 2, the material is rather homogenous. Hydraulic conductivity is typically a function of the clay content. But the 10% difference in the clay content has not even made any considerable impact on hydraulic conductivity. Therefore it is clear that the clay content of one of the sample types is incorrect.

Sample Processing by a Soil Grinder

During the next phase of the research, the laboratory processing of S-samples were performed using a soil grinder (hence SG-samples). The mechanical soil grinder utilized is the one typically used for preparing samples for Atterberg limits and specific gravity tests. Air dried samples were placed into the soil grinder which consists of an electric motor driving a ½ inch shaft with ¼ inch spinning steel bars. The grinder "broke the soil down" to clods to the approximate size of a number 10 sieve or smaller. This method was designated as G-processing (hence processed samples designated as SG-samples). An example of a sample prepared by this method is shown in Figure 4(a) and the apparatus for grinding can be seen in Figure 4(b). Visual appearance of the ground samples appeared similar to the samples collected through the reverse circulation drilling exploration (D-samples).

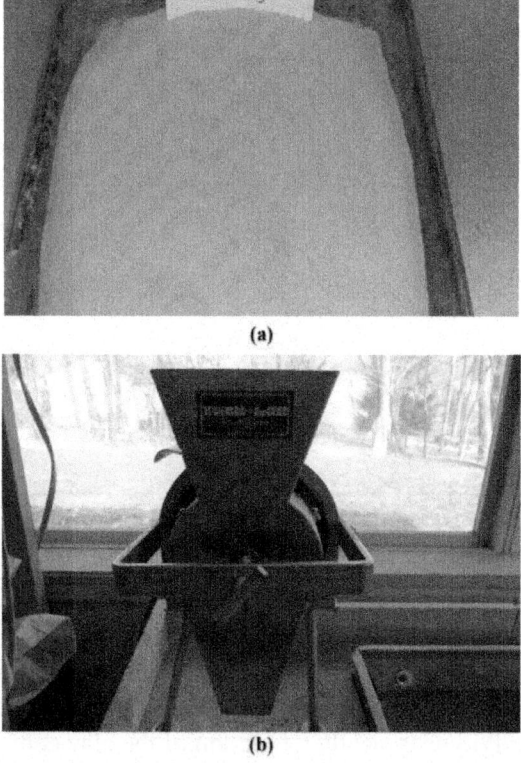

Figure 4: (a) A stockpiled sample processed through grinding (SG); (b) the soil grinder used for processing.

The results of particle size analysis, Atterberg limits, modified Proctor and hydraulic conductivity tests conducted on the SG-samples are displayed in Table 3. Figure 5 compares the grain-size distributions for SM-1 and SG-1 samples (from the same location). The plot shows a clear difference in the clay content in the gradation curves obtained from manual processing and grinder processing of the same S-samples.

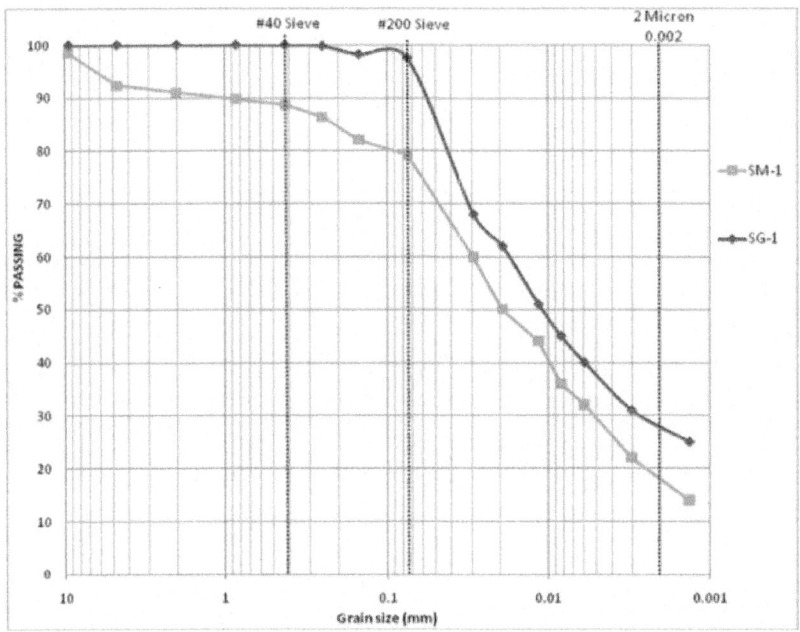

Figure 5: Comparison of grain size distributions: manually processed stockpiled sample (SM-1) *vs.* grinder processed stockpiled sample (SG-1).

However, referring back to Tables 1–3, little to no change is observed in other geotechnical characteristics. The small variations noted are well within the typical expected test margins of error for Atterberg limits, modified Proctor and hydraulic conductivity tests. On the other hand, the grain size distributions of SG-samples (Table 3) nearly mimic the results for the DM-samples (Table 1). Figure 6 compares the grain-size distributions of DM-1 and SG-1 samples.

Table 3: Geotechnical characteristics of SG-samples

Sample Number	Grain Size Distribution						Atterberg Limits[1]			USCS Soil Type	Compaction Characteristics Based On Modified Proctor Test[2]		Hydraulic Conductivity (cm/sec)
	% Finer	% Finer	% Finer	% Finer	% Finer	% Finer	LL	PL	PI		γ_{dmax} (kN/m³)	OMC (%)	
	12.5 mm	9.5 mm	#4	#40	#200	.002 mm							
SG-1	100.0	100.0	100.0	100.0	97.5	28.9	29	15	14	CL	21.9	7.27	8.4×10^{-6}
SG-2	100.0	100.0	100.0	90.0	80.0	20.2	27	19	8	CL	21.0	7.20	8.1×10^{-6}
SG-3	100.0	100.0	100.0	98.4	92.4	29.2	27	18	9	CL	22.0	7.08	8.8×10^{-6}
SG-4	100.0	100.0	100.0	91.8	89.5	28.1	28	18	10	CL	21.4	7.85	6.9×10^{-6}
SG-5	100.0	100.0	100.0	96.5	95.4	29.1	29	18	11	CL	–	–	–
SG-6	100.0	100.0	100.0	98.4	92.0	28.0	25	15	10	CL	–	–	–
SG-7	100.0	100.0	100.0	96.1	88.7	25.0	26	16	10	CL	–	–	–
SG-8	100.0	100.0	100.0	97.2	94.5	27.5	29	15	14	CL	–	–	–
SG-9	100.0	100.0	100.0	96.4	95.2	27.4	29	15	14	CL	–	–	–
SG-10	100.0	100.0	100.0	98.5	94.2	29.5	30	18	12	CL	–	–	–
SG-11	100.0	100.0	100.0	96.5	91.0	26.0	31	20	11	CL	–	–	–
SG-12	100.0	100.0	100.0	91.0	82.5	22.4	26	16	10	CL	–	–	–
SG-13	100.0	100.0	100.0	99.1	96.9	29.2	28	17	11	CL	–	–	–

[1] LL = liquid limit, PL = plastic limit, PI = plasticity index;
[2] γ_{dmax} = maximum dry unit weight, OMC = optimum moisture content.

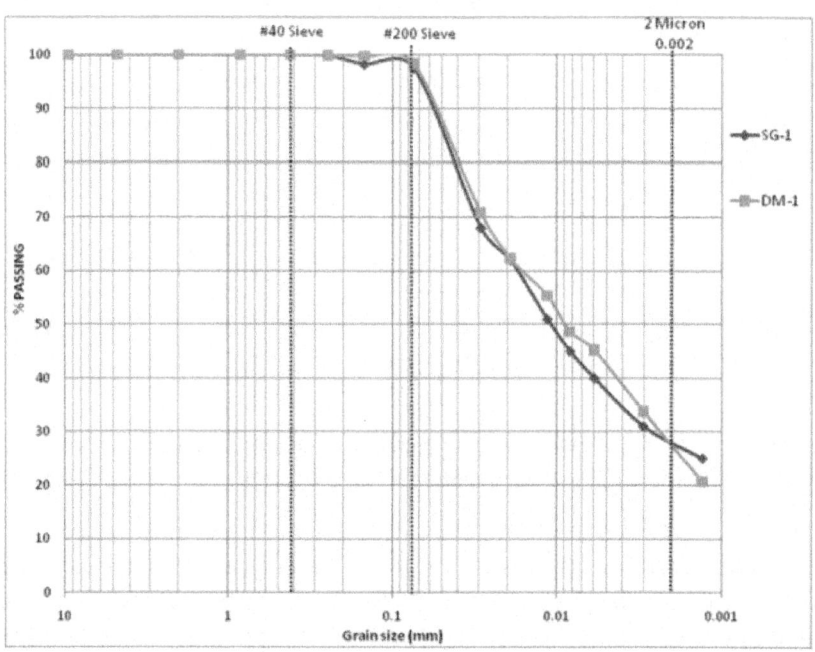

Figure 6: Comparison of grain size distributions: Manually processed drilled *vs.* grinder processed stockpiled samples (DM-1 & SG-1).

The results of the laboratory samples processed using a soil grinder, indicate the impact of the mechanical process to have a direct affect on the grain-size distribution. Since the same S-samples showed a clay content similar to that of DM when processed using a grinder (SG), it is clear that the discrepancy could be due to the soil clods that are not easily broken during processing.

Standard ASTM test protocols for sieve and hydrometer tests have been in use for some time and they have an excellent track record for producing repeatable and reliable data. The question that needs an answer is that why mined clay samples collected at the stockpiles have become an exception.

Stockpile Mined Clay Samples from Other Locations in the Midwest

To see if the use of a soil grinder has any impact on the grain size distribution, comparable clayey soil samples obtained from two other locations in the Midwest were analyzed. Four samples from southwestern Michigan and another four from central Kentucky were used for this purpose. Samples were split in half prior to processing. One set of samples was first processed by hand through a number 4 screen. The other set of samples was then processed using the soil grinder after air drying. Both sets were then subjected to gradation analysis, Atterberg limits, modified Proctor, and hydraulic conductivity testing.

Included in Table 4 and Table 5 are the results from the tests conducted on samples from southwestern Michigan and central Kentucky respectively. No significant change is noted in the Atterberg limits, modified Proctor, hydraulic conductivity as well as the gradation of the test specimen based on the processing method. Figure 7 compares the gradation curves of the mined clay from southwestern Michigan collected at the stockpiles but processed manually (MI SM-1) and using a soil grinder (MI SG-1). Figure 8 presents the same comparison for the samples from Central Kentucky (KY SM-1 and KY SG-1). Figure 7 and Figure 8 provide examples of how well the gradation curves from two methods compare with each other. The conclusion these results provide is that the discrepancy observed in mined clay from Appalachian Ohio is an exception and may not be due to a deficiency of the standard testing methods.

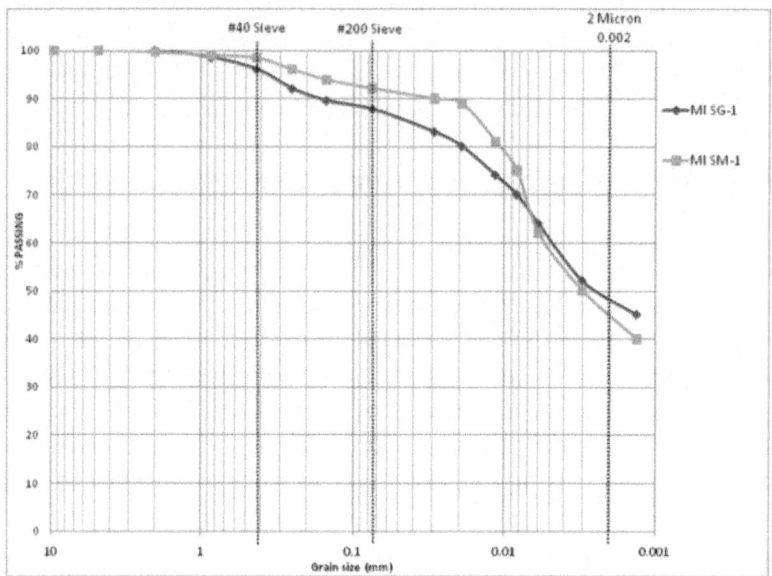

Figure 7: Comparison of grain size distributions: manually processed *vs*. grinder processed stockpiled samples of mined clay from Southwest Michigan (MI SM-1 & MI SG-1).

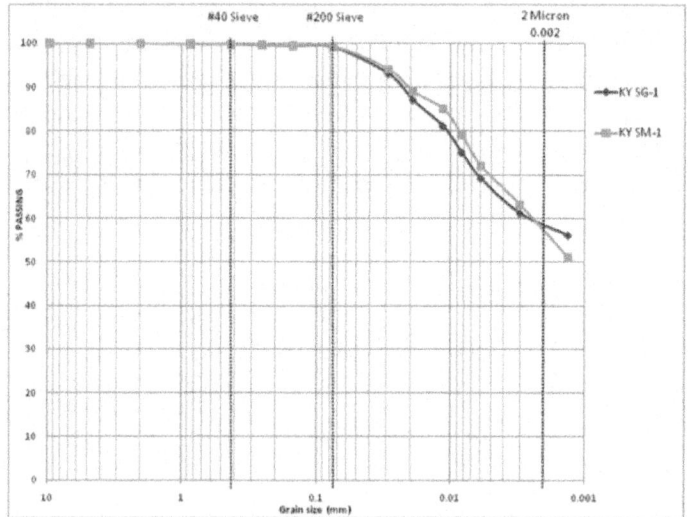

Figure 8: Comparison of grain size distributions: manually processed *vs*. grinder processed stockpiled samples of mined clay from Central Kentucky (KY SM-1 & KY SG-1).

Table 4: Geotechnical characteristics of stockpiled mined clay-samples from Southwest Michigan processed manually and using a soil grinder

Sample Number	Grain Size Distribution						Atterberg			USCS Soil Type	Compaction Characteristics Based On Modified Proctor Test[2]		Hydraulic Conductivity (cm/sec)
	% Finer	% Finer	% Finer	% Finer	% Finer	% Finer	Limits[1]				γ_{dmax} (kN/m³)	OMC (%)	
	12.5 mm	9.5 mm	#4	#40	#200	.002 mm	LL	PL	PI				
MI SM-1	100.0	100.0	99.6	98.4	92.1	45.5	38	24	14	CL	18.8	12.45	6.8 × 10⁻⁸
MI SM-2	100.0	100.0	100.0	99.2	93.9	31.9	34	16	18	CL	19.1	12.36	5.4 × 10⁻⁸
MI SM-3	97.0	96.5	94.2	86.5	62.8	21.1	24	16	8	CL	–	–	–
MI SM-4	100.0	100.0	99.0	98.2	96.7	50.7	39	24	15	CL	–	–	–
MI SG-1	100.0	100.0	100.0	96.0	87.8	48.9	38	24	14	CL	18.6	12.55	7.0 × 10⁻⁸
MI SG-2	100.0	100.0	100.0	100.0	96.5	38.2	35	16	19	CL	19.0	12.44	5.3 × 10⁻⁸
MI SG-3	100.0	100.0	98.5	88.2	63.5	25.2	27	16	11	CL	–	–	–
MI SG-4	100.0	100.0	99.2	98.2	96.9	51.0	40	22	18	CL	–	–	–

[1] LL = liquid limit, PL = plastic limit, PI = plasticity index;
[2] γ_{dmax} = maximum dry unit weight, OMC = optimum moisture content.

Table 5: Geotechnical characteristics of stockpiled mined clay-samples from Central Kentucky processed manually and using a soil grinder

Sample Number	Grain Size Distribution						Atterberg			USCS Soil Type	Compaction Characteristics Based On Modified Proctor Test[2]		Hydraulic Conductivity (cm/sec)
	% Finer	% Finer	% Finer	% Finer	% Finer	% Finer	limits[1]				γ_{dmax} (kN/m³)	OMC (%)	
	12.5 mm	9.5 mm	#4	#40	#200	.002 mm	LL	PL	PI				
KY SM-1	100.0	100.0	99.9	99.7	99.2	58.6	47	21	26	CL	18.0	15.27	3.2 × 10⁻⁸
KY SM-2	100.0	100.0	99.9	99.8	99.2	42.5	50	22	28	CH	18.5	13.89	4.1 × 10⁻⁸
KY SM-3	100.0	100.0	99.3	97.9	96.6	54.6	53	24	29	CH	–	–	–
KY SM-4	100.0	100.0	100.0	99.8	99.4	50.9	52	27	25	CH	–	–	–
KY GR-1	100.0	100.0	100.0	99.8	99.1	58.0	46	21	25	CL	18.3	14.91	3.4 × 10⁻⁸
KY GR-2	100.0	100.0	99.8	99.6	99.0	43.0	51	24	27	CH	18.5	14.01	4.1 × 10⁻⁸
KY GR-3	100.0	100.0	99.8	98.0	96.8	54.2	54	25	29	CH	–	–	–
KY GR-4	100.0	100.0	100.0	99.9	99.3	50.1	50	25	25	CH	–	–	–

[1] LL = liquid limit, PL = plastic limit, PI = plasticity index;
[2] γ_{dmax} = maximum dry unit weight, OMC = optimum moisture content.

Geological Background of Clay Deposits in Appalachian Basin

Information on the regional geology and the composition of mined clay from Appalachian Ohio was taken into consideration to investigate the discrepancies it shown in the grain size analysis. As explained before, soil found beneath coal layers at the base of the Appalachian basin is comprised of shale enriched clay deposits [1]. Shale in this region is mostly of lightly indurated silt and clay that can be broken with little force applied with the addition of moisture. The shale, depending on moisture content, will either appear as a gravel sized particle or as a clay clod. With the cyclothemic affects in Appalachian Ohio, one would expect a large variety and discontinuity within and between the sedimentary layers. Natural moisture content makes a direct influence on the visual classification of this soil. The higher the natural moisture content, the more likely the material will be visually classified as clayey soil and not as shale.

This specific composition of mined clay from Appalachian Ohio reveals that the soil clods observed in the stockpile samples are mostly the fragments of shale. Manual processing does not break all shale clods. The dispersing agent used during the hydrometer test is only able to separate some of the fines into individual particles, but not strong enough to break the shale fragments into the respective parent materials (*i.e.* clay and silt). Therefore, it is not surprising to see a lower clay fraction in SM-samples, because the small shale fragments may have behaved similar to coarse grained soils during the mechanical analysis. When a grinder was used to process soils, it was able to break the shale fragments into smaller clods. These smaller clods must have allowed the dispersing agent to more readily separate the shale fragments into its parent particles and increase the clay fraction in SG-samples. This explains why SG-samples behaved similar to DM-samples during mechanical analysis (Figure 6).

On the other hand, the mined clay from southwestern Michigan was from a glacial deposit and samples from central Kentucky were from a cyclothemic deposit. These samples from Michigan and Kentucky were high in clay content with traces of sand. Unlike the soil samples from Appalachian Ohio, none of these other samples would be considered to contain any shale fragments. As a result, they did not exhibit a considerable difference in the grain size distributions based on the processing method. Therefore, the presence of shale fragments should be the primary cause for the discrepancies observed in the grain-size analysis.

Impact on Clay Mining Industry

Hydraulic conductivity is the most important characteristic of RSL material and The State of Ohio has used 20 percent as a requirement for borrow soils used in RSL systems [2]. As per the results shown in Table 1 through Table 5, SM samples do not meet the 20% clay content requirement, but SG and DM-samples do meet this requirement. However, it is interesting to note that all three types of samples are from the same origin and they share the same hydraulic conductivity. Hydraulic conductivity values obtained for all three types of samples are compared in Figure 9. As can be seen in Figure 9 all hydraulic conductivity values are bordering along 1.0×10^{-7} cm/sec regardless of the sample collection/processing method.

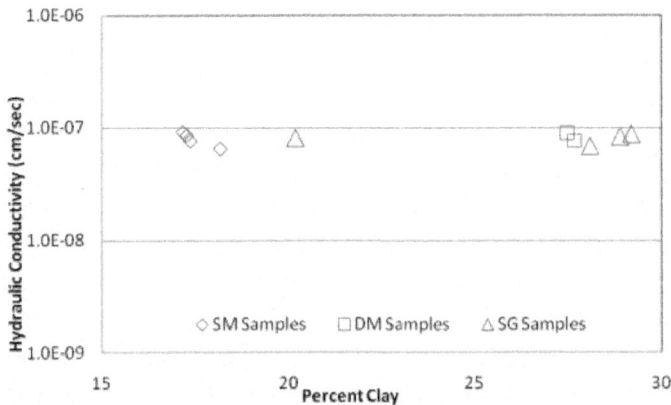

Figure 9: Hydraulic conductivity *vs.* percent clay content obtained for DM, SM and SG samples.

When soil samples are delivered to a geotechnical laboratory for testing, geological history, sampling methods, and the expectations of soil evaluation results are seldom provided to the laboratory staff. The geotechnical laboratory is typically expected to act as an independent third party in the soil evaluation process even if the laboratory is within the same company as the sampling technician. This independent behavior works well when the testing standards prescribed within the industry include limited error margins. However, this intentional lack of communication between the geotechnical laboratory and the geotechnical engineer can cause problems when soils with unique characteristics such as the mined clay from Appalachian Ohio are tested.

If not instructed specifically, a laboratory may or may not use a grinder for processing soils, and depending on that decision, even a suitable soil may very well be rejected. It was reported that in 2003, over 2.1 million tons of clay were mined in Ohio [6]. With prices averaging $20/ton, selling mined clay for RSL construction is significant to Ohio's mining industry. Therefore, decision making purely based on the apparent clay content without paying much attention to the hydraulic conductivity, can make a considerable negative impact on the clay mining industry.

CONCLUSIONS

The following conclusions were from the material discussed in this research.

It should be noted that all conclusions are exclusively applicable to mined clay from Appalachian Ohio and may not be applicable to other general type of clayey soils.

- Grain size distributions of mined clay from Appalachian Ohio exhibit sensitivity to sampling and/or laboratory processing methods. Other geotechnical characteristics such as Atterberg limits, compaction (based on modified Proctor method) and hydraulic conductivity were not sensitive to sampling and or laboratory processing methods.

- The Presence of shale fragments in mine clay from Appalachian Ohio was found to be the reason for observing sampling and laboratory processing dependant results in particle size analysis of mined clay from Appalachian Ohio.

- Manual processing is not sufficient to separate shale fragments into individual silt and clay particles in the soil samples collected at mined clay stockpiles. Samples collected through drilling or samples collected at the stockpiles but processed using a soil grinder, seem to be producing more reliable grain size distribution data for mined clay from Appalachian Ohio.

- Use of clay content as the sole criteria to accept or reject Mined clay from Appalachian Ohio for RSL applications could produce misleading results. If clay content is the only criteria to be used, the samples for mechanical analysis should be collected through drilling or else samples collected at stockpiles should be processed using soil grinder.

REFERENCES

1.   Coogan, AH. Ohio Surface Rocks and Sediments. In *Fossils of Ohio*; Rodney, MF, Ed.; Ohio Geological Survey: Columbus, OH, USA, 1996; Bulletin 70; Volume Chapter 3, pp. 31–50.

2.   *Sanitary Landfill Facility Construction, OAC-3745-27-08*; Ohio Environmental Protection Agency: Columbus, OH, USA, 2003.

3.   Das, BM. *Principles of Geotechnical Engineering*, 6th ed; Thomson Canada Limited: Toronto, Ontario, Canada, 2006.

4.   *ASTM Annual Book of Standards, Section Four, Construction. Soil and Rock (i)*; ASTM International: West Conshohocken, PA, USA, 2008; pp. D420–5876.

5.   Moran, A. Characterization of Mined Clay from Appalachian Ohio: Implications of Sampling and Laboratory Processing Techniques on Local Regulations. M.D. thesis, Department of Civil Engineering, Lawrence Technological University, Southfield, MI, USA, 2009.

6.   Wolfe, M. *Rocks and Minerals Mined in Ohio and Their Uses, GeoFacts No. 11*; The Division of Geological Survey, Ohio Department of Natural Resources: Columbus, OH, USA, 2005.

Chapter 9

NEW DEVELOPMENTS IN GEOTECHNICAL EARTHQUAKE ENGINEERING

Yang Changwei,[1] Su Tianbao,[2] Zhang Jianjing,[1] and Du Lin[1]

[1]School of Civil Engineering, Key Laboratory of Transportation Tunnel Engineering, Ministry of Education, Southwest Jiaotong University, Chengdu 610031, China

[2]Henan University of Urban Construction, Pingdingshan, Henan 467036, China

ABSTRACT

Based on the review on the advances of several important problems in geotechnical seismic engineering, the authors propose the initial analysis theory of time-frequency-amplitude (known as TFA for short), in an effort to realize the organic combination of time and frequency information and develop a groundbreaking concept to the traditional idea in the geotechnical seismic engineering area.

REVIEW ON THE PRESENT MAIN DEFECT IN THE GEOTECHNICAL EARTHQUAKE ENGINEERING

Geotechnical seismic engineering is an important area in geotechnical engineering, and its topic is to resolve the seismic problems related to geotechnical engineering. With development for more than 50 years, the time domain analysis theory and the frequency analysis theory have been generally established in the geotechnical seismic engineering [1–6]. However, a common problem still exists in the time domain analysis theory and the frequency domain theory, which is that the research methods for geotechnical seismic engineering are still the separate application of time domain or frequency domain, and no combination of above two is considered. [7–11]. For example, time domain analysis theory can only consider the time histories of acceleration, velocity, and displacement but not the frequency contents of ground motions. Frequency domain analysis theory can only consider the frequency contents of ground motions, while time histories of acceleration, velocity, and displacement are not included. Seismic wave is however a very complex nonstationary signal, whose amplitudes and frequencies change with time. Present outcomes in the time domain theory and the frequency domain theory cannot reflect synthetically the characteristics of geotechnical seismic engineering, very close to an old Chinese saying "The blind man feels an elephant—to take a part for the whole." Therefore, research on the time-frequency-amplitude analysis theory considering the time, frequency, and amplitude can be an important task, which will be a new direction of geotechnical seismic engineering.

BRIEF INTRODUCTION OF THE SIGNAL ANALYSIS TECHNOL-OGIES

Fourier transform and wavelet transform are two seismic signal analysis technologies [12]. Fourier transform is a steady-state analysis technology and suitable for frequency domain analysis. Wavelet transform has some time-frequency domain resolution ability, but it is hard to carry out an accurate time-frequency analysis [13] because of uncertainty principle. Based on the Hilbert transform, Huang et al. proposed a new signal analysis method for nonstationary signal [14] called Hilbert-Huang transform (HHT). The approach can perform linearized and stabilized analysis for nonlinear and nonstationary signals, initial data in the analysis can be retained, and energy leaking can be avoided. Studies in the past showed that the results from HHT can reflect true cases. The concrete calculation principle of Hilbert-Huang transform is shown as follows: generally, a nonstationary signal can be decomposed into a series of intrinsic mode function (known as IMF for short) components by using ensemble empirical mode decomposition (known as EEMD for short) method. The instantaneous frequency of each IMF will be derived by the Hilbert-Huang transform, and then the Hilbert spectrum will be obtained by integrating all the instantaneous frequency spectrums. The analytical signal, amplitude function, phase function, and instantaneous frequency function can be obtained by using the Hilbert-Huang transform. The computational formulas are shown in

$$Z(t) = c(t) + jH[c(t)] = a(t)e^{j\phi(t)},$$

$$a(t) = \sqrt{c^2(t) + H^2[c(t)]},$$

$$\phi(t) = \arctan\frac{H[c(t)]}{c(t)},$$

$$f(t) = \frac{1}{2\pi}\frac{d\phi(t)}{d(t)}.$$

(1)

From the above formulas, it is known that both amplitude and frequency are the function of time. If the amplitude is calculated by using a combined time and frequency domain approach, the Hilbert spectrum can be obtained, as shown in (2). The marginal spectrum and the instantaneous frequency spectrum can be obtained by using (3) and (4). The Hilbert spectrum can be obtained by integrating amplitude squared, as shown in (5).

Consider

$$H(\omega, t) = \mathrm{Re} \sum_{i=1}^{n} a_i(t) e^{j\phi_i(t)},$$
(2)

$$h(\omega) = \int_0^T H(\omega, t)\, dt,$$
(3)

$$IE(t) = \int_\omega H(\omega, t)\, dt,$$
(4)

$$ES(\omega) = \int_0^T H^2(\omega, t)\, dt.$$
(5)

New Development in the Geotechnical Earthquake Engineering

Up to now, our team has conducted a field investigation along the road and railway nearly 3000 km in 5.12 Wenchuan earthquake-stricken areas (see Figure 1), a number of field monitoring of typical slopes, more than twenty shaking table tests, a number of numerical simulations and theoretical studies [15, 16]; some shaking tabletest models are shown in Figures 2, 3, 4, 5, and 6.

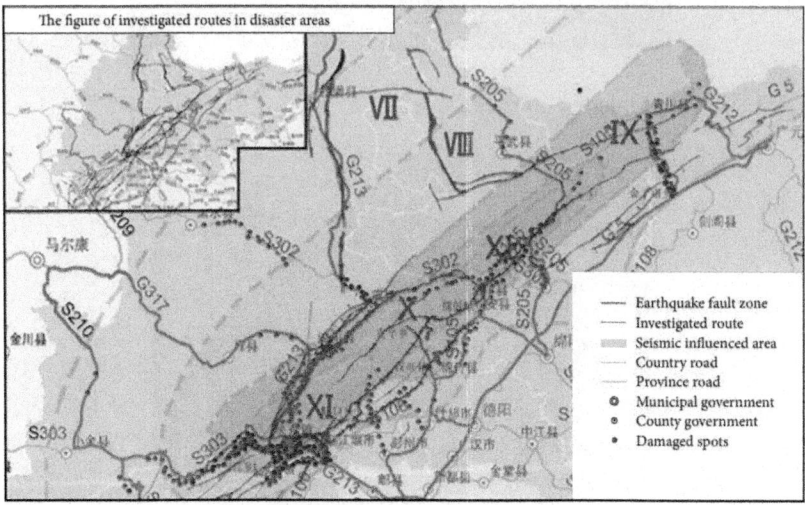

Figure 1: Roads investigated and seismic intensity.

Figure 2: Shaking table test of site under nonuniform seismic excitation.

Figure 3: Shaking table test of rock slope.

Figure 4: Shaking table test of retaining wall.

Figure 5: Shaking table test of antisliding pile.

Figure 6: Simulation test of fault slip.

On the basis of extensive studies, several important problems in the geotechnical seismic engineering are studied, including analysis methods of seismic responses of the site, analysis methods of seismic responses and seismic stability of the retaining structures, analysis methods of seismic responses of the slope, analysis methods of the landslide mechanism, and the analysis method of seismic stability of the slope. With the help of above-mentioned signal analysis skills in the aerospace area and signal analysis area (e.g., Hilbert-Huang transform), our team has conducted a series of analyses on civil structure depending on the proposed time-frequency-amplitude (known as TFA for short) analysis theories, which are comprised of horizontal layers, inclined layers, gravity retaining wall, reinforced retaining wall, slopes, and landslide mechanism [17, 18]. Our team gives the initial prototype of TFA analysis theories in geotechnical seismic engineering, which realizes the organic combination of time and frequency information, and develops a groundbreaking concept to the traditional idea in the geotechnical seismic engineering area.

ESSENCE OF THE TIME-FREQUENCY-AMPLITUDE ANAL-YSIS THEORIES

The time-frequency-amplitude analysis theories give reasonable considerations to the time-frequency characteristics of seismic waves, mainly reflected in the elastic displacement magnitude and the frequency of the wave vector. This time, this paper selects Wolong earthquake wave (as shown in Figure 7) to make an explanation; procedures are as follows: first, Wo-Long wave is decomposed into intrinsic mode functions (IMF) using the empirical mode decomposition (EMD) method. The acceleration time histories of each IMF can be seen in Figure 8 and the instant frequency time histories of each IMF can be seen in Figure 9. Based on the acceleration time histories and instant frequency time histories of each IMF, bring them into the related formulas in order to get the stress time histories of each IMF; finally, sum the stress time histories together in order to obtain the total results.

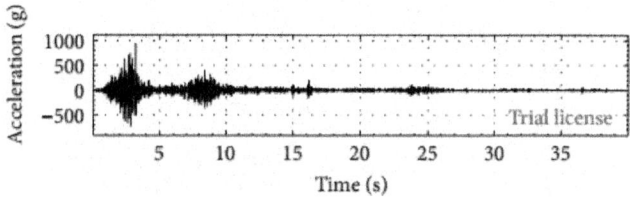

Figure 7: Time history of the Wenchuan-Wolong seismic wave.

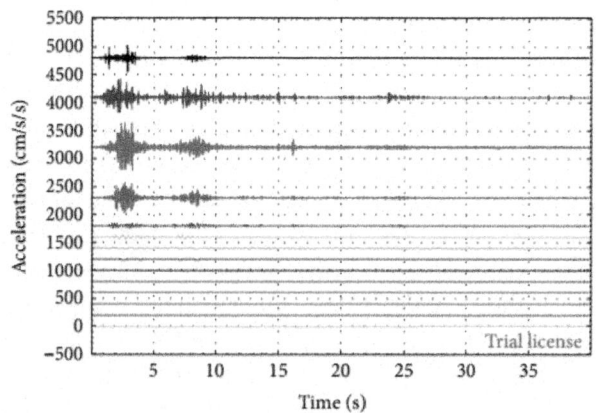

Figure 8: Acceleration time history of IMF.

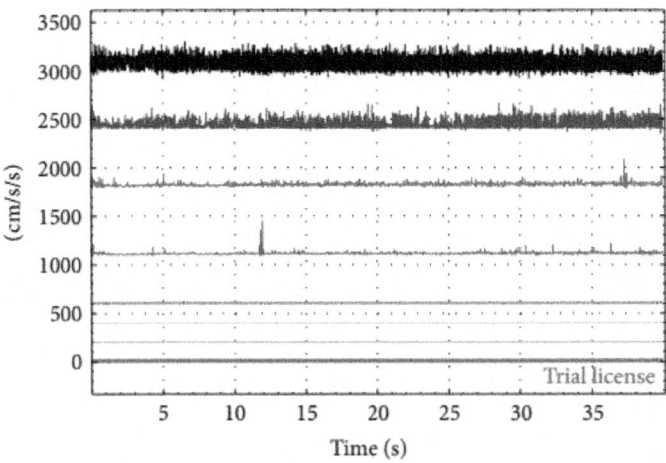

Figure 9: Instant frequency time history of IMF.

Advantages of the Time-Frequency Analysis Method of Seismic Safety of the Rock Slope

Comparisons with pseudostatic methods using the various countries' earthquake resistant design codes, TFA analysis theory of seismic stability of rock slope is capable of considering the time, frequency, and amplitude of seismic wave, and then a brief introduction of the above-mentioned theory is made as follows.

In the TFA analysis theory of seismic stability of rock slope, the calculated model is shown in Figures 10 and11. In the model, the sliding mass was divided into seven slices, and an angle of θ between the tangent and horizontal directions at any point A was obtained at the middle of a slice. Figure 11 shows the reflection and refraction model at point A. The mechanical parameters in the upper medium (bedrock) are ρ_1, C_{p1}, C_{s1}, and G_1 (density, P wave speed, S wave speed, and shear modulus) and ρ_2, C_{p2}, C_{s2}, and G_2 in the lower medium (colluvial soils), and the I-I stands for the sliding surface. When transverse wave S^1 reaches the I-I surface, some new waves are produced, for example, reflected transverse wave S^2 , reflected longitudinal wave S^3 , refracted transverse wave S^4, and refracted longitudinal wave S^5. α_1, α_1', β_1, and β_1' stand for the incident angle of the incident transverse wave, the reflected angle of the reflected longitudinal wave, the refracted angle of the refracted transverse wave, and the refracted angle of the refracted longitudinal wave. Note that the reflected angle of the reflected transverse wave is the same as the incident angle of the incident transverse wave.

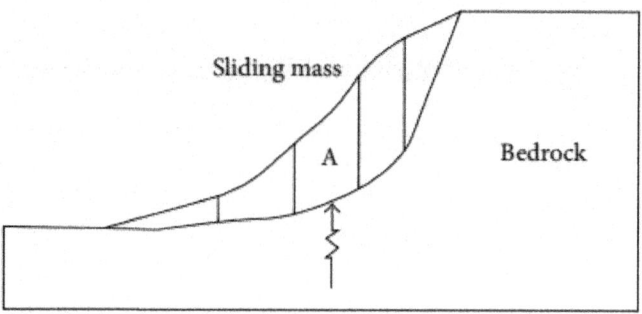

Figure 10: Generalized model of the bedrock-regolith slope.

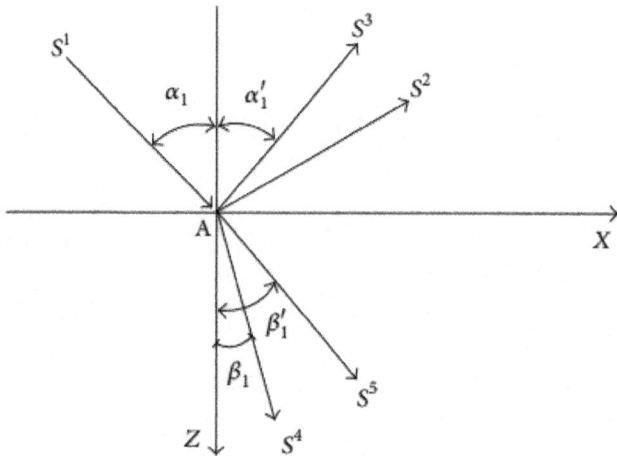

Figure 11: Analysis model of reflection and refraction at any point A.

After the derivation, the stress state at any point can be calculated according to (6) and (7). And then, the stress state at whole sliding surface can be calculated according to (8). At last, the seismic stability of slope can be estimated according to (9).

Consider

$$\tau_s = \tau_0 + \mu_1 \left[S_0^1 \cdot k_z^{11} \cdot \cos \alpha_1 - S_0^2 \cdot k_z^{21} \cdot \cos \alpha_1 \right.$$
$$+ S_0^3 \cdot k_z^{31} \cdot \sin \alpha_1' - S_0^1 \cdot k_x^{11} \cdot \sin \alpha_1 - S_0^2 \cdot k_x^{21}$$
$$\left. \cdot \sin \alpha_1 - S_0^3 \cdot k_x^{31} \cdot \cos \alpha_1' \right],$$

$$(6)$$

$$\sigma_0 \tan \varphi + C$$
$$= \left\{ \lambda_1 \left[S_0^1 \cdot k_x^{11} \cdot \cos \alpha_1 - S_0^2 \cdot k_x^{21} \right. \right.$$
$$\left. \cdot \cos \alpha_1 + S_0^3 \cdot k_x^{31} \cdot \sin \alpha_1' \right] + (\lambda_1 + 2\mu_1)$$
$$\times \left[-S_0^1 \cdot k_z^{11} \cdot \sin \alpha_1 - S_0^2 \cdot k_z^{21} \cdot \sin \alpha_1 \right.$$
$$\left. \left. - S_0^3 \cdot k_z^{31} \cdot \cos \alpha_1' \right] + \sigma_0 \right\} \tan \varphi + C.$$

$$(7)$$

In (6) and (7), S_0^1 stand for the displacement of incident transverse wave, S_0^2 stand for the displacement of reflected transverse wave, S_0^3 stand for the displacement of reflected longitudinal wave, S_0^4 stand for the displacement of refracted transverse wave, and S_0^5 stand for the displacement of refracted longitudinal wave; k_x^{i1}, k_z^{i1} stand for the wave vectors in the X and Z directions of the incident, reflected, and the refracted waves, respectively; C stand for the cohension of the sliding mass and φ stand for the internal friction angle. Consider

$$F_s = \sum_{i=1}^{n} \tau_{si} \cdot dA_i;$$

$$(8)$$

$$F_r = \sum_{i=1}^{n} (\sigma_{0i} \cdot \tan \varphi_i + C_i) \cdot dA_i,$$

$$K = \frac{F_r}{F_s} < K_0 \quad \text{risk;}$$

$$K = \frac{F_r}{F_s} > K_0; \quad \text{safety.}$$

$$(9)$$

In (8) and (9), τ_{si} represents the sliding shear stress at point $A_i = 0.5 (L_{i-1} + L_{i+1}) \times 1$ is the slice area, L_{i-1} is the slice area, L_{i-1} is the width of the $i-1$th slice and L_{i+1} is the width of the $i+1$th slice; σ_{0i} represents the normal stress at point i; φ_i represents the frictional angle at point i; Ci represents the cohesion at point i. According to the above-mentioned method, a simple example is used to illustrate the advantage of the TFA analysis theory: model for calculation is shown in Figure 12. Figures 13 and 14 give the adopted sine wave whose instantaneous frequencies gradually increase with time and the calculated parameters can be seen in Table 1.

Table 1: Physical and mechanics parameters of bedrock, soil layer, and structural plane.

Model	Gravity	Shear wave velocity	Longitudinal wave velocity	Lame coefficients	Shear modulus
Bedrock	30 kN/m³	826.19 m/s	3240 m/s	881.89 MPa	2047.24 MPa
Slip mass	19 kN/m³	99.21 m/s	145.34 m/s	6.30 MPa	6.30 MPa
Sliding surface	Normal stiffness	Shear stiffness	Internal friction angle	Tensile strength	Cohesion
	4500 MPa	2300	32°	23 kPa	50 kPa

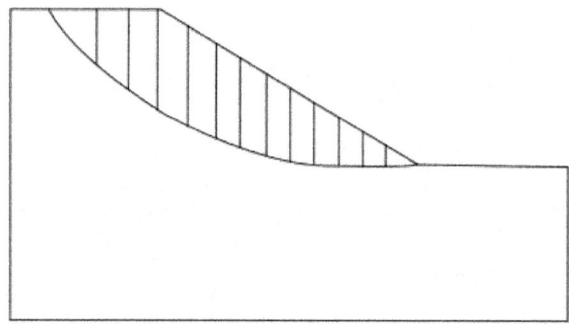

Figure 12: Calculated model.

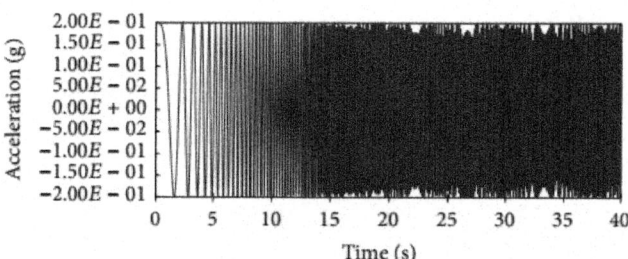

Figure 13: Time history of sine wave.

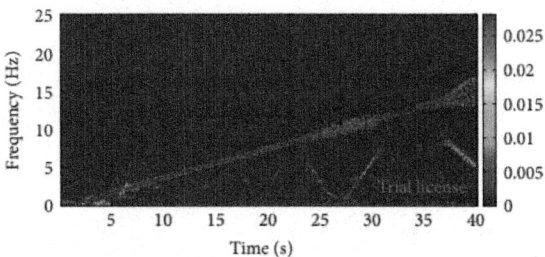

Figure 14: Hilbert spectrum of sine wave.

After the calculation is done, results of TFA analysis method are shown in Figure 15, indicating that the slope would fail after T = 10.0 s. The result of pseudostatic method shows that the surplus sliding force is −47 kN and means that the slope is stable. While the slope is proven to fail in the numerical simulation, which is consistent with the result of TFA analysis method shown in Figure 16, this phenomenon can be elucidated as follows

(i)   At T=10.0s, the instantaneous frequency of the seismic wave is 4.9 Hz and the natural frequency calculated according to the literature [17] is 6.8 Hz, with both being very close to each other, leading to a further resonance. However, the instantaneous frequency of the input seismic wave reaches 6.8 Hz at T=18.0s, and the resonance shows the strongest value, which leads to the maximum surplus shear force. Then the resonance gradually decreases with difference between the frequencies of the input wave and the natural frequency goes up, and the surplus shear force gradually decreases in response.

(ii)   Time-frequency analysis method and numerical method both not only can consider the effect of PGA on the seismic stability of slope, but also consider the effect of frequency on the seismic stability of slope. But pseudo-static method only can consider the effect of PGA on the seismic stability of slope, which leads to the difference between the calculated results.

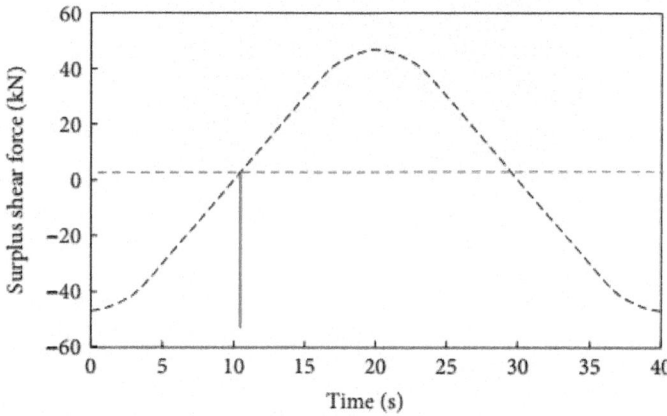

Figure 15: Result of TFA analysis method.

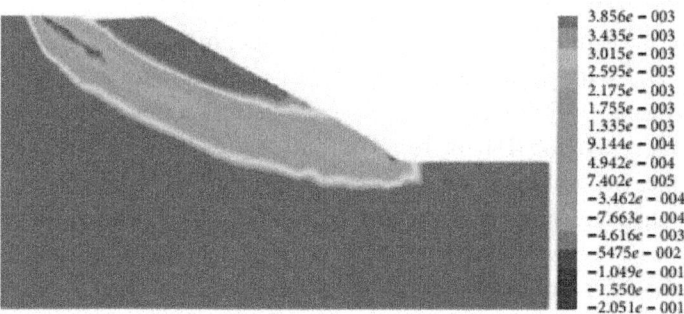

Figure 16: Results of numerical simulation.

Synthesizes the above analysis, we can know that, therefore, the TFA analysis method can solve the fundamental flaw existing in the analysis method using the codes.

APPLICATION IN PRACTICE

Up to now, the TFA analysis theory has been already used to direct the aseismic designs of geotechnical engineering, containing several kilometers of YaLe expressway, and the contribution to several traffic lines opening.

CONCLUSIONS AND FUTURE WORKS

The main conclusions of this study are as follows.(i)Based on the above analysis, this paper gives an initial prototype of time-frequency-amplitude analysis theory in the geotechnical seismic engineering. This theory realizes the organic combination of time information and frequency information and gives reasonable considerations to the time-frequency characteristics of seismic waves, which mainly is reflected in the elastic displacement magnitude and the frequency of the wave vector and that is the core content of TFA analysis theories. (ii)Comparing to the results of TFA analysis theory of rock slope, pseudostatic method, and numerical analysis method shows that the TFA analysis theory can consider the effect of the time and frequency of the input wave on the seismic stability of slope, which solves the fundamental flaw existing in the analysis method using the codes. At the same time, according to the derivation results, we can know that the time-frequency analysis method of rock slope has some universalization and comprehensiveness, which can consider some factors reasonably as follows: (1) initial geostress field; (2) the inner frictional angle, the cohesion, the reflection coefficients, and the refraction coefficients; (3) the physical and mechanical properties of the bedrock and the regolith; and (4) the instantaneous frequency, the incidence angle, the time histories of the acceleration, the velocity, the displacement, and so forth.(iii)The time-frequency-amplitude analysis theory is built in compliance with the elastic wave theory and geotechnical seismic engineering in which the elastic state of the system is considered. The next works are making the further studies based on the present outcomes; time-frequency-amplitude analysis theory can be thus proposed to consider the nonlinear of the geotechnical engineering and earthquake wave.

CONFLICT OF INTERESTS

The authors declare that there is no conflict of interests regarding the publication of this paper.

ACKNOWLEDGMENTS

This study is supported in part by the National Science Foundation of China (Contract no. 41030742), NBRP of China (973 Program, 2011CB013605), and Construction and Science and Technology Project of the Ministry of Communications (Contract no. 2013318800020).

REFERENCES

1.   D. K. Keefer, "Landslides caused by earthquakes," Geological Society of America Bulletin, vol. 95, pp. 406–421, 1984.

2.   J. N. Hutchinson, "Mechanism producing large displacements in

landslides on preexisting shears,"Memoir of the Society of Geology of China, vol. 9, pp. 175–200, 1987.

3. K. Sassa, "Prediction of earthquake induced landslides," in Proceedings of the 7th International Symposium on Landslides, K. Senneset, Ed., vol. 1, pp. 115–132, Balkema, Rotterdam, The Netherlands, 1996.

4. Prestininzi and R. Romeo, "Earthquake-induced ground failures in Italy," Engineering Geology, vol. 58, no. 3-4, pp. 387–397, 2000.

5. M. Lin and K. Wang, "Seismic slope behavior in a large-scale shaking table model test," Engineering Geology, vol. 86, no. 2-3, pp. 118–133, 2006.

6. M. Chigira, X. Wu, T. Inokuchi, and G. Wang, "Landslides induced by the 2008 Wenchuan earthquake, Sichuan, China," Geomorphology, vol. 118, no. 3-4, pp. 225–238, 2010.

7. J. Wasowski and V. Del Gaudio, "Evaluating seismically induced mass movement hazard in Caramaico Terme (Italy)," Engineering Geology, vol. 58, no. 3-4, pp. 291–311, 2000.

8. H. B. Havenith, D. Jongmans, E. Faccioli, K. Abdrakhmatov, and P. Bard, "Site effect analysis around the seismically induced Ananevo rockslide, Kyrgyzstan," Bulletin of the Seismological Society of America, vol. 92, no. 8, pp. 3190–3209, 2002.

9. H.-B. Havenith, A. Strom, D. Jongmans, K. Abdrakhmatov, D. Delvaux, and P. Tréfois, "Seismic triggering of landslides, part A: field evidence from the Northern Tien Shan," Natural Hazards and Earth System Science, vol. 3, no. 1-2, pp. 135–149, 2003. ·

10. S. A. Sepúlveda, W. Murphy, R. W. Jibson, and D. N. Petley, "Seismically induced rock slope failures resulting from topographic amplification of strong ground motions: the case of Pacoima Canyon, California," Engineering Geology, vol. 80, no. 3-4, pp. 336–348, 2005.

11. L. Che and X. R. Ge, "Earthquake-induced toppling failure mechanism and its evaluation method of slope in discontinuous rock mass," International Journal of Applied Mechanics, vol. 4, no. 3, Article ID 1250036, 15 pages, 2012.

12. Z. Zuwu and Y. Lingkan, "Seismic waves scattering in rock interface and energy dissipation characteristics—taking M8.0 Wenchuan Earthquake as an example," Journal of Catastrophology, no. 1, pp. 5–9, 2011.

13. S. Xu, W. Zheng, Y. Liu, D. Xi, and G. Li, "A preliminary analysis of scale effect of elastic wave propagation in rock mass," Chinese Journal of Geotechnical Engineering, vol. 33, no. 9, pp. 1348–1356, 2011.

14. N. E. Huang, Z. Shen, S. R. Long et al., "The empirical mode decomposition and the Hilbert spectrum for nonlinear and non-stationary time series analysis," The Royal Society of London A, vol. 454, pp. 903–995, 1998.

15. J. Zhang, H. Qu, Y. Liao, and Y. Ma, "Seismic damage of earth structures of road engineering in the 2008 Wenchuan earthquake," Environmental Earth Sciences, vol. 65, no. 4, pp. 987–993, 2012.

16. Y. Changwei and Z. Jianjing, "A prediction model for horizontal run-out distance of landslides triggered by Wenchuan Earthquake," Earthquake Engineering and Engineering Vibration, vol. 12, no. 2, pp. 201–208, 2013.

17. C. Yang, J. Zhang, and D. Zhou, "Research on time-frequency analysis method for seismic stability of rock slope subjected to SV wave," Chinese Journal of Rock Mechanics and Engineering, vol. 32, no. 3, pp. 483–491, 2013.

18. C. Yang and J. Zhang, "Landslide responses of high steep hill with two-side slopes under ground shaking," Journal of Southwest Jiaotong University, vol. 48, no. 3, pp. 415–422, 2013.

Chapter 10

SOFT SCHEMES FOR EARTHQUAKE-GEOTECH-NICAL DILEMMAS

Silvia García

Geotechnical Department, Institute of Engineering, National University of Mexico, Mexico

ABSTRACT

Models of real systems are of fundamental importance in virtually all disciplines because they can be useful for gaining a better understanding of the organism. Models make it possible to predict or simulate a system's behavior; in earthquake geotechnical engineering, they are required for the design of new constructions and for the analysis of those that exist. Since the quality of the model typically determines an upper bound on the quality of the final problem solution, modeling is often the bottleneck in the development of the whole system. As a consequence, a strong demand for advanced modeling and identification schemes arises. During the past years, soft computing techniques have been used for developing unconventional procedures to study earthquake geotechnical problems. Considering the strengths and weaknesses of the algorithms, in this work a criterion to leverage the best features to develop efficient hybrid models is presented. Via the development of schemes for integrating data-driven and theoretical procedures, the soft computing tools are presented as reliable earthquake geotechnical models. This assertion is buttressed using a broad history of seismic events and monitored responses in complicated soils systems. Combining the versatility of fuzzy logic to represent qualitative knowledge, the data-driven efficiency of neural networks to provide fine-tuned adjustments via local search, and the ability of genetic algorithms to perform efficient coarse-granule global search, the earthquake geotechnical problems are observed, analyzed, and solved under a holistic approach.

INTRODUCTION

There are significant challenges for the future development and application of earthquake-geotechnical engineering that requires innovative approaches within a multidisciplinary framework. Very useful and up-to-date information on the occurrence frequency and impact of earthquake disasters is being assessed and analyzed by a number of organizations around the world. The earthquake-geotechnical engineering is an important bridge between geology, geomorphology, seismology, and civil engineering and serves as the environment where integrated and multidisciplinary approaches can be developed. In such applications, regarding specialized geotechnical engineering merely as a subset of civil engineering will lead to incomplete understanding of problems and the development of inadequate or incomplete solutions. Narrow perspectives can also suffocate progress and innovation. Links between the geosciences, seismology, mathematics, computing, and geotechnical engineering in terms of common concerns and needs such as obtaining, organizing, validating, displaying, and interpreting surface and subsurface must be considered.

One must look at the big picture for understanding past developments and present practices and for developing valid perspectives of the future. This should not mean simply following well-worn paths and considering progress only in terms of improvements, adjustments, and modifications of the current elements of what is regarded as good practice. Adopting new paradigms may be desirable or even necessary. The inclusion of innovative archetype of knowledge and skills concerning computing, creativity, and globalization in addition to factual and analytical knowledge seems mandatory. Innovative computing refers to the importance of exploiting the emerging processing resources to improve basic concepts and logic rather than the present emphasis only on faster applications and the blind using of software developed by others.

A competent modeling of engineering systems, when they are affected by seismic activity, has to be flexible enough to handle various degrees of complexity and uncertainty, and at the same time be sufficiently powerful to deal with situations in which the input signal may or may not be controllable. Mathematically based models are developed using scientific theories and concepts that just apply to particular conditions. Thus, the core of the model comes from assumptions that for complex systems usually lead to simplifications (perhaps oversimplifications) of the problem phenomena. It is fair to argue that the representativeness of a particular theoretical model largely depends on the degree of comprehension the developer has on the behavior of the actual engineering problem. Predicting natural-phenomena characteristics like those of earthquakes, and thereupon their potential effects at particular sites, certainly belongs to a class of problems we do not fully understand. Accordingly, analytical modeling often becomes the bottleneck in the development of more accurate procedures. Soft computing (SC) technologies have provided us with a unique opportunity to establish coherent seismic analysis environments in which uncertainty and partial data knowledge are systematically handled.

By seamlessly combining learning, adaptation, evolution, and fuzziness, SC complements current engineering approaches allowing us to develop a more comprehensive and unified framework to the effective management of earthquake phenomena. Each SC algorithm has well-defined labels and could usually be identified with specific scientific communities. Lately, as we improved our understanding of these algorithms' strengths and weaknesses, we began to leverage their best features and develop hybrid algorithms that indicate a new trend of coexistence and integration between many scientific communities to solve a specific task. In this paper, geotechnical aspects of earthquake engineering under a soft examination are covered. Via the development of reinterpretations of the following selected topics: (i) spatial variation of soil dynamic properties, (ii) attenuation laws for rock sites (seismic input), (iii) generation of artificial-motion time histories, and (iv) evaluation of liquefaction susceptibility, SC techniques are presented as appealing alternatives for integrated data-driven and theoretical procedures to generate reliable and intelligent geoseismic models.

The author of this document is well aware that standards for geotechnical seismic design are under development worldwide. While there is no need to "reinvent the wheel", there is a requirement to adapt such initiatives to fit the emerging safety philosophy and demands. This investigation also strongly endorses the view that "guidelines" are far more desirable than "codes" or "standards" disseminated all over seismic regions. Flexibility in approach is a key ingredient of geotechnical engineering and the cognitive technology in this area is rapidly advancing. The science and practice of geotechnical earthquake engineering is far from mature and needs to be expanded and revised periodically in the coming years. It is important that readers and users of the computational models presented here familiarize themselves with the latest advances and amend the recommendations herein appropriately.

The following sections are not intended to be a detailed treatise of the latest research in geotechnical earthquake engineering, but to provide sound guidelines to support rational cognitive approaches. While every effort has been made to make the material useful in a wider range of applications, applicability of the material is a matter for the user to judge. The main aim of this guidance document is to promote consistency of cognitive approach to everyday situations and, thus, improve geotechnical-earthquake aspects of the performance of the built safe environment.

COMPUTATIONAL INTELLIGENCE: SOFT COMPUTING TECHNOLOGIES

The computational intelligence is a synergistic integration of essentially three computing paradigms, namely, neural networks, fuzzy logic, and evolutionary computation entailing probabilistic reasoning (belief networks, genetic algorithms, and chaotic systems) [1]. This synergism provides a framework for flexible information processing applications designed to operate in the real world and is commonly called soft computing (SC) [2]. Soft computing technologies are robust by design and operate by trading off precision for tractability. Since they can handle uncertainty with ease, they conform better to real world situations and provide lower cost solutions.

The three components of soft computing differ from one another in more than one way. Neural networks operate in a numeric framework and are well known for their learning and generalization capabilities. Fuzzy systems [3] operate in a linguistic framework, and their strength lies in their capability to handle linguistic information and perform approximate reasoning. The evolutionary computation techniques provide powerful search and optimization methodologies. All the three facets of soft computing differ from one another in their time scales of operation and in the extent to which they embed a priori knowledge.

Figure 1 shows a general structure of soft computing technology. The following main components of SC are known by now: fuzzy logic (FL), neural networks (NN), probabilistic reasoning (PR), genetic algorithms (GA), and chaos theory (ChT) (Figure 1). In SC, FL is mainly concerned with imprecision and approximate reasoning, NN with learning, PR with uncertainty and propagation of belief, GA with global optimization and search, and ChT with nonlinear dynamics. Each of these computational paradigms (emerging reasoning technologies) provides us with complementary reasoning and searching methods to solve complex, real-world problems. In large scope, FL, NN, PR, and GA are complementary rather that competitive [4, 5]. The interrelations between the components of SC, shown in Figure 1, make the theoretical foundation of hybrid intelligent systems. As noted by Zadeh: "… the term hybrid intelligent systems is gaining currency as a descriptor of systems in which FL, NC, and PR are used in combination. In my view, hybrid intelligent systems are the wave of the future" [6]. The use of hybrid intelligent systems leads to the development of numerous manufacturing system, multimedia system, intelligent robots, and trading systems, which exhibits a high level of MIQ (machine intelligence quotient).

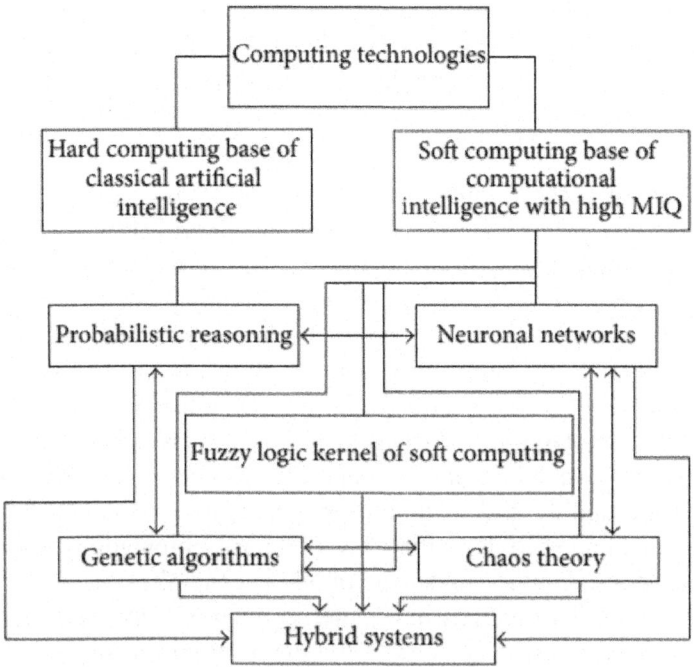

Figure 1: Soft computing components.

The constituents of SC can be used independently (fuzzy computing, neural computing, evolutionary computing, etc.) and more often in combination [7–11]. Based on independent use of the constituents of soft computing, fuzzy technology, neural technology, chaos technology, and others have been recently applied as emerging technologies to both industrial and nonindustrial areas.

Fuzzy logic is the leading constituent of soft computing. In soft computing, fuzzy logic plays a unique role. FL serves to provide a methodology for computing [7]. It has been successfully applied to many industrial spheres, robotics, complex decision making and diagnosis, data compression, and many other areas. To design a system processor for handling knowledge represented in a linguistic or uncertain numerical form, we need a fuzzy model of the system. Fuzzy sets can be used as a universal approximator, which is very important for modeling unknown objects. If an operator cannot tell linguistically what kind of action he or she takes in a specific situation, then it is quite useful to model his/her control actions using numerical data. However, fuzzy logic in its so called pure form is not always useful for easily constructing intelligent systems. For example, when a designer does not have sufficient prior information (knowledge) about the system, development of acceptable fuzzy rule base becomes impossible. As the complexity of the system increases, it becomes difficult to specify a correct set of rules and membership functions for describing adequately the behavior of the system. Fuzzy systems also have the disadvantage of not being able to extract additional knowledge from the experience and correcting the fuzzy rules for improving the performance of the system.

Another important component of soft computing is neural networks. Neural networks (NN), viewed as parallel computational models, are parallel fine-grained implementation of nonlinear static or dynamic systems. A very important feature of these networks is their adaptive nature, where "learning by example" replaces traditional "programming" in problems solving. Another key feature is the intrinsic parallelism that allows fast computations. Neural networks are viable computational models for a wide variety of problems including pattern classification, speech synthesis and recognition, curve fitting, approximation capability, image data compression, associative memory, and modeling and control of nonlinear unknown systems [11, 12]. NN are favorably distinguished for efficiency of their computations and hardware implementations. Another advantage of NN is generalization ability, which is the ability to classify correctly new patterns. A significant disadvantage of NN is their poor interpretability. One of the main criticisms addressed to neural networks concerns their black box nature [6].

Evolutionary computing (EC) is a revolutionary approach to optimization. One part of EC—genetic algorithms—are algorithms for global optimization. Genetic algorithms (GA) are based on the mechanisms of natural selection and genetics [13]. One advantage of genetic algorithms is that they effectively implement parallel multicriteria search. The mechanism of genetic algorithms is simple. Simplicity of operations and powerful computational effect are the two main advantages of genetic algorithms. The disadvantages are the problem of convergence and the absence of strong theoretical foundation. The requirement of coding the domain of the real variables into bit strings also seems to be a drawback of genetic algorithms. It should be also noted that the computational speed of genetic algorithms is low. Table 1 presents the comparative characteristics of the components of soft computing. For each component of soft computing, there is a specific class of problems, where the use of other components is inadequate.

As it was shown above, the components of SC complement each other, rather than compete. It becomes clear that FL, NC, and GA are more effective when used in combinations. Lack of interpretability of neural networks and poor learning capability of fuzzy systems are similar problems that limit the application of these tools. Neurofuzzy systems are hybrid systems which try to solve this problem by combining the learning capability of connectionist models with the interpretability property of fuzzy systems. As it was noted above, in case of dynamic work environment, the automatic knowledge base correction in fuzzy systems becomes necessary. On the other hand, artificial neural networks are successfully used in problems connected to knowledge acquisition using learning by examples with the required degree of precision.

Table 1: Central characteristics of soft computing technologies

	Fuzzy sets	Artificial neural networks	Evolutionary computing, GA	Probabilistic reasoning	Chaotic computing
Weaknesses	(i) Knowledge acquisition (ii) Learning	Black box interpretability	(i) Coding (ii) Computational speed	(i) Limitation of the axioms of probability theory (ii) Lack of complete knowledge (iii) Computational complexity	(i) Computational complexity (ii) Chaos identification complexity
Strengths	(i) Interpretability (ii) Transparency (iii) Plausibility (iv) Graduality (v) Modeling (vi) Reasoning (vii) Tolerance to imprecision	(i) Learning (ii) Adaptation (iii) Fault tolerance (iv) Curve fitting (v) Generalization ability (vi) Approximation ability	(i) Computational efficiency (ii) Global optimization	(i) Rigorous framework (ii) Good understanding	(i) Nonlinear dynamics simulation (ii) Discovering chaos in observed data (with noise) (iii) Determining the predictability (iv) Prediction strategies formulation

The cooperation between these formalisms gives a useful tool for modeling and reasoning under uncertainty in complicated real-world problems. Such cooperation is of particular importance for constructing perception-based intelligent information systems. We hope that the mentioned intelligent combinations will develop further, and the new ones will be proposed. These SC paradigms will form the basis for creation and development of computational intelligence.

COGNITIVE MODELS OF GROUND MOTIONS

The existence of numerous databases in the field of civil engineering, and in particular in the field of geotechnical earthquake, has opened new research lines through the introduction of analysis based on soft computing. Three methods are mainly applied in this emerging field: the ones based on the neural networks (NN), the ones created using fuzzy sets (FS) theory, and the ones developed from the evolutionary computation [14].

The SC hybrids used in this investigation are directed to tasks of prediction (classification and/or regression). The central objective is obtaining numerical and/or categorical values that mimic input-output conditions from experimentation and in situ measurements and, then, through the recorded data and accumulated experience to predict future behaviors.

The examples presented herein have been developed by an engineering committee that works for generating useful guidance to geotechnical practitioners in geotechnical seismic design. This effort could help to minimize the perceived significant and undesirable variability within geotechnical earthquake practice. Some urgency in producing the alternative guidelines was seen, after the most recent earthquakes disasters, as being necessary with a desire to avoid a long and protracted process. To this end, a two-stage approach was suggested with the first stage being a cognitive interpretation of the well-known procedures with appropriate factors for geotechnical design, with a posterior step identifying the relevant philosophy for a new geotechnical seismic design.

Spatial Variation of Soil Dynamic Properties

When using considerable volumes of 3D geotechnical information, with nonlinear and multidimensional relations, its proper management and analysis with conventional tools are limited, reducing its exploitation in natural phenomena modeling. The spatial variability of subsoil properties represents a major challenge in both the design and construction phases of most geoengineering projects. Subsoil investigation is an imperative step in any civil engineering project and it is also considered a prerequisite to the economical design of substructures. In general, the purpose of an exploratory investigation is to infer accurate information about actual soil and rock conditions at the site.

The geotechnical investigation stage of any project should be carefully planned to obtain the most reliable soil parameters with the minimum expense. A detailed planned program of boring and sampling, as well as the utilization of an accurate interpretation technique (i.e., modeling), is the cornerstone of any reliable and precise exploratory investigation. It is impossible to determine the optimum spacing of borings before an investigation begins because the spacing depends not only on type of structure but also on uniformity or regularity of encountered soil deposits. Even the most detailed soil maps are not efficient enough for predicting a specific soil property because it might exhibit variations from place to place, even for the same soil type. Consequently, interpolation techniques have been extensively applied. Unfortunately, the most common used methods, kriging and cokriging, require a significant number of measurements for each soil type to obtain assertive estimations, which is generally unmanageable. Based on the high cost of collecting soil attribute data at many locations across landscape, new interpolation methods have to be tested. The tested methodologies must be able to properly organize the historical geotechnical into multiple databases for establishing spatial/chronological predictive models through interpreting properties (soil exploration) and behaviors (in situ measured).

In the following, the spatial variation of the cone tip resistance q_c and shear wave velocity V_s defined using NN and GA and a georeferenced 3D model of the soils underlying Mexico City is presented. The classification/prediction criterion for this very complex urban area is established according to two much related soils properties, a mechanical property and a dynamic property.

Cone penetration testing CPT and shear wave velocity information are frequently used to determine the vertical succession of different layers of soft soil and parameters related to these materials. Huijzer [15] worked first on the automatic interpretation of cone penetration tests—using linear regression—but a satisfactory recognition process based exclusively on resistance value turned out to be impossible. A later study by Coerts [16] started the interpretation from a predefined number of geotechnical units (number of boreholes) based on profiles published in geological maps whose parameter values determination was contaminated by many vague and uncertain sources. Additionally, a general regression neural approximation for site characterization, in terms of soil strengths, was developed by Juang et al. [17]. The authors presented the benefits of using NNs; however, their estimations exhibited very low confidence since the total number of boreholes available was used to train the network and no test patterns were presented for evaluating the capabilities for generalization of the network.

In this investigation, a proved methodology [18] was used to estimate V_s and q_c at locations on the half space where no measurements were performed. Therefore, V_s and q_c were generalized in any subdomain of the studied space. The estimations at any point within the half space represent a virtual boring so that an intricate grid of values has to be created to support the neural contour maps. Although of an NN model is not the most powerful interpolator, it represents an excellent classification/prediction function (parametric and structural related to geographical position and mechanical soil properties). In this investigation, the García and Romo's methodology [18] is reinterpreted and improved by using a mechanical and a dynamical property coupled with linguistic and historical information to better reproduce the half-space of the lacustrine soil deposits of the Mexico Valley. In addition, the suggested NN hidden structure tuned by GA represents novelty.

Database. The application of the NN-GAs methodology is illustrated by an example dealing with the spatial variability of the V_s and q_c in the lacustrine zone of Mexico valley. The result is a 3D model of the soils underlying the city area. Cone-tip penetration resistances and shear wave velocities have been measured along 19 bore holes spread throughout the clay deposits of Mexico City. Borehole locations are marked with dots in Figure 2. This information was used as the set of examples inputs, which are georeferenced in latitude, longitude, and depth and the output will be the ratio CPT resistance on shear wave velocity. It is important to point out that 25% of these patterns (sample points and complete depth-property information) are not used in the training stage; they will be presented for corroborating the generalization capabilities of the closed system components. Once the training is stopped, the ratio test/ validation set is presented and the correlation between measured and NN-estimated values is checked and the error is assessed.

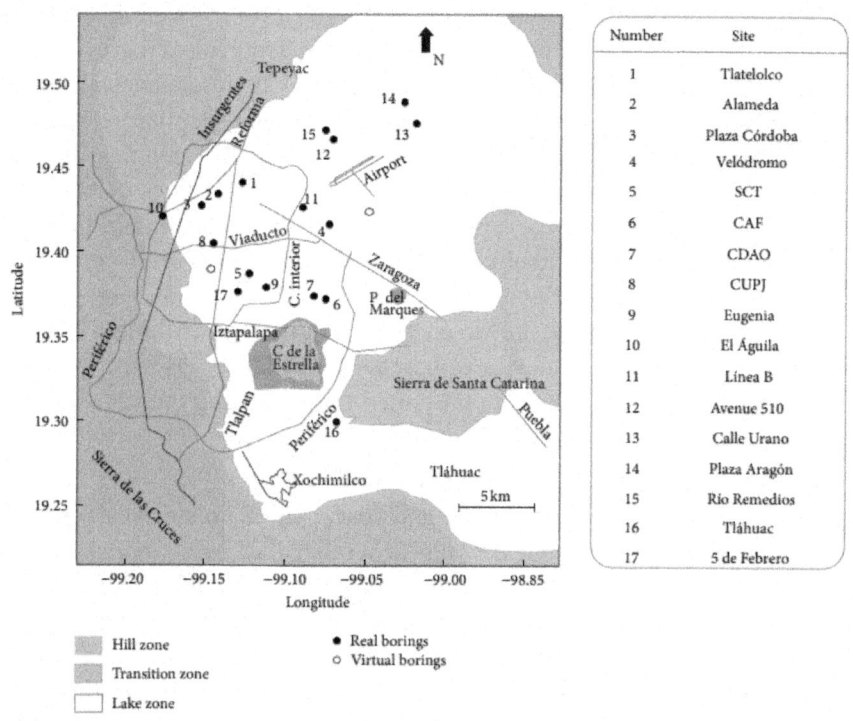

Figure 2: Geotechnical zones and exploration sites, Mexico City.

Spatial Model Construction. The figure of merit used to compare the NN architectures trained (those trails for acquiring the optimum neural design) was percent root mean squared error defined as

$$RMSE\% = \sqrt{\frac{1}{N}\sum_{i=1}^{N}\left(\frac{y_i - x_i}{\max(x)}\right)^2},$$

(1)

where x_i are the correct values on the test set and   are the values obtained by interpolation. It is important to note that the definition of spatial variation of a soil property using neurogenetic tools does not need a priori identification or selection of homogeneous soil layers, which is mandatory in many procedures. In the 3D neurogenetic analysis, the functions $q_c = \{q_c(X, Y, Z)\}/V_s = \{V_s(X, Y, Z)\}$ are to be approximated, with which the q_c/V_s values at any half-space point can be determined. For predicting and defining soil properties and layers geometries, the 3D neuroenvironment is generated using the procedure outlined below.

(1) The generation of the database including identification of the boring of the site (X and Y geographical coordinates, Z depth, and a CODE-ID number), elevation reference (meters above sea level, m.a.s.l.), thickness of predetermined structures (layers), and additional information related to geotechnical zoning that could be useful for results interpretation. The database is organized in three tables. The first contains the CODE, X, Y, and Z data for each sounding (Figure 3). In a second table (Figure 4) any previous knowledge of strata structure is included for all soundings (parametric, geometrical, or/and linguistic descriptions). This "strata" table contains the properties arranged as columns related to depth. It can be extended through additional columns (mechanical, physical, or geometrical soil parameters) to generate a wide-ranging characterization. A third table (Figure 5) contains the NN results (crisp evaluations) and it is constructed once the NN model is labeled as "optimum." This table contains the training/testing information (the "real" in situ measurements) plus the "virtual" borings, the neural information that supports the 3D grid and completes the media with materials and properties neuroinferred.

(2)The first two tables (Figures 3 and 4) are used to train an initial neural topology whose weights and layers are adjusted by an evolutive algorithm, until the minimum RMSE between calculated and measured q_c/V_s values is achieved (Figure 6). In the case of 2D prediction, the q_c/V_s vectors separated for testing the generalization capabilities of the NN are compared with those obtained during the direct phase of the NN (Figure 7). In three dimensional analyses, all (training) data points from whole volume were considered at once. Through examining the neurogenetic results for unseen measurements, it can be concluded that the procedure works extremely well in identifying the general trend in materials resistance (stiffness).

(3)The construction of the 3D soils environment is done using the real/virtual parametric vectors from Table 3 (Figure 5) but it is tuned considering the additional information from Table 2. A clustering fuzzy subroutine is applied to the database in order to find the incongruences, analyze them, and decide about its possible occurrence or not. This 3D view of the studied zone represents an easier and more understandable engineering frame. Some examples of the kind of information obtained from this assembly are shown in Figure 8.

Table 2: Database used to construct the ML model

Dataset	Input attributes	Number of selected instances
1	Z_{NAF} (m), $Z_{TOP\ LAYER}$ (m), H (m), σ'_V (KPa), σ_V (KPa), S_T, V_S (m/s), M, a_{max} (g)	181 (Andrus and Stokoe, 1997, 2000) [20, 21]
2	$Z_{TOP\ LAYER}$ (m), R_f (%), σ'_V (KPa), q_C (MPa), M, a_{max}(g)	226 (Juang et al.,1999; Juang, 2003) [9, 22]

Table 3: Seismic events contained in the database

Earthquake	Magnitude M_w	PGA* (g)	Epicentral distance (km)	Number of patterns
1906 San Francisco	7.9	0.28–0.26	13–27	4
1964 Alaska	9.2	0.21–0.39	35–100	7
1964 Niigata, Japan	7.5	0.19	21	299
1971 San Fernando	6.4	0.55	1	28
1979 Imperial Valley	6.5–6.6	0.21–0.51	2–6	32
1983 Nihonkai-Chubu, Japan	7.7	0.25	27	72
1987 Superstition Hills	6.6	0.21	23	6

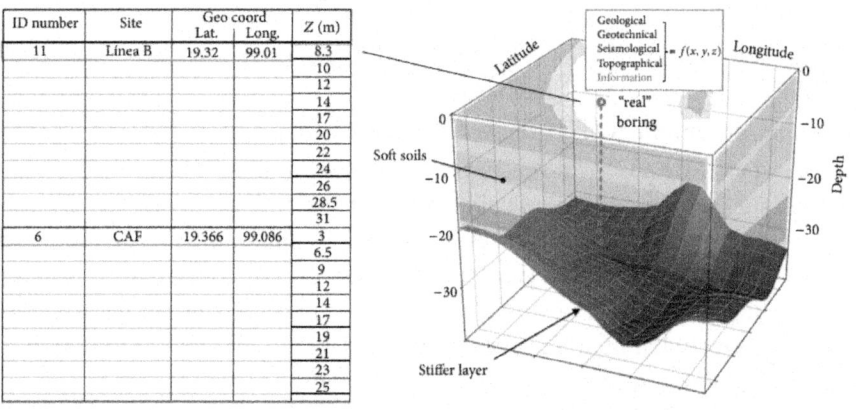

Figure 3: Schematic representation of Table 1: "real" borings situation.

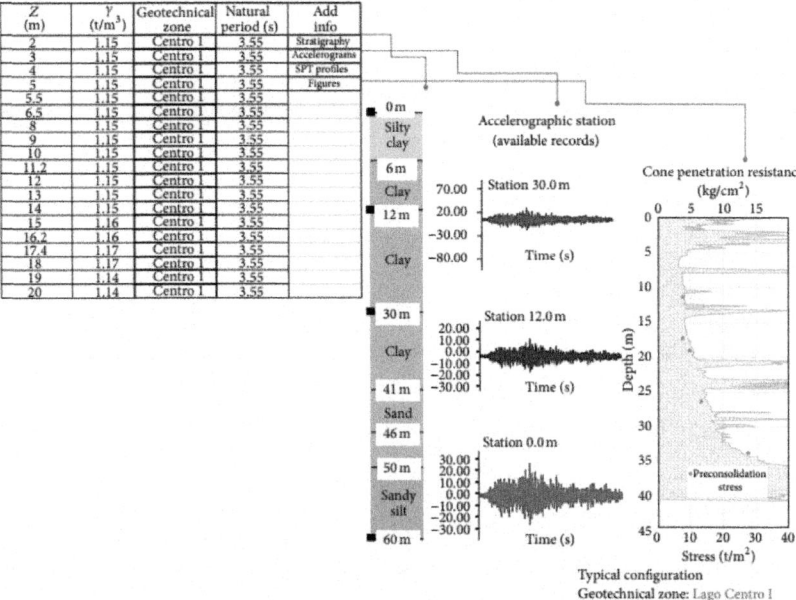

Figure 4: Schematic representation of Table 2: "real" borings geo- and seismic measurements.

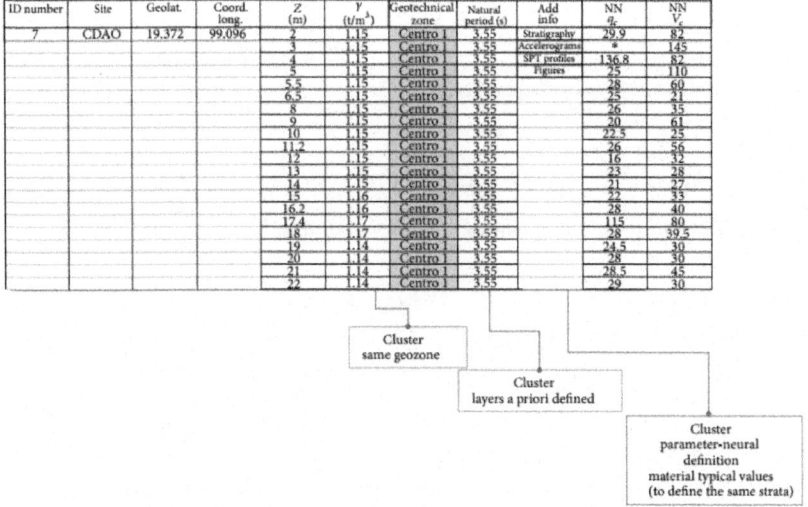

Figure 5: Schematic representation of Table 3: "real" and "virtual" borings to construct the stratigraphies or the 3D environment.

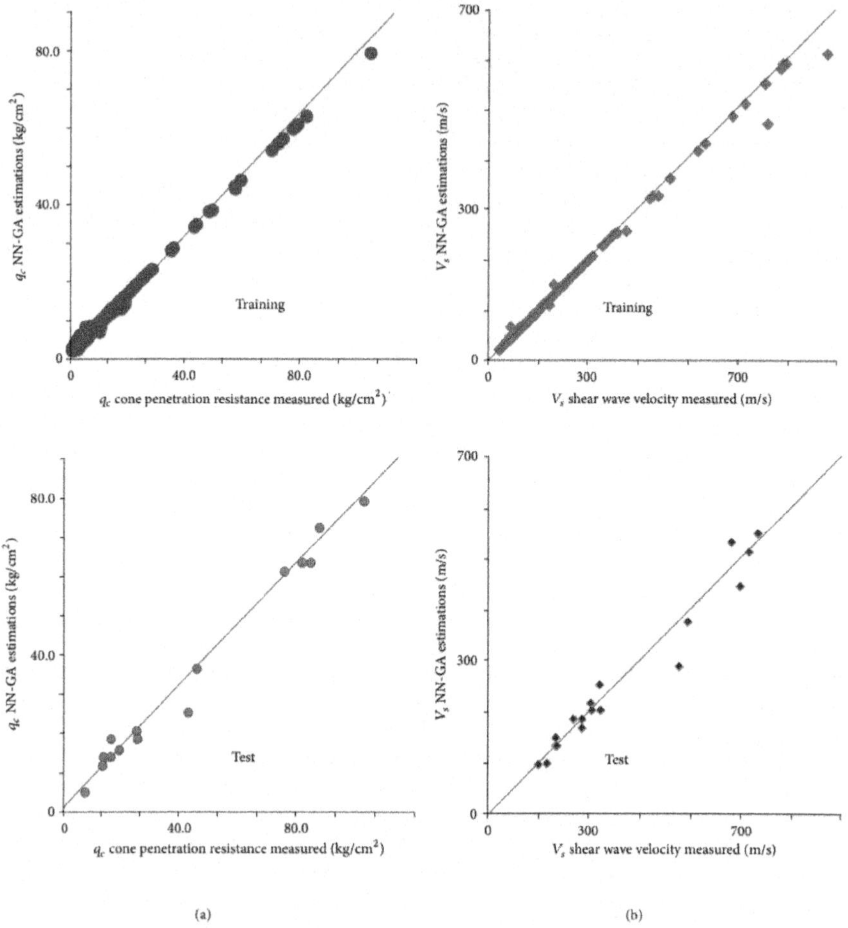

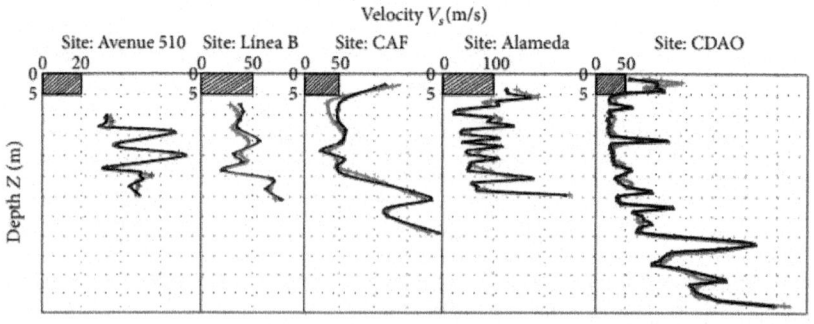

Figure 6: Measured versus evaluated values of q_c and V_s.

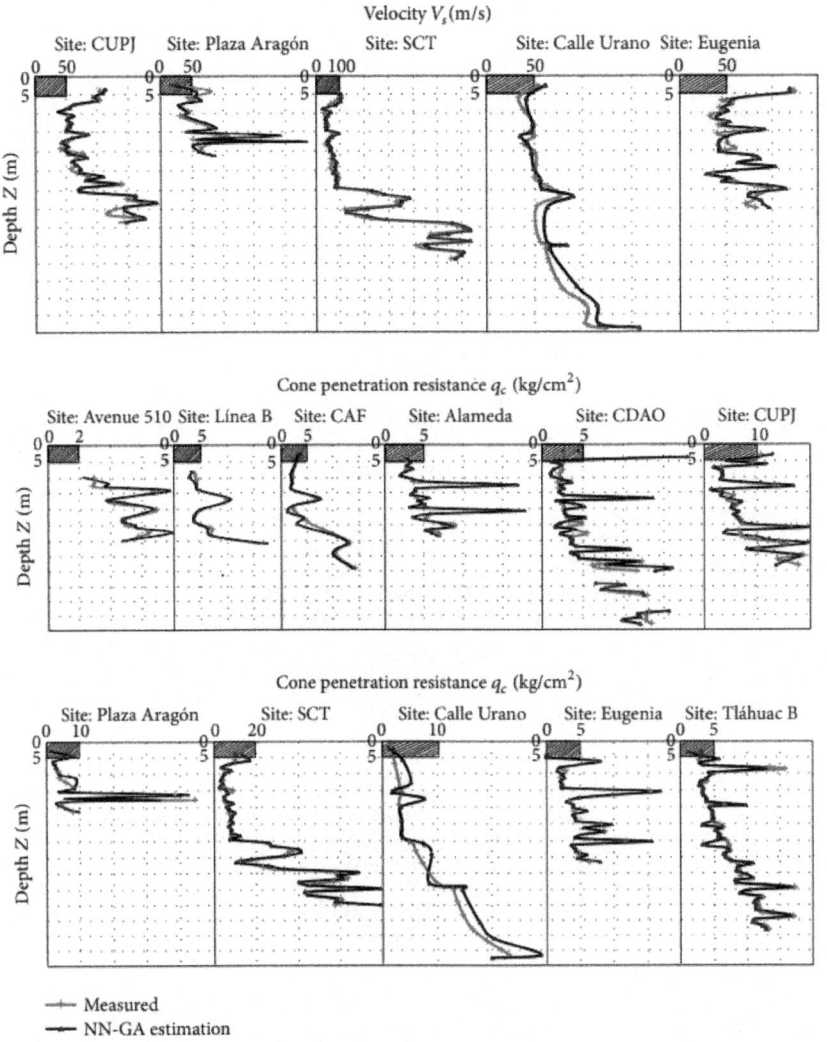

Figure 7: Neural V_s and q_c vectors compared with in situ profiles.

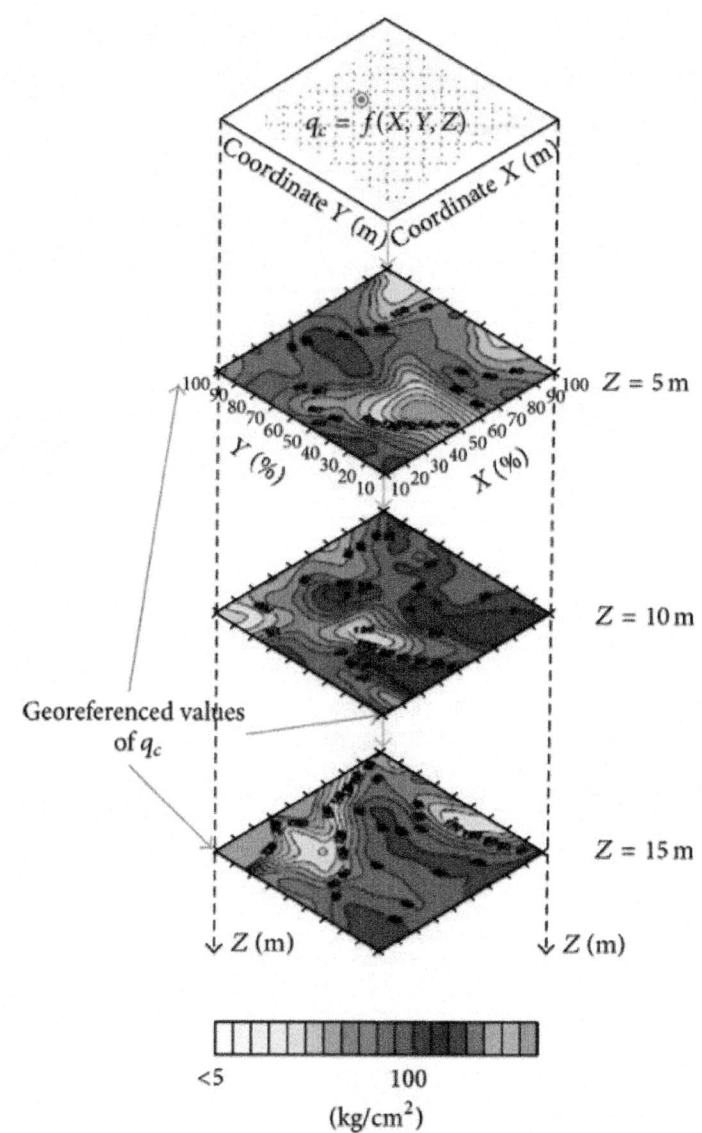

(a)

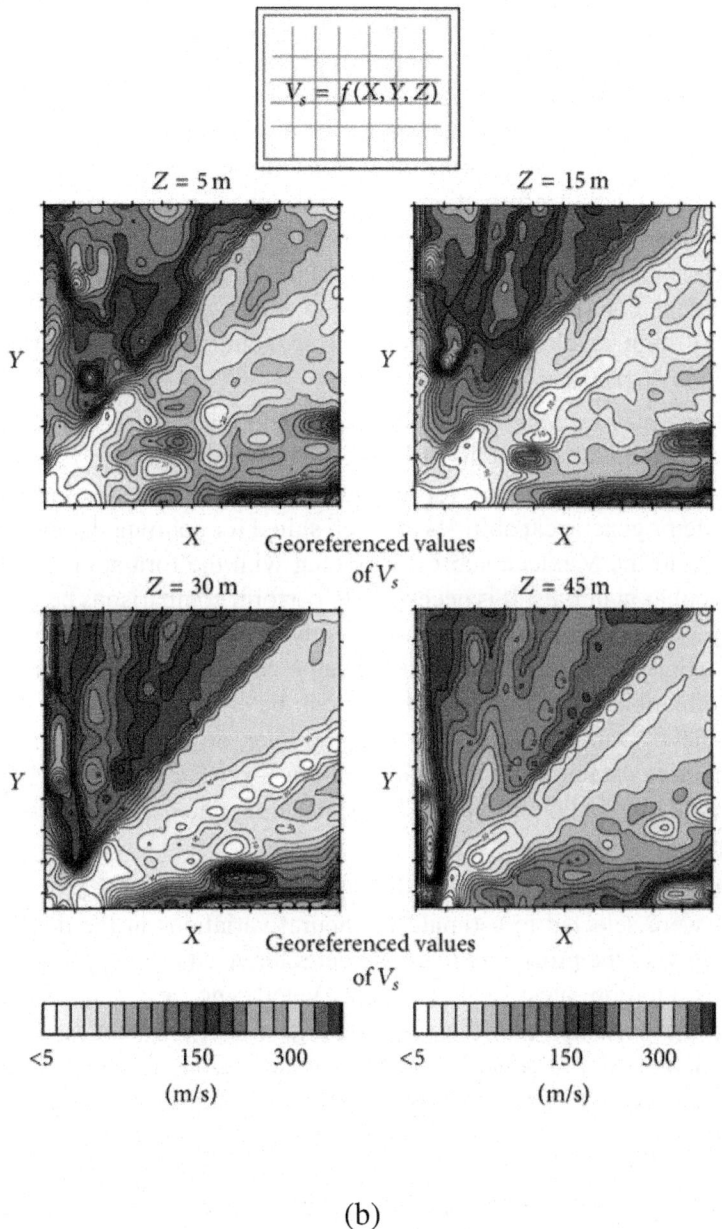

(b)

Figure 8: Isovalues curves for q_c and V_s.

Many advantages has the 3D NN exploration, for example, determining drillhole spacing requires, for obvious reasons, to obtain the maximum benefit at minimum cost, meaning that the number of boreholes has to be sufficient to ensure continuity, without costing more than necessary. The inevitable question about the optimum drillhole spacing and the location of the boreholes can be answered using the 3D neural representation taking into account the prospections, the experience, and the engineers' criteria. The level of risk that the management is willing to accept controls how rigorous should be the data collection at the start of a project as well as during subsequent developments. The cost of collecting information has to be weighed up against the potential cost of uncertainty. Neuronal estimations combined with geological/geotechnical evidences permit to infer and in some cases to verify the geological and/or geotechnical continuity. Therefore, in situ exploration could be conducted with very high level of confidence and economic efficiency during further exploration steps (new borings).

To generate a reliable model of the soils underlying, it is necessary to review the horizontal continuity of the layers defined, from sounding to sounding. The definition of the stratigraphy for Mexico City clays is presented for showing that the neurogenetic capabilities are well suited for defining the intrusions and the lenses in the Mexican subsoil, agreeing with the formation and origin of these peculiar materials. It is necessary to perform comparisons between layers at every site, in order to classify differently those layers that fulfill the same classification but do not belong to the same continuous layer. The procedure of identification and selection of layers can be easily conducted by using the neurogenetic tool that permits to manage, inter, and extrapolate information from massive databases. If a specific pattern is repeated in nearby soundings, then it is likely the condition of a layer intruding another. Contrarily, it might be the presence of a lens.

In this study, soundings on a line along the East-West direction (E-W) of the city were selected to estimate the neural variations in the distribution of soils following the procedure to create georeferenced profiles (Figure 9). Four "real" exploration sites: SCT, Eugenia, Velódromo, and Línea B constitute the line of soundings and the schematic grid suggests that is the "virtual" information used to complete the layers successions. There are four well-defined soil layers but the shallowest one is not considered for the analyses. The variation of q_c and V_s (in Figure 9 the q_c profile is not included to make it more understandable) with depth from about 5 to 35 m is presented at Eugenia, Velódromo, and Línea B sites, while the maximum depth is 50 m at the site SCT. Clearly, Velódromo site exhibits the lowest values of q_c and V_s, and shear wave velocities range approximately from 25 to 50 m/s, while varies from 24.5 to 49 MPa. Otherwise, the site with the highest q_c and V_s correspond to SCT site. q_c and V_s are in the range from about 50 to 625 m/s and 49 and 613 MPa, respectively.

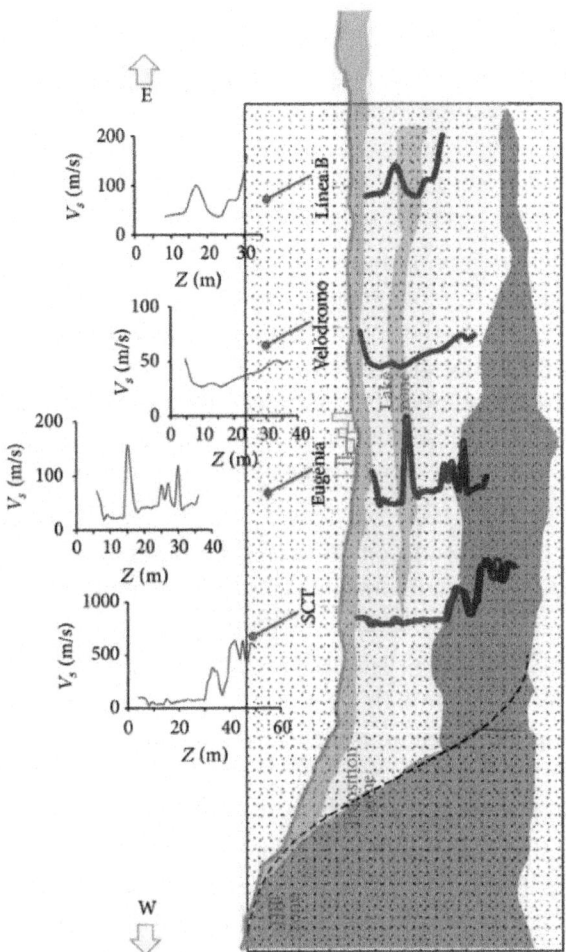

Figure 9: Layer definition using "virtual" profiles (every vertical dotted line is a "virtual" axis boring) and "real" exploration sites.

The neurogenetic model has been proven to be useful in the interpretation of natural resource information. The above methodology can be automated to produce geotechnical maps as a software system: a package of programs for conducting analyses of the spatial variability of one or more interrelated variables. Based on the results presented here, it can be concluded that a soft system would be menu driven and simple to use and visualize, requiring a minimum number of input data with dynamic allocation of memory so that datasets with a wide range of variables and positions can be used without altering the basic structure's program.

Estimation of Peak Ground Accelerations for Mexican Subduction Zone Earthquakes

Earthquake ground motions are affected by several factors including source, path, and local site response. These factors should be considered in engineering design practice using seismic hazard analyses that normally use attenuation relations derived from strong motion recordings to define the occurrence of an earthquake with a specific magnitude at a particular distance from the site.

These relations are typically obtained from statistical regression of observed ground motion parameters. Because of the uncertainties inherent in the variables describing the source (e.g., magnitude, epicentral distance, focal depth, and fault rupture dimension), the difficulty to define broad categories to classify the site (e.g., rock or soil), and our lack of understanding regarding wave propagation processes and the ray path characteristics from source to site, commonly the predictions from attenuation regression analyses are inaccurate.

As an effort to recognize these aspects, multiparametric attenuation relations have been proposed by several researchers (e.g., [19, 23–28]). However, most of these authors have concluded that the governing parameters are still source, ray path, and site conditions.

In this investigation, an empirical NN formulation that uses the minimal information about magnitude, epicentral distance, and focal depth for subduction-zone earthquakes is developed to predict the three components (two horizontal, one vertical) of peak ground acceleration PGA at rock sites (consisting of at most a few meters of stiff soil over weathered or sound rock). The NN model was obtained from existing information compiled in the Mexican strong motion database.

Events with poorly defined magnitude or focal mechanism, as well as recordings for which site-source distances are inadequately constrained, or recordings for which problems were detected with one or more components were removed from the data. It uses earthquake moment magnitude M_w, epicentral distance ED, and focal depth FD. The obtained results indicate that the proposed NN is able to capture the overall trend of the recorded PGAs.

This approach seems to be a promising alternative to describe earthquake phenomena despite of the limited observations and qualitative knowledge of the recording stations geotechnical site conditions, which leads to a reasoning of a partially defined behavior. Based on the procedure for achieving PGAs, spectral ordinates for any particular period can be estimated in the same manner with similar confidence levels.

Database. The database used in this study consists of 1058 records. These events were recorded at rock and rock-like sites during Mexican subduction earthquakes (Figure 10). Event dates range from 1964 to 2006. Events with poorly defined magnitude or focal mechanism, as well as recordings for which site-source distances, are inadequately constrained, or recordings for which problems were detected with one or more components were removed from the data. To test the predicting capabilities of the neuronal model, 186 records were excluded from the dataset used in the learning phase. One was the September 19, 1985, M_w =8.1 earthquake, and the other 185 events were randomly selected, making sure that a broad spectrum of cases were included in the testing database.

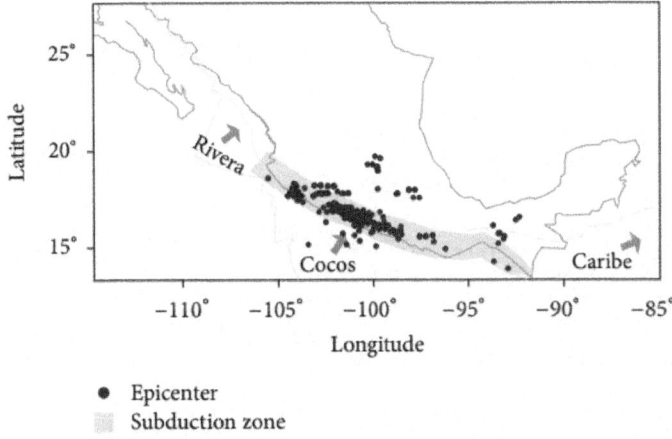

Figure 10: Motion array and events included in the data base.

The moment magnitude scale M_w is used to describe the earthquakes size, resulting in a uniform scale for all intensity ranges. If the user has another magnitude scale, the empirical relations proposed by Scordilis [29] can be used. In this paper the used ED is considered to be the length from the point where fault rupture starts in the recording site, as indicated in Figure 11. The third input parameter, FD, does not express mechanism classes; it is declared as a nominal variable which means that the NN identifies if there is an event type (broad class) or the value impacts on the result through its crisp quantity. Some studies have led to consider that subduction-induced earthquakes may be classified as interface events (FD < 50 km) and intraslab events (FD > 50 km) ([19, 30]). This is a rough classification because crustal and interface earthquakes would be mixed [31], and therefore it was considered not relevant for this model. The dynamic range of M_w goes from 3 to 8.1 approximately and the events were recorded at near (a few km) and far field stations (about 690 km). The depth of the zone of energy release ranged from very shallow to about 360 km.

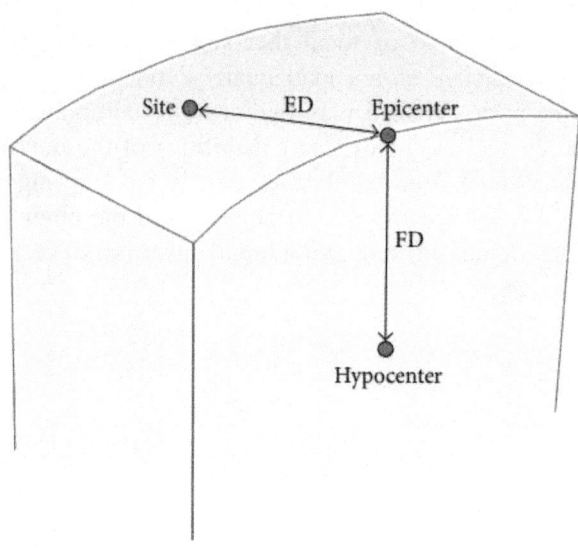

Figure 11: Schematic representation of parameters: focal depth, FD, and epicentral distance, ED.

Neural-Attenuation Model Construction. Modeling of the database has been performed using the quick propagation QP learning algorithm [32]. Horizontal (mutually orthogonal PGAh1, N-S component, and PGAh2, E-W component) and vertical components (PGAv) are included as outputs for neural mapping into three units. The neural modules that met the convergence criterion (mean square error $\leq 5\%$) have a total of 72 and 126 hidden nodes for PGAh1 and PGAh2, respectively, while the PGAv module behavior was quite acceptable using a simple alternative (QP, 2 layers/15 units or nodes each). Details of the topology-selection process can be found in [33].

The neuronal attenuation model for $\{M_w, \text{ED}, \text{FD}\} \rightarrow \{\text{PGAh1}, \text{PGAh2}, \text{PGAv}\}$ was evaluated by performing testing analyses. The predictive capabilities of the NNs were verified by comparing the PGAs estimated to those induced by the 186 events excluded from the original database that was used to develop the NN architectures (training stage). In Figure 12(a), the PGA's computed during the training and testing stages are compared to the measured values. The relative correlation factors $(R^2 \approx 0.93)$, obtained in the training phase, indicate that those topologies selected as optimal behave consistently within the full range of intensity, distances, and focal depths depicted by the patterns. Once the networks converge to the selected stop criterion, learning is finished and each of these black boxes becomes a nonlinear multidimensional functional. Every functional is then assessed (testing stage) by comparing their predictions to the separated 186 PGA values of the database. As new conditions of M_w, ED, and FD are presented to the neural functional, R^2 decreases. This drop is more appreciable for the horizontal components. Nonetheless, as indicated by the upper and lower boundaries included in the Figure 12(b), forecasting of all three seismic components is reliable enough for practical applications.

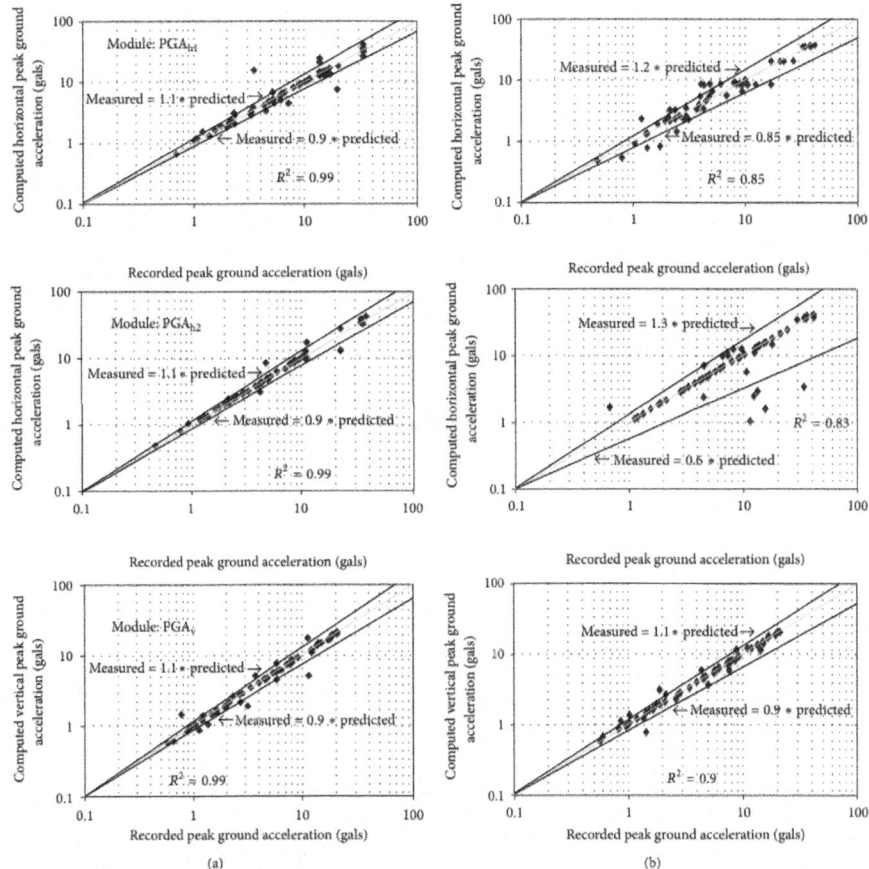

(a)   (b)

Figure 12: Comparisons between estimated and recorded peak ground accelerations for each component independently. (a) Left: training phase and (b) right: testing phase.

A sensitivity study for the input variables was conducted for the three neuronal modules. The results are strictly valid only for the data base utilized; nevertheless, after several sensitivity analyses conducted changing the database composition, it was found that the following trend prevails; the M_w would be the most relevant parameter (presents larger relevance) then would follow the epicentral distance, ED, and the less influential parameter was the focal depth,

FD. However, for near site events the epicentral distance could become as relevant as the magnitude, particularly, for the vertical component. The selected functional forms incorporate the results of analyses into specific features of the data, such as the PGAh dependence of the geometrical ED-FD description.

At this stage it is convenient to note that traditionally the PGAs that are used to develop attenuation relationships are defined as randomly oriented (e.g., [26]), mean (e.g., [34]), and the larger value from the two horizontal components (e.g., [35]). Accordingly, considering the variety of PGA definitions used in most existing attenuation relations, it was deemed fairer to use the PGAh1-h2 prediction modules for comparison purposes. Figure 13 compares five fitted relationships to PGA data from interface earthquakes recorded on rock and rock-like sites. The two case histories correspond to a large and a medium size events (the September 19, 1985, Michoacán earthquake and the July 4, 1994, event, resp.).

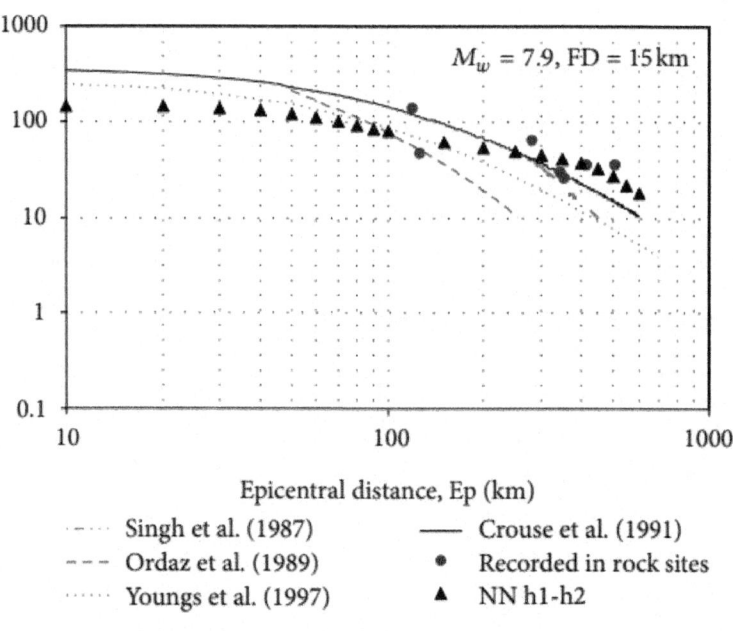

Epicentral distance, Ep (km)

- - - - Singh et al. (1987)	—— Crouse et al. (1991)	
- - - Ordaz et al. (1989)	• Recorded in rock sites	
.... Youngs et al. (1997)	▲ NN h1-h2	

(a)

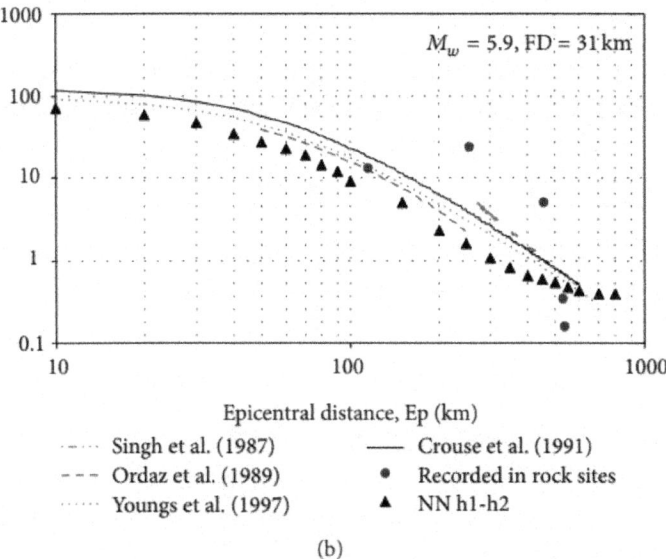

$M_w = 5.9, FD = 31\ km$

Epicentral distance, Ep (km)

Singh et al. (1987)   ——— Crouse et al. (1991)

– – – Ordaz et al. (1989)   ● Recorded in rock sites

·········· Youngs et al. (1997)   ▲ NN h1-h2

(b)

Figure 13: Attenuation relationship obtained with the proposed neural attenuation model PGAh1-h2 for two subduction-related events: (a) September 19, 1985, earthquake and (b) July 4, 1994, Event.

The estimated values obtained for these events using the relationships proposed by Gómez et al. [36], Youngs et al. [19], Atkinson and Boore [20]—proposed for rock sites—and Crouse [24]—proposed for stiff soil sites—and the predictions obtained with the PGAh1-h2 modules are shown in Figure 13. It can be seen that the estimation obtained with Gómez et al. [36] seems to underestimate the response for the large magnitude event. However, for the lower magnitude event both the measured responses and NN predictions follow closely. Youngs et al. [19] attenuation relationship follows closely the overall trend but tends to fall sharply for long epicentral distances. Although, as mentioned previously, the PGAh1-h2 modules yielded important differences in the testing phase, its predictions follow closely the trends and yield a better behavior, in the full range of epicentral distances included in the data base, than traditional attenuation relations applied to the Mexican subduction zone.

Furthermore, it should be stressed the fact that the September 18, 1985, earthquake was not included in the database used in the development of the neural networks and that this event falls well outside the range of values in such database, and hence it is an example of the extrapolation capabilities of the networks developed in this paper. It is worth to note that while the NN trend follows the general behavior of the measure data, the traditional functional approaches have predefined extreme boundaries. On the other hand, when the intensity of the earthquake is moderate, most of the PGAs measured in rock sites are within a narrow band, and thus generally the NN and traditional functionals follow similar patterns. The generalization capabilities of the PGAh1-h2 module can be explored even more by simulating other subduction zones events. Measured random horizontal PGAs taken from Youngs et al. [19] belonging to Japan and North America for two magnitude intervals (M_w: 7.8–8.2 and M_w: 5.8–6.2) were compared to the NN predictions. These results are plotted in Figure 14. It can be seen that the NN prediction agrees well with the general trend even considering averages of both earthquake magnitude and focal depth.

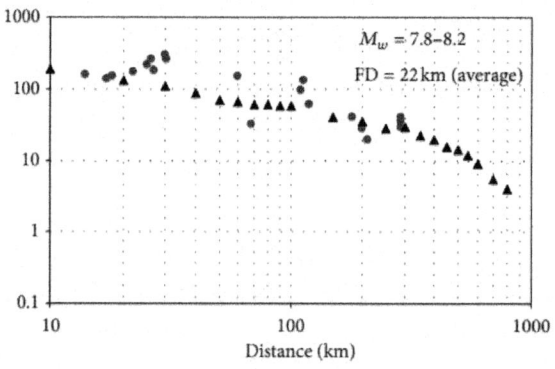

• Recorded (taken from Youngs et al. 1997)
▲ NN h1-h2

(a)

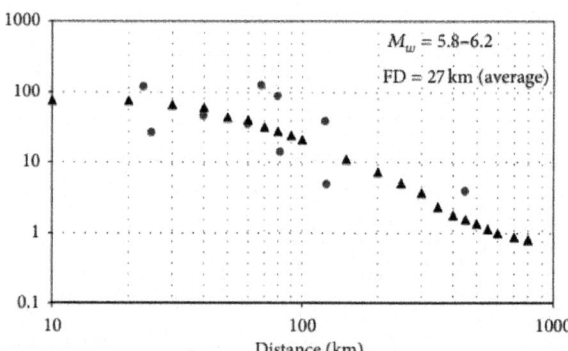

• Recorded (taken from Youngs et al. 1997)
▲ NN h1-h2

(b)

Figure 14: Comparison of NN model PGAh1-h2 with data published by Youngs et al. (1997) [19].

Artificial Generation of Time Series: Accelerograms Application

For nonlinear seismic response analysis, where the superposition techniques do not apply, earthquake acceleration time histories are required as inputs. Virtually all seismic design codes and guidelines require scaling of selected ground motion time histories so that they match or exceed the controlling design spectrum within a period range of interest (e.g., [21]). After many years of strong motion recording programs, there are now more significant accelerograms that have been recorded, digitized, processed, and analyzed. While the available data represents a unique and invaluable collection for studies and research of strong earthquake ground motion, it does not cover all the need ranges of the parameters commonly used in empirical scaling laws (e.g., earthquake magnitude and focal depth, source to station distance, percentage of rock along the wave paths, and recording geological and soil conditions).

Considerable variability in the characteristics of the recorded strong motions under similar conditions may still require a characterization of future shaking in terms of an ensemble of accelerograms rather than in terms of just one or two "typical" records. This situation has thus created a need for the generation of synthetic (artificial) strong-motion time histories that simulate realistic ground motions from different points of views and/or with different degrees of sophistication. To provide the ground motions for analysis and design, various methods have been developed: (i) frequency-domain methods where the frequency content of recorded signals is manipulated (e.g., [22, 37–39]) and (ii) time-domain methods where the recorded ground motions amplitude is controlled (e.g. [40, 41]). Regardless of the method, it is separated the selection of the representative earthquake and the scaling to match the design spectrum. First, one or more time histories are selected subjectively, and then scaling mechanisms for spectrum matching are applied.

In this research a Genetic Generator of Signals GENES is presented. GENES is a tool for finding the coefficients of a pre-specified functional form, which fit a given sampling of values of the dependent variable associated with particular given values of the independent variable(s). When GENES is applied to synthetic accelerograms construction, the proposed tool is capable of (i) searching, under specific soil and seismic conditions, between thousands of earthquake records and recommending a desired subset that better match a target design spectrum, and (ii) through processes that mimic mating, natural selection, and mutation, of producing new generations of accelerograms until an optimum individual is obtained. The procedure is fast and reliable and results in time series that match any type of target spectrum with minimal tampering and deviation from recorded earthquakes characteristics.

The objective of GENES, when applied to synthetic earthquakes construction, is to produce compatible artificial signals with specific design spectra. GENES introduces specific seismic and site characteristics taking into consideration that (i) a typical strong motion record consists of a variety of waves whose contribution depends on the earthquake source mechanism (wave path) and its particular characteristics are influenced by the distance between the source and the site, some measure of the size of the earthquake, and the surrounding geology and site conditions; and (ii) the design spectra can be an envelope or integration of many expected ground motions that are possible to occur in certain period of time, or the result of a formulation that involves earthquake magnitude, distance and soil conditions.

The input data consist of the ordinates of the target acceleration design spectrum, the period range for the matching, lower- and upper-bound acceptable values for scaling signal shape, and a set of seismic/soil and GA parameters. The output is the more success chromosome in terms of an accelerations vector (or a set of). Additionally some GAs parameters are required: a population size, number of generations, crossover ratio, and mutation ratio.

The algorithm (see Figure 15) is started with a set of solutions (each solution is called a chromosome). A solution is composed of thousands of components or genes (accelerations a_i recorded at the time t_j), each one encoding a particular trait. The initial solutions (original population) are selected based on seismic parameters at a site (defined by the user): moment magnitude, epicentral distance, geotechnical and geological site classification, depth of sediments and component direction. If the designer does not have a priori seismic/site knowledge, GENES selects the initial population randomly (Figure 16). One of the GENES advantages is the possibility of modifying on line the image of the expected earthquake. While the GAs is running the user interface shows the individual per epoch and its response spectra in the same window, if the duration time, the highest intensities interval or the Δt are not convenient for the designer's interests, these values can be modified without GENES retraining or a change on its structure (Figure 17).

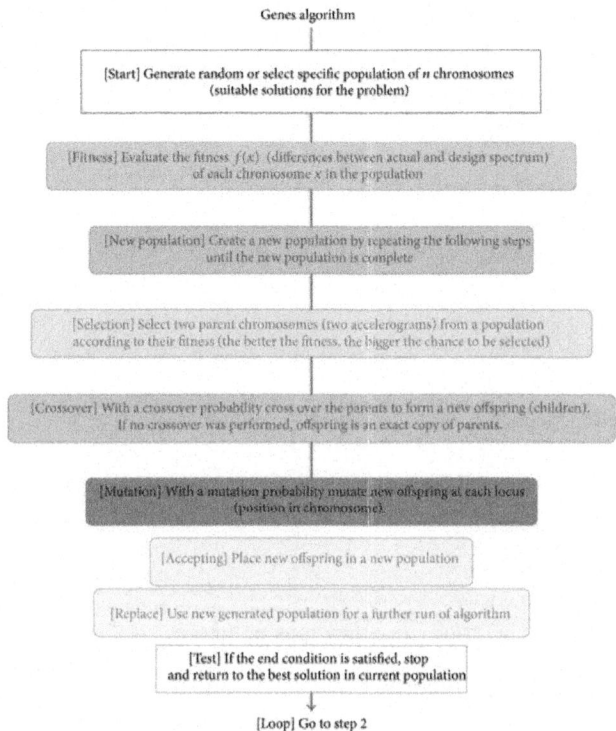

Figure 15: Genetic Generator: flow diagram.

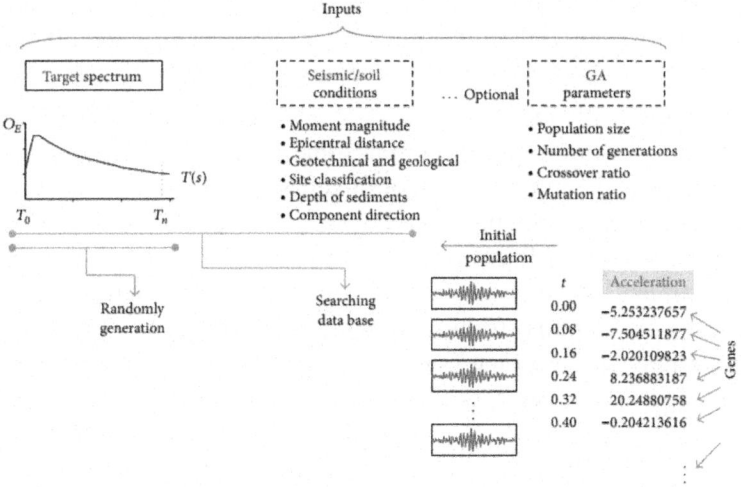

Figure 16: Genetic Generator: working phase diagram.

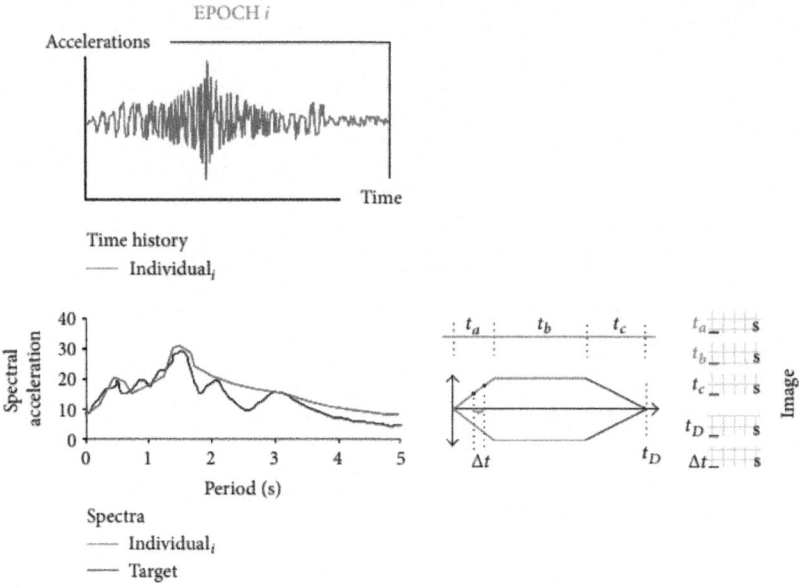

Figure 17: Genetic generation: modifications on line.

Proposing an entire recorded accelerogram as a chromosome, the space of all feasible solutions can be calledaccelerograms space (state space). Each point in this search space represents one feasible solution and can be "marked" by its value or fitness for the problem. The looking for a solution is then equal to a looking for some extreme (minimum or maximum) in the space. The accelerograms space can be whole known by the time of solving a problem, but usually only a few points from it are known and other points as the process of finding solution continues are generated.

According to the individuals' fitness, expressed by difference between the target design spectrum and thechromosome response spectrum, the problem is formulated as the minimization of the error function, Z, between the actual and the target spectrum in a certain period range. Solutions with highest fitness are selected to form new solutions (offspring). During reproduction, the recombination (or crossover) and mutation permits to change the genes (accelerations) from parents (earthquake signals) in some way that the whole new chromosome (synthetic signal) contains the older organisms attributes that assure success. This is repeated until some user's condition (e.g., number of populations or improvement of the best solution) is satisfied (Figure 18). The program is very fast and it takes only few minutes to converge to an optimum solution on a PC.

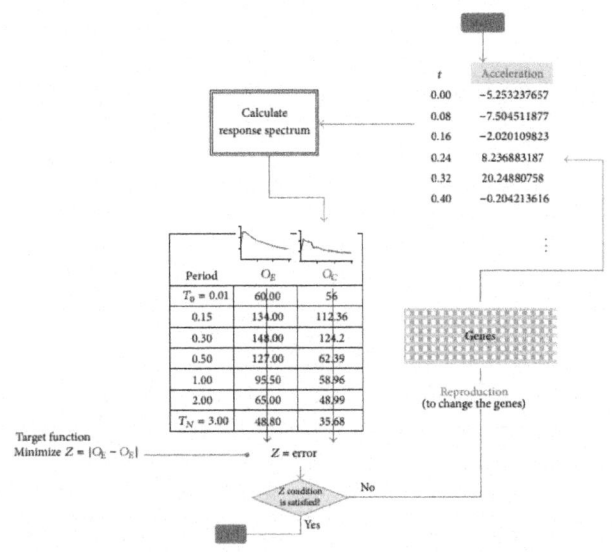

Figure 18: Iteration process of the genetic generator GENES.

In Figure 19 three examples of signals recovered following this methodology are shown. The examples illustrate the application of the GENES to select any number of records to match a given target spectrum (only the more successful individuals for each target are shown in the figure). It can be noticed the stability of the genetic algorithm in adapting itself to smooth, code, or scarped spectrum shapes.

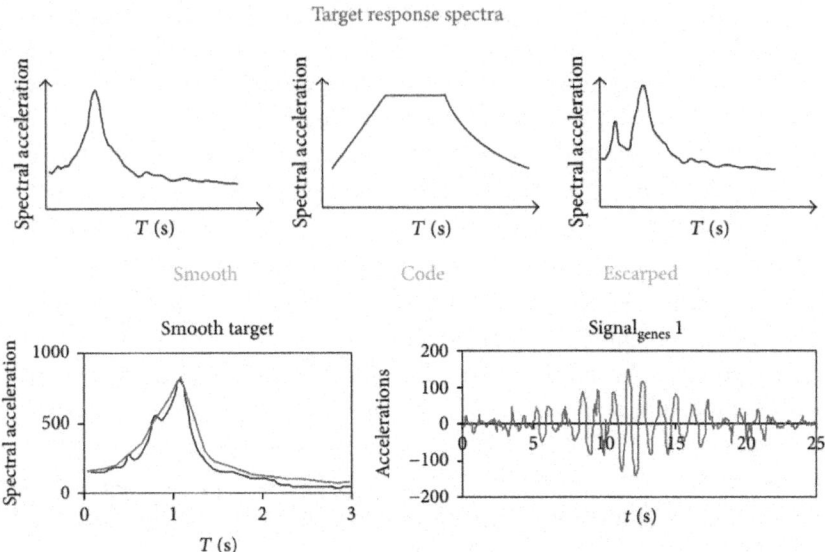

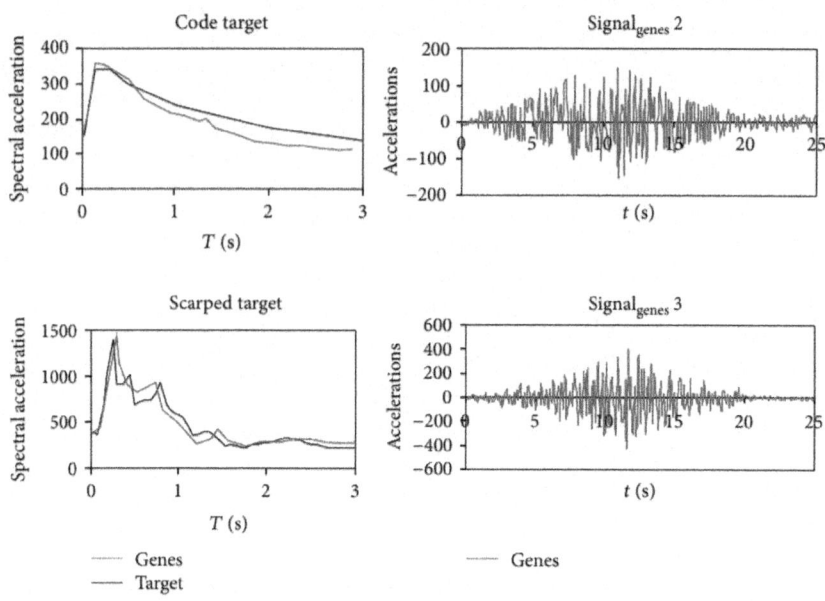

Figure 19: Some generator results: accelerograms application.

In GENES, contrary to other techniques, the characteristics of seismic source, path attenuation, and local soil conditions have been taken into account explicitly when generating synthetic ground motions. Given seismic/ site conditions and a target (design response spectra in this accelerograms application), the processes inspired in Darwin's theory about evolution (GENES mimics mating, natural selection, and mutation) generate accelerations time histories following a very simple method. The procedure is fast and reliable as the results in records match the target spectrum with minimal deviation. GENES has been applied successfully to generate synthetic ground motions having different amplitudes, duration, and combinations of moment magnitude and epicentral distance. Although the variations in the target spectra, the resulted signals preserve the nonlinear and nonstationary characteristics of real earthquakes.

As an advantage, in GENES it is possible to incorporate the uncertainties related with geotechnical and/or seismological effects on the earthquake wave forming process. An additional toolbox is still under development that will permit the use of advanced signal analysis instruments because, as it has been demonstrated (e.g., [42,43]), monitoring nonstationary signals through Fourier or response spectra is not the most convenient feat. Supplementary guidelines for practicing engineers will be implemented in order to make GENES a self-contained kit for risk seismic analyses.

Liquefaction Phenomena

Soil liquefaction and related ground failures are commonly associated with large earthquakes. In common usage, liquefaction refers to the loss of strength in saturated, cohesionless soils due to the build-up pore water pressures during dynamic loading. The losses are attributed to the earthquake-induced liquefaction phenomenon bill of up to hundreds of millions dollars all over the world each year. Therefore, the assessment of the liquefaction potential and the associated damages is an imperative task in earthquake geotechnical engineering. In this section, two models for analyzing this important topic are presented: (i) a classification tree for predicting the occurrence of liquefaction and (ii) a neurofuzzy model for estimating the lateral spreading induces by liquefaction.

Classification Tree for Liquefaction Occurrence Prediction

Over the past forty years, scientists have conducted extensive research and have proposed many methods to predict the occurrence of liquefaction. In the beginning, undrained cyclic loading laboratory tests had been used to evaluate the liquefaction potential of a soil [44] but due to difficulties in obtaining undisturbed samples of loose sandy soils, many researchers have preferred to use in situ tests [45]. Empirical field-based procedures for determining liquefaction potential have two critical constituents: (i) the analytical framework to organize past experiences, and (ii) an appropriate in situ index to represent soil liquefaction characteristics. In a semiempirical approach the theoretical considerations and experimental findings provide the ability to make sense out of the field observations, tying them together, and thereby having more confidence in the validity of the approach as it is used to interpolate or extrapolate to areas with insufficient field data to constrain a purely empirical solution.

The original simplified procedure [46] for estimating earthquake-induced cyclic shear stresses continues to be an essential component of the analysis framework. The refinements to the various elements of this context include improvements in the in situ index tests (e.g., standard penetration test SPT, cone penetration test CPT, self-boring pressure meter tests BPT, and shear wave velocities V_s). Unfortunately, as new liquefaction cases from recent earthquakes have become available, these empirical plots have to be modified for calibrating the boundary between states [47]. In recent years, a powerful computing tool, artificial neural networks (NN), has been introduced for solving the problem of assessing the liquefaction potential (two classes pattern recognition). Many researchers have reported similar or superior accuracy to that of simplified methods using NN in discriminating between liquefaction and nonliquefaction cases [48–54]. They adopted different types of NN architectures and various combinations of input variables for evaluating liquefaction potential from field records (both CPT-q_c and SPT-N datasets), concluding that NN are simpler than and as reliable as conventional simplified methods. Despite the notable performing of these neural approximations, these "black box" models have central shortcomings: (i) their impractical knowledge interpretation, (ii) their slight power to determine the strategic parameters (relative importance), and (iii) the existing NN models cannot be used for others than the model designers.

Based on the recognized weaknesses when using NN, this study presents an empirical machine learning ML model for evaluating liquefaction potential. ML, a branch of soft computation, is a scientific discipline concerned with the design and development of algorithms that allow computers to evolve behaviours based on empirical data (numerical/categorical). Between the ML representations, a classification tree (CT) was selected to determine the liquefaction occurrence and to uncover the character of driving parameters, including earthquake and soil conditions [55–58], being a crucial concern the representation as the more direct way to understand the information structures and relate them to the data from which they came.

The CT for liquefaction prediction establishes a natural connection between experimental and theoretical findings. The CT is presented as a feasible tool for discovering unknown, valid patterns and relationships between geotechnical, seismological, and engineering descriptions through the relevant available information about liquefaction occurrence around the world (expressed as empirical prior knowledge or input-output data).

Database. The database used in this study was constructed using the information compiled by Juang and Chen [51], Juang and Jiang [59], and Andrus et al. [60]. A summary of the parameters included in these datasets is presented in Table 1. From the 407 patterns, the 53% are cases were liquefaction occurred and the other 47% cases are nonliquefied ones. The 80% of the lines were selected as training patterns (used during the model construction) and the 20% was separated for testing the CLIP generalization capabilities. The information is derived from CPT and V_s measurements and different seismic conditions (USA, China, Taiwan, and Japan). The soils types range from clean sand and silty sand to silt mixtures (sandy and clayey silt). Diverse geological and geomorphological characteristics are included. The reader is referred to the citations in Table 2 for details.

In Table 2, according to nomenclature in each original database, $Z_{TOP\ LAYER}$ is the top layer depth, Z_{NAF} is the water table depth, H is the layer thickness, a_{max} is the maximum registered acceleration (peak ground acceleration (PGA)), q_c is the cone penetration resistance, R_s is the fine content, σ'_v is the effective vertical stress and σ_v is the total one, M is the magnitude, V_s is the shear wave velocity, and S_T is the soil type.

CT Construction. The general CSR and CRR formulations by Seed and Idriss [61] and Youd et al. [62] are adopted in the CT proposal. The cyclic stress ratio definition includes

$$CSR = f_{ML}\left(M, \sigma_v, \sigma'_v, a_{max}\right)$$

(2)

while the cyclic resistance ratio the CT follows the expression

$$CRR = f_{ML}\left(\frac{q_c}{V_s}, \sigma'_v, S_T\right).$$

(3)

For the CT construction, none of these variables are adjusted, preclassified, or transfigured as in other approximations that must be done (i.e., the correction of q_c for overburden stress). This implies that any restrictive or subjective hypothesis needs to be included to predict the liquefaction occurrence. It should be noted that the soil type S_T is a parameter that has not been properly defined [63] and it is still a debate point between researchers. In the CT pursuing the classification reported in the original data base is included as categorical instance. Although there are more sophisticated soil definitions, they require multiple steps of subjective calculation and the proposed instance is easier to use and is more suitable for training the machine learner model and does not need to be calibrated along with other formulas, methods, or the model results.

The final schematic representation of the liquefaction tree model is shown in Figure 20. The following input variables were booked.

(1) Geotechnical: cone penetration resistance "q_c", shear wave velocity "V_s", soil type "S_T", fines content R_f, and total and effective stresses "σ_v", "σ'_v".

(2) Geometrical: layer thickness "H", water level depth "Z_{NAF}", and top of layer depth "$Z_{TOP\ LAYER}$".

(3) Seismic: maximum magnitude "M", and peak ground acceleration "a_{max}".

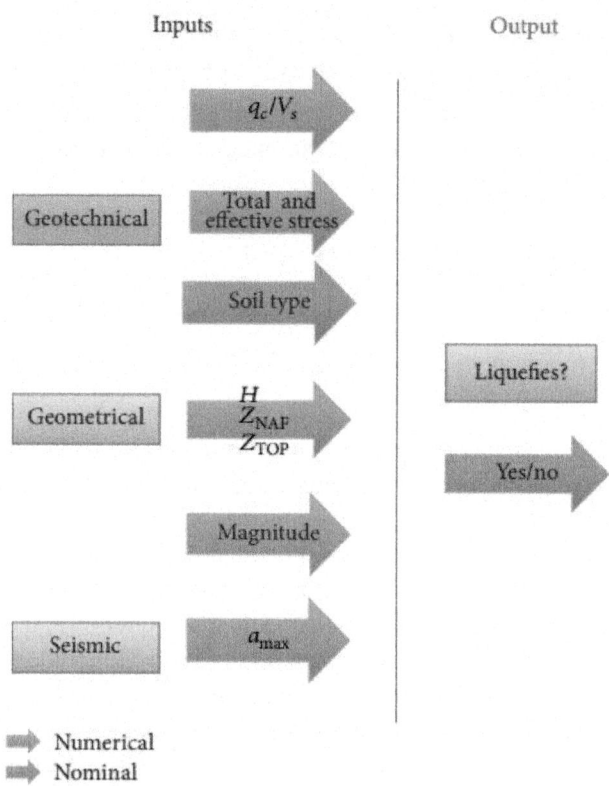

Figure 20: Inputs-output schema.

The output variable is "Liquefies?" and it can take the categorical linguistic values "YES"/"NO."

The objective of the tree training is to determine the more suitable representation for acquiring knowledge and structures and relate them to the data from which they came. The discovered patterns are accepted as successful as the prediction from the tree for the given set of inputs $(M, \sigma_v, \sigma_v', a_{\max}, q_c/V_s, S_T)$ matches the target (Liquefies? YES/NO), based on field observations. The CT was built through a process known as binary recursive partitioning. After the iterative process of splitting the data into partitions, and splitting it up further on each of the branches, all of the records in the training set (the preclassified records that are used to determine the structure of the tree) initially put together in one big box are breaking up using every possible binary split on every field. The resulting classification tree is a connected acyclic graph with remarkable advantages over regression, discriminant analysis, and other procedures based on algebraic models.

The CT for liquefaction analysis was pruned to minimize the sum of (1) the output variable variance in the validation data, taken a terminal node at a time, and (2) the product of the cost complexity factor and the number of terminal nodes. After working with global liquefaction data, an automatic procedure for detecting interactions among variables (related to classical cluster analysis) generates the bulky trees shown in Figures 21 and 22.

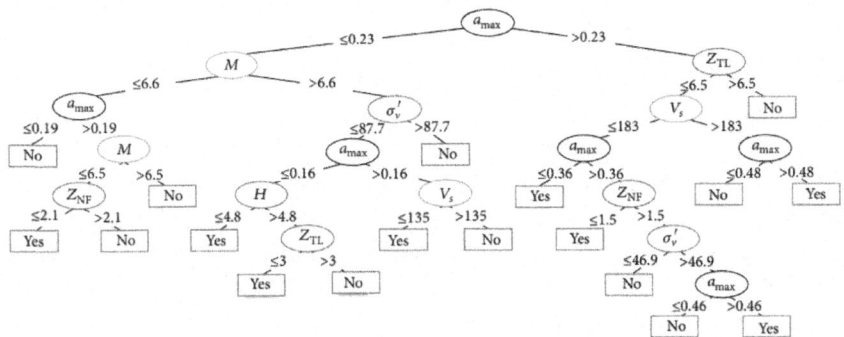

Figure 21: CLIP: Geotechnical description V_s.

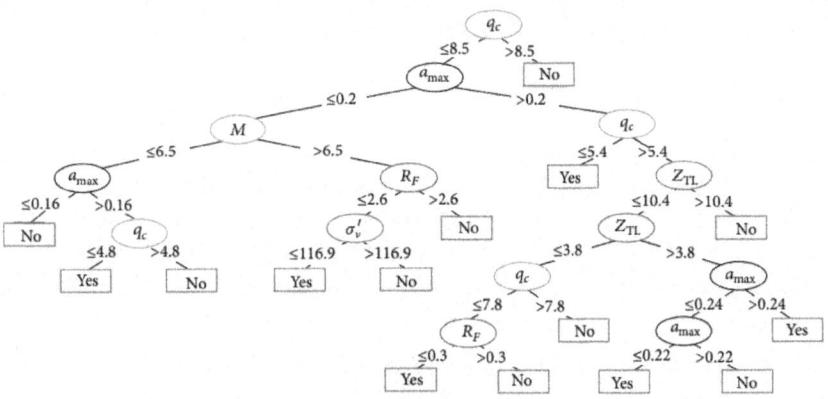

Figure 22: CLIP: Geotechnical description q_c.

CT Interpretation. The input data for the liquefaction occurrence is complex and contains many different categories and many possible predictors for performing the classification (deficient taxonomy) and then the resulting tree is large. This is not so much a computational problem as it is a problem of presenting the trees in a manner that is easily accessible to the analyst or for presentation to the "consumers" of the research. Figures 21 and 22 show the pruned model trees (after CCF operation) using q_c and V_s as geotechnical inputs, respectively, and sharing the terminal class nodes (YES/NO). The trees for predicting the liquefaction occurrence by concerns about seismic, geotechnical, and geometrical parameters can serve as a basis for structuring the discussions about phenomena-parameterization strategy. Notice that the classes predicted are at the bottom of the tree and any further step is not necessary. The practical exploit of this tool is straight forward: the user comes into the CT and presents basic parameters for defining the event and site conditions (even if there are missed attributes a CT can be exploited) then each branch and node of the tree is tracked for offering, in the terminal node, a simple conclusion about liquefaction occurrence.

To use the tree structure, the analyst has to tag on the branches in line with the instance being analyzed, and when it reaches a terminal node a simple class expression is given for determine the concept according to the attributes and values contained in the example. In engineering practice, it is very uncommon to get a full description of soil conditions (in situ testing results); therefore, the CT was developed for evaluating the liquefaction occurrence having q_c and V_s data information.

When the dynamic variable V_s is being used, the CT has the following configuration (Figure 21): the driven variable is a_{max} and the partitioning algorithm found a behavior boundary in 0.23 g. For seismic events where the intensity is under this value, the following driving parameter is M; lower magnitudes (<6.6) need the definition of a_{max} and Z_{NAF} for offering the prediction, while higher magnitudes (>6.6) require the numerical description of σ'_v, H, $Z_{TOP\ LAYER}$, and V_s. This structure and the relations contained are very important for understanding the "occurrence" phenomena. For minor cyclic loading, the CT uses information about water level depth and the maximum accelerations as its driven parameters, while for greater loads, the criteria for selecting the unsafe situations is based on practically the whole set of geotechnical, geometrical, and seismological variables.

If the geotechnical description available is the mechanical variable q_c, the CT has the configuration depicted in Figure 22. The primary categorical variable is q_c with a boundary around 8.5 MPa for the first split. When the analyzed soil layer has a $q_c > 8.5\ MPa$, there is no possibility of liquefaction under the dynamic range of a_{max} and M included in the training examples, and this indicates that the differences in the amplitudes, magnitudes, and geometries are critical when analyzing soils with resistances under 8.5 MPa. Then, to predict the liquefaction occurrence for soils with $q_c < 8.5\ MPa$, it is necessary (i) to declare M value, (ii) to define the fine content percent, and (iii) to declare de σ'_v value. As a consequence of this tree separation, appealing conclusions can be achieved, for example, $Z_{TOP\ LAYER}$ related to Z_{NAF} explanation is important for accelerations above 0.2 g while for those under this frontier the percent of fine particles contained in the soil mass is more important.

As it was exposed, besides the huge size of the CT, important features about the physics of the problem can be easily detected. For example, between the databases used in this investigation, many input soil classes are declared (gravel, sand, gravel and sand, sandy silt, silty sand, and clayey silt) but when these labels are studied in a multidimensional environment, the relationships derived from experience registered in the databases eliminate some of them (broader categories are detected) and still having satisfactory prediction results. The CT shows that the refinement in numerous labels for S_T is not in benefit of assertive prediction of liquefaction occurrence; it seems that the in situ soil properties (V_s or q_c) are better classifiers for soil masses and its responses. The pruned classification trees use a total geometrical description (H, $Z_{TOP\ LAYER}$, Z_{NAF}) as a driven condition to get the final nodes in the set of examined conditions. Observing the seismological descriptions, to use the CT it is not necessary to separate the instances from minor and severe magnitudes to develop two models (as is required in many published models). The whole dynamic range of M and a_{max} is integrated in the training set being a general conclusion of the learning algorithm; the behavior boundary is at M=6.5 and a_{max} has a direct impact in the liquefaction occurrence only for M values above 6.5.

The impact of R_f on the number and direction of tree branches evidences the distracting effect of using a fusion of specialist and numerical criteria for defining this "indefinable" parameter. The fines content expression seems very subjective, between the selected inputs and output parameters, when trying to discover the numerical relationships. R_f coupled with q_c generates a better (smaller and efficient) tree and consequently improved knowledge; on the other hand, when the soil description is through V_s, its inclusion is not in profit of efficient predictions. This can be explained based on the property nature: one is mechanical (q_c) and needs more information for a response under seismic forces and the other one is directly a dynamic property (V_s) and it seems sufficient for elaborate useful relations.

Without potentially confusing regression strategies, the attenuation tree validation results (cases not used during the model construction and used for proving the CT generalization capabilities) shown in Figures 23 and 24 can be seen as an indication of how effectively the CT maps the assigned predictor variables to the response parameter. The success rate of the trained CT in predicting the occurrence of liquefaction (i.e., distinguishing liquefied cases from nonliquefied cases) is 99% during the training phase of network development. The success rate of the trained model in predicting the occurrence of liquefaction in the cases in the testing data subset and the overall success rate of the trained network in predicting liquefaction in all cases in the entire dataset is 96%. Using data from many regions and broad ranges of inputs, the prediction capabilities of the tree are superior to many other approximations used in common practice, but the most important remark is the generation of meaningful clues about the reliability of physical parameters, measurement and calculation process, and practice recommendations.

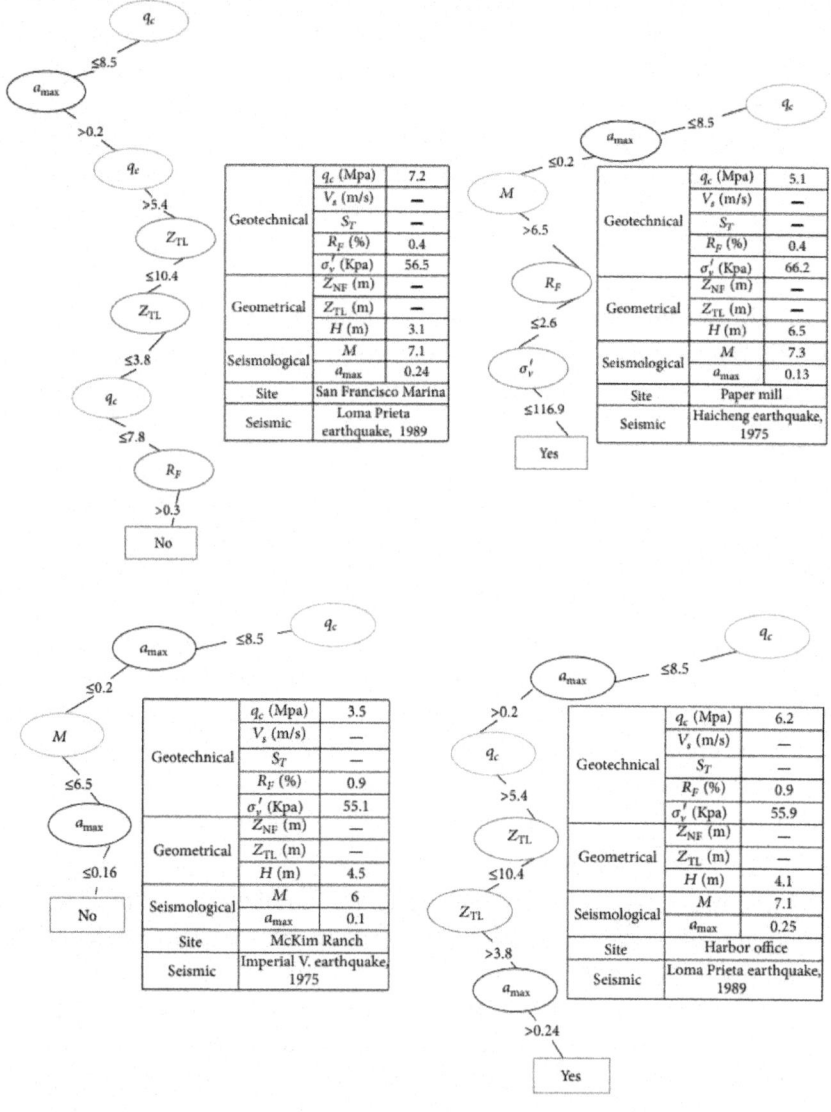

Figure 23: Validation tests: using q_c as geotechnical descriptor.

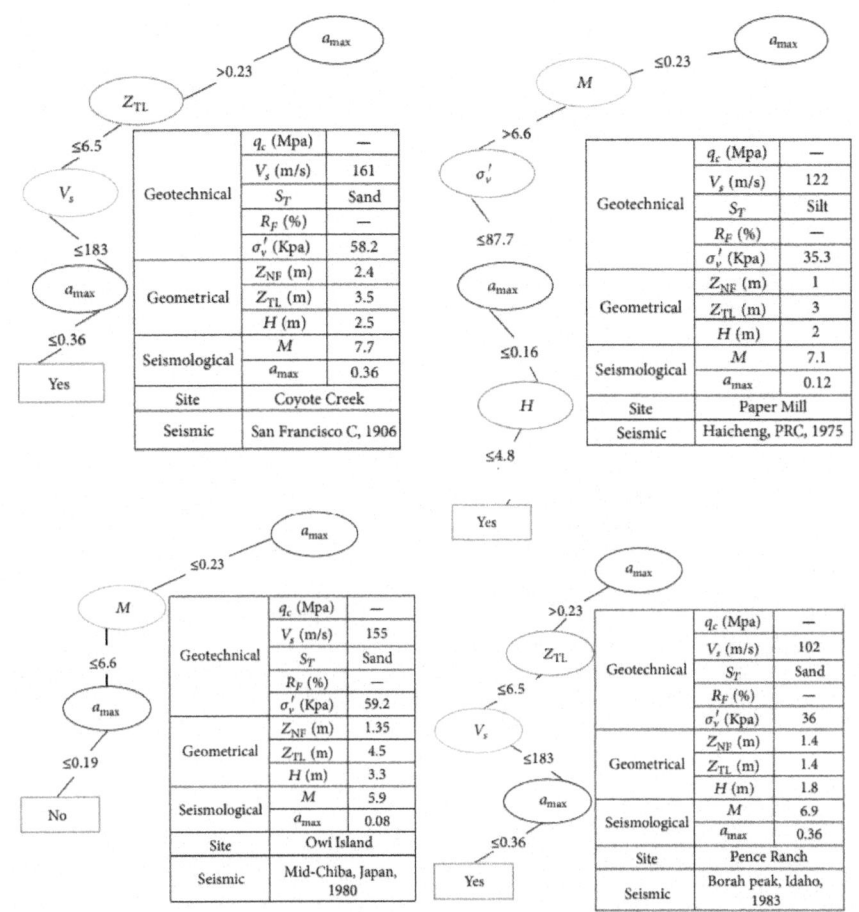

Figure 24: Validation tests: using V_s as geotechnical descriptor.

In 1999, two major earthquakes, namely, Chi-Chi, Taiwan, earthquake (magnitude $M_w=7.6$) and Kocaeli, Turkey, earthquake, (magnitude $M_w=7.4$) triggered ground failure principally induced by soil liquefaction throughout the city of Adapazari (Turkey) and the cities of Wufeng, Nantou, and Yuanlin (Taiwan). These case records are used for validating the proposed tree model. The multidimensional CT permits a very efficient prediction of the liquefaction occurrence (Figures 25 and 26). If the analyst has uncertainties in the inputs definition, or even some of them are missed, CLIP can be used to understand the effect of the fuzziness on behavior path and to use intelligently the information in the branches and leaves.

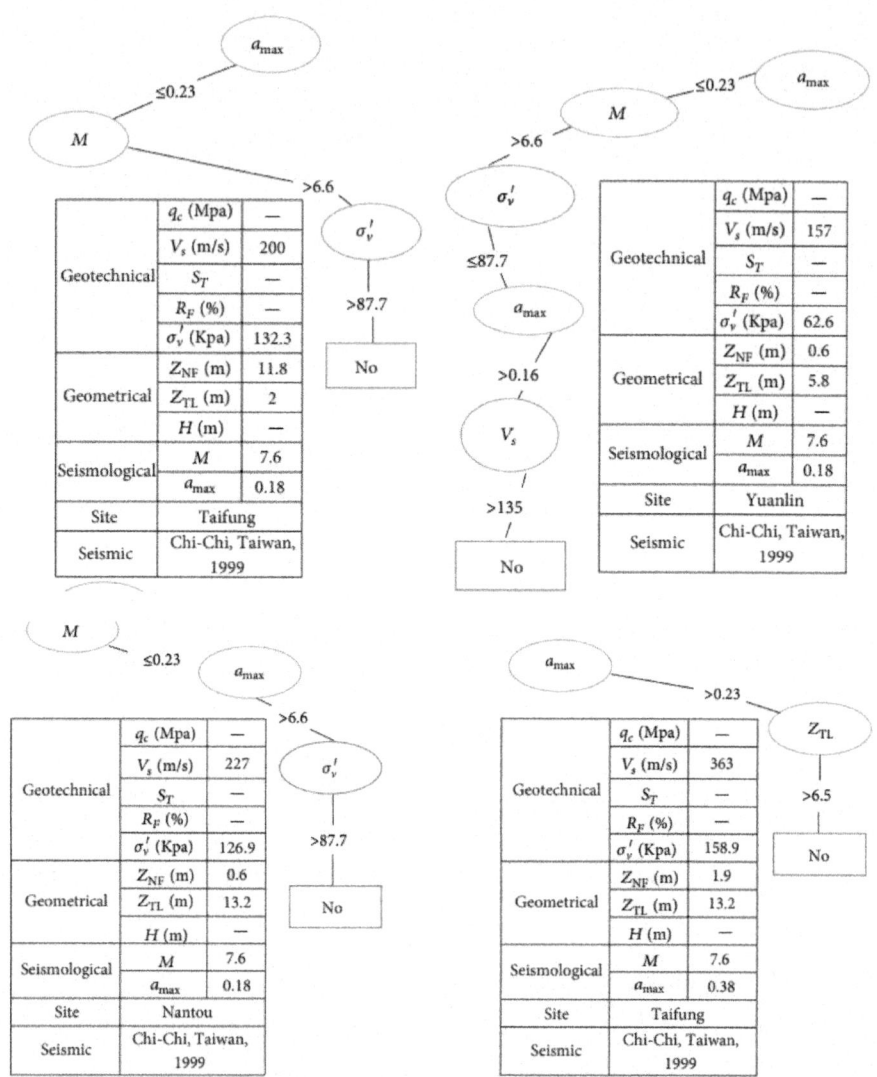

Figure 25: CLIP results for Chi-Chi, Taiwan, earthquake.

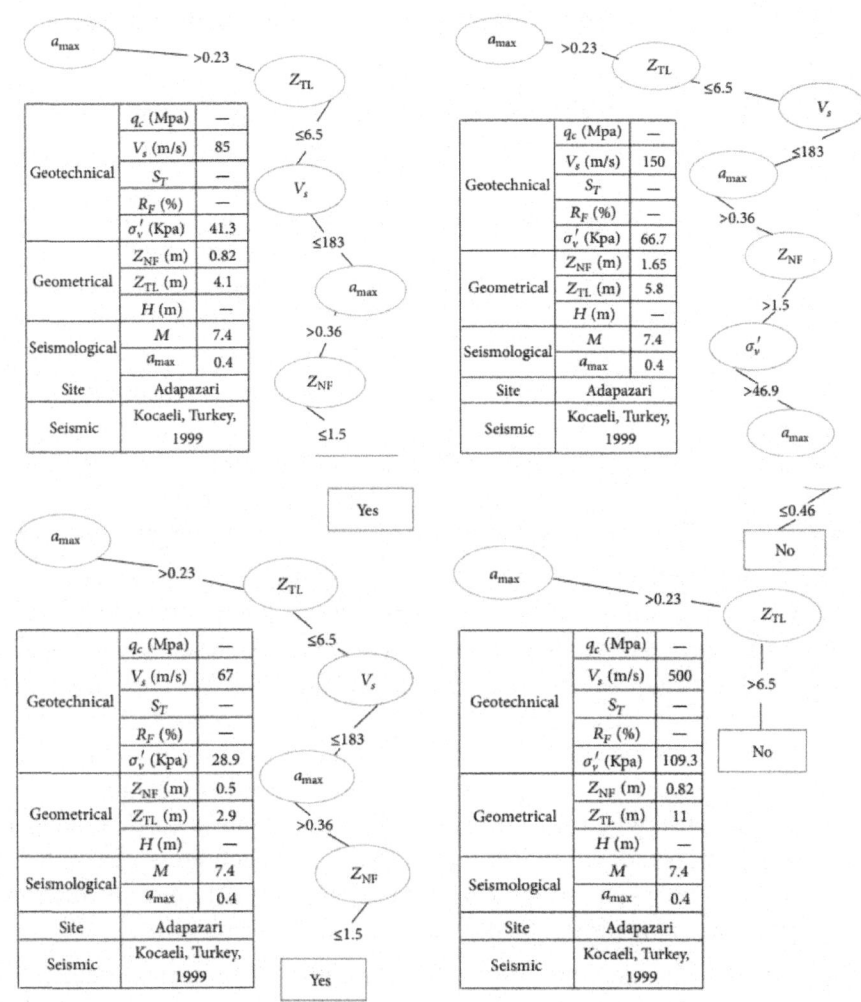

Figure 26: CLIP results for Kocaeli, Turkey, earthquake.

The classification tree is clearly an advantageous tool: practical, free to use, easy to understand, and with straightforward behaviors interpretation. The CT simplicity is useful not only for purposes of rapid classification (or prediction) of new observations but also for yielding a much simpler model for explaining why observations are classified or predicted in a particular manner (e.g., to analyze input-output parameters importance, to present simple statements to management, or to eliminate elaborate and inaccurate equations). There is no doubt that ML represents a powerful alternative in predicting the liquefaction occurrence and phenomenon related.

A Neurofuzzy System to Analyze Liquefaction-Induced Lateral Spread

Lateral spreading is conceivably the most common type ground failure induced by the liquefaction of saturated fine granular materials. Horizontal spreads cause basically two loading mechanisms upon engineered structures. One is due to drag forces exerted, mainly on piles and piers, and the thrust that may be caused by a crust of nonliquefied soil riding on the liquefied layer as it is driven against buried structures. Indeed, piles and piers can be subjected to appreciable loading by the soil flowing past them, causing severe damage. Also, during lateral spreads, blocks of intact, superficial soil displace along a shear zone within the liquefied layer, either down slope or toward a free face (i.e., river, channel, or an abrupt topographical depression) driven by gravity or earthquake forces. The resulting ground deformation usually has extensional fissures at the head of the failure as well as shear deformations at the flanks and compression of the soil at the toe [64].

Several researchers (e.g., [65, 66]) have studied drag forces exerted on piles by liquefied soil and have found that such forces are often too small to inflict any damage. However, other studies of centrifuge tests have shown that large forces were applied by the flowing ground (e.g., [67]). This discrepancy on the potential effects may be due mainly to the amount of lateral spread computed by these researchers and thereupon the importance of having means to estimate as accurately as possible the magnitude of lateral spreads caused by seismic events. When the stiffer unliquefied stratum is carried along the underlying, the spreading sand may produce high enough pressures on buried structures to cause them severe damage and even their failure. In the limit, these forces reach the passive resistance of the unliquefied soil.

Displacements ranging from a few centimeters to several meters are commonly developed by this phenomenon [68]. Thus, it is not difficult to infer that engineered structures be severely marred when enduring lateral soil displacements of such a magnitude. Some examples that support this assertion are given by the recorded effects of the 1906 (M_w 7.9) San Francisco earthquake on buildings, bridges, roads, and pipelines. Similarly, the 1964 Alaska (M_w 9.2) and Niigata (M_w 7.5) earthquakes caused extensive damage to engineered structures in cities such as Valdez and Anchorage, Alaska, and Niigata, Japan, respectively. In [69], it caused severe damage to pile bridge foundations induced by the 1987, M_w 6.3, Edgecumbe, New Zealand, earthquake has been reported. The San Pedrito dock-pile foundation was also damaged during the 1995, M_w 8.0, Manzanillo earthquake [70, 71]. Similarly, extensive destruction was induced during the Hyogoken-Nambu earthquake (M_w 7.2) of 1995 to buildings and bridges in the reclaimed land areas along the coastline of Kobe, Japan.

These examples and many others reported in the technical literature clearly show that lateral spreads on gently sloping ground can severely impair engineered structures. Acknowledgement of the chaotic response of liquefying soil masses to earthquakes and having a clear evidence (as shown later) that the available neural and empirical methods do not adequately forecast lateral spreads, it was deemed important to develop alternate procedures of analysis. In the following, a powerful computational paradigm, a neurofuzzy NF empirical model, is presented as a tool for the estimation of liquefaction-induced lateral spreads due to seismic loading. It takes into account parameters related to general earthquake characteristics, topographical, regional, geologic, and soil data.

Database. The information compiled by Bartlett and Youd [68] and extended later by Youd et al. [72] includes 448 entries corresponding to seven earthquakes. A summary of this database is given in Table 3. The data was divided (classified) in four main categories according to the known qualitative and quantitative information: (a) ground displacement amplitude; (b) borehole data; (c) boundary conditions, that is, ground-slope and free-face topographical data; (d) seismic knowledge. Since the raw database contains redundant and conflicting information, before being used in the training stage of the neurosystem development, it was preprocessed using a fuzzy clustering technique (see García and Romo [73] for technical details). The input-output data considered in the neurofuzzy system may be represented by the following function:

$$D_h = f\left(M_w, R^*, A, S, W, L, T_{15}, D50_{15}, F_{15}\right).$$

(4)

Notice that there are some differences with respect to the traditional equations published to calculate D_h. In the neurofuzzy case, the maximum ground acceleration, A, in units of gravity, g, and the length from the free-face to the point of displacement, L, in meters are included. On the other hand, N160S is not involved because it was found that its dynamic range was too narrow, thus its influence as classifier was found to be negligible [73]. In Figure 27 the main geometric aspects of the lateral spreading problem are depicted, and some of the neurofuzzy parameters S, and L, are depicted. The rest of the parameters in (4) are defined in Figure 27. According to (4), the general approach used in determining the horizontal displacements, D_h, caused by earthquake-induced liquefaction assumes that the pattern of ground displacements can be classified by their topographical variables (S, W, L), geological and soil conditions $(T_{15}, D50_{15}, F_{15})$, and the earthquake characteristics (M_w, R^*, A). Therefore, the fuzzy system training process was carried out by considering the coupled effect of all dependent parameters (right hand side of (4)) on the independent variable D_h.

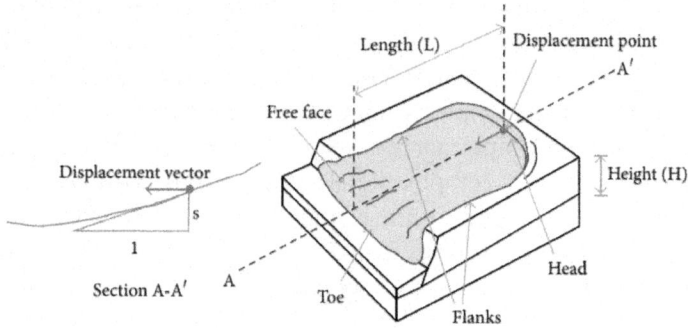

Figure 27: Sketch of lateral spread phenomenon.

Prior to the training stage, all data points of the input-output function that showed some redundancy or antagonism were lumped with others of the database employing the fuzzy clustering technique described in García and Romo [73]. In this way, a single data point can replace each of the members of such a cluster to reduce the number of points. These remaining points, called typicals, constitute the patterns that combined with raw data (information points that did not fit any of the classifications defined by the fuzzy clustering technique used) and were utilized in the training of the NF system (see the example of S-D_h in Figure 28; these are the originals examples distribution in the training set). After fuzzy preprocessing, the number of patterns was reduced to 337 (see Figure 29 where the typical and raw data for S-D_h are depicted). These patterns were found after many trials for searching the optimum size of the database (i.e., it represents the dataset having lower incongruence, thus leading to a better distribution in the input variables space).

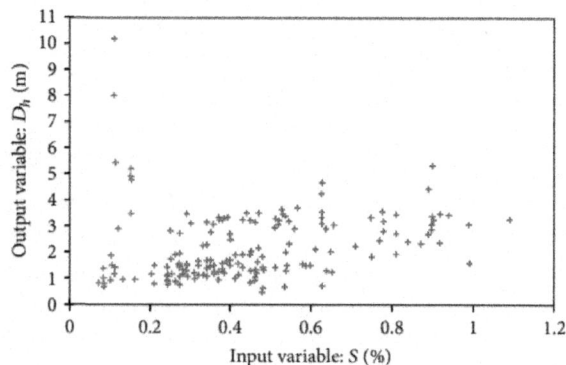

Figure 28: Lateral spreading sample dataset. One input-one output.

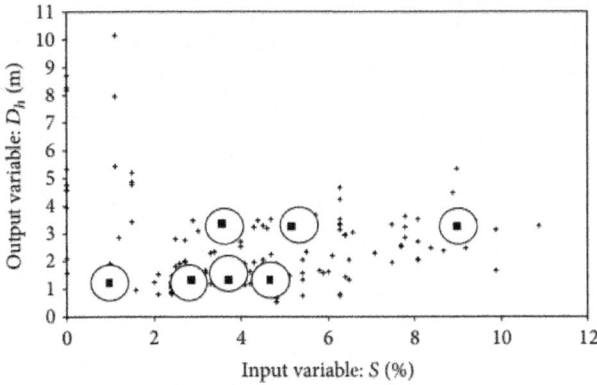

Figure 29: Clusters and typicals after fuzzy clustering (example with one input-one output).

Out of the 337 patterns, 257 were randomly selected for the training stage and the rest for testing. The training process used the "new" pattern distribution associated with the membership functions (here the popular triangles were considered) initially defined by experts. Now, to proceed with the training process, the membership degree of each membership function has to be obtained. To illustrate the procedure to define the membership degree (fuzzification), assume (see Figure 30) that S: 6.5%, thus HIGH: 0.22, MEDIUM: 0.40, and LOW: 0.00. Once all the fuzzification process is carried out for all the dependent variables (M_w, R^*, A, etc.), all possible fuzzy rules (if-then) are developed similarly. Afterwards, the training process begins and the fuzzy rules defined by the experts are modified according to the relationships between all the data patterns.

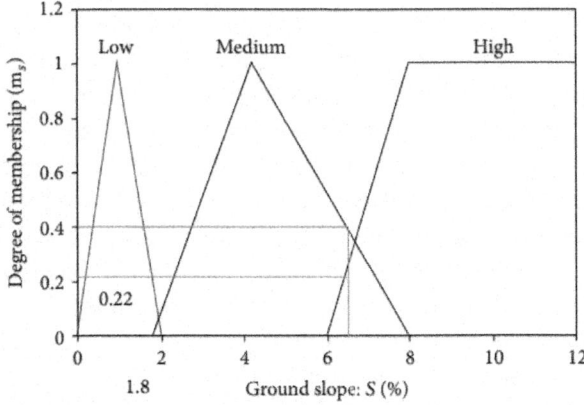

Figure 30: Membership functions for S.

The membership functions defined by the experts are modified until the system response is optimized, which is achieved when the mean-squared error threshold specified is not exceeded. Once the optimization is reached, the system obtained is the fuzzy system that minimizes the $D_h\text{-}f$ (M_w, R^*, $A, S, L, W, D50_{15}, T_{15}, F_{15}$) error mapping.

One of the many resulting if-then rules that optimizes the $D_h \rightarrow f(-)$ mapping is included below:

	IF
Earthquake	M_w is HIGH and R^* is MEDIUM and A is LOW
Topography	and S is LOW and W is MEDIUM and L is LOW
Soil	and T_{15} is LOW and $D50_{15}$ is LOW and F_{15} is MEDIUM
	THEN
	D_h is LOW

Considering the character of components that influence the lateral-spreading problem, it was decided to build up the neurofuzzy system using the following three modules.

(i)   Reg-neurofuzzy: appropriate for predicting horizontal displacements in geographic regions where seismic hazard surveys have been identified.

(ii) Site-neurofuzzy: proper for predictions of horizontal displacements for site-specific studies with minimal data on geotechnical conditions.

(iii) Geotech-neurofuzzy: more refined predictions of horizontal displacements when additional data is available from geotechnical soil borings.

Since the training process is carried out in parallel for all parameters, the resulting fuzzy system is integrated by all the membership functions, given in the column farthest to the right in Figure 31. This figure also includes the experts' membership functions (fourth column in Figure 31), as well as the input/soft variables and their corresponding units. It may be seen that their base and slope are appreciably modified, depicting the importance of taking into account the numerical interrelationships among all the variables that are thought to dominate the physical phenomenon. This aspect is very important for nonlinear problems having high dimensionality, such as the lateral-spreading problem. The procedures for carrying out this training process and a detailed sketch of the neurofuzzy system are given in [74]. Briefly, this system is incorporated in a model to perform the fuzzy clustering of the raw data and to obtain the improved database which is fed into a NF module, which then performs the NF training that yields the output, D_h. The schematic representation of the fuzzy structure is depicted in Figure 32. The organization of the model components is such that it allows the prediction of lateral spreads, D_h, even when no full descriptions of the seismic, topographical, or soil information are available. Lexical assumptions about any input group can be implemented in terms of linguistic data for getting an approximate estimation. But, as would be expected, better predictions can be made as the quality (i.e., less vague and contradictory influences of M_w, R^*, A, y, F_{15}) of the input variables to the fuzzy system model improves.

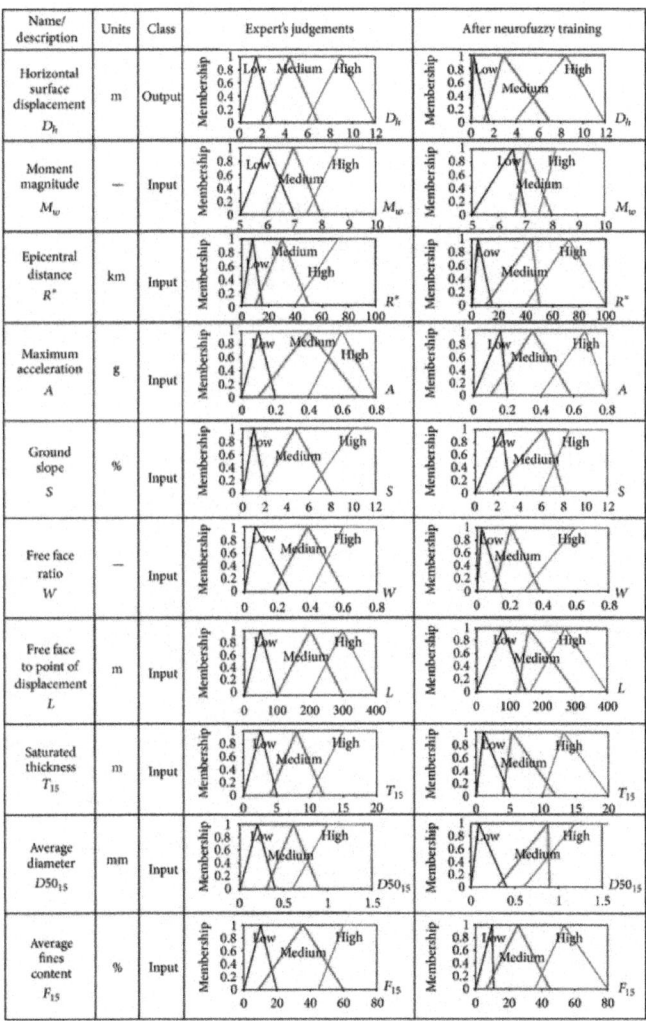

Figure 31: Membership functions used as neurofuzzy system variables.

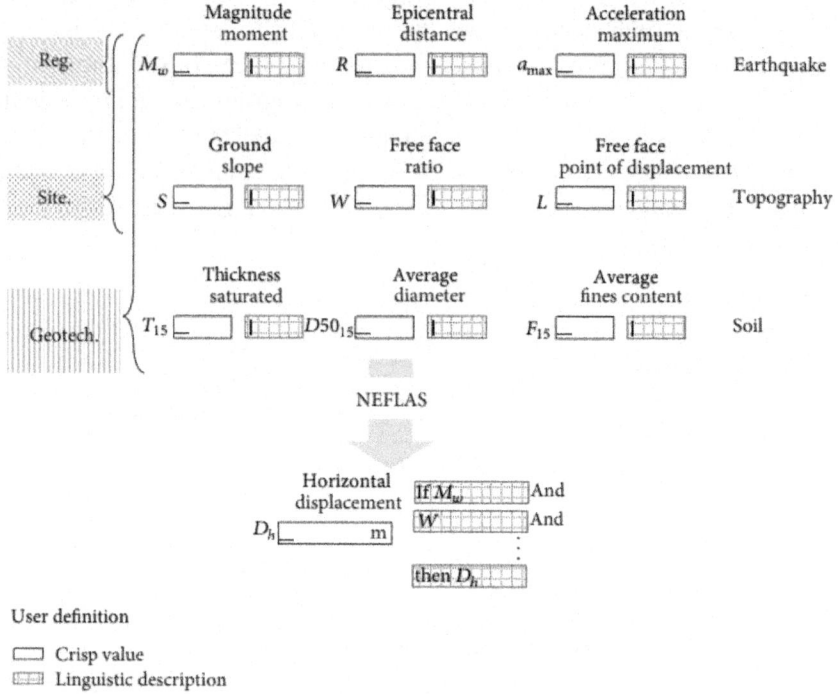

Figure 32: Neurofuzzy structure (NEFLAS is for neurofuzzy system to estimate lateral spread displacement).

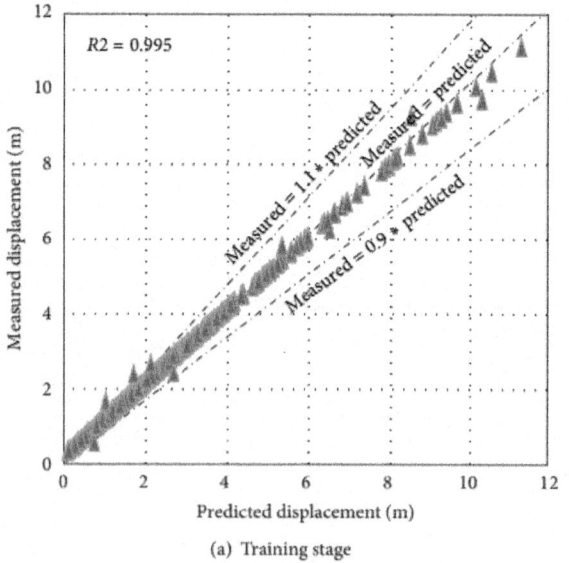

(a) Training stage

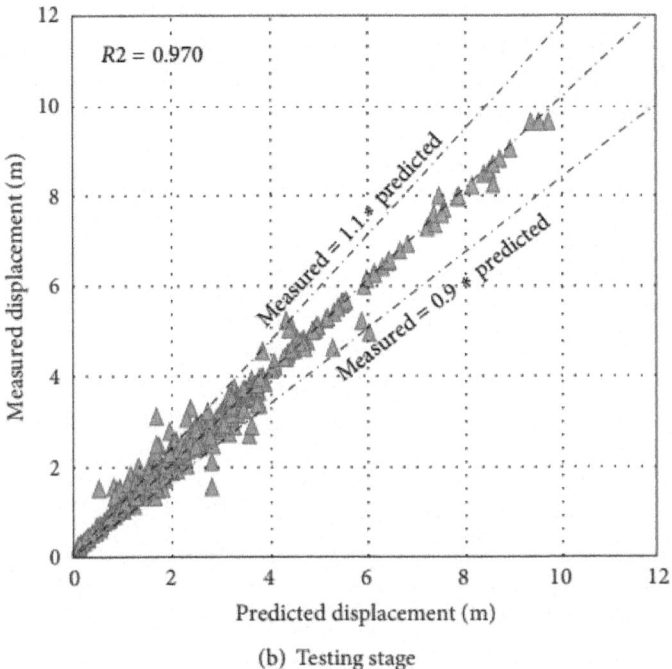

(b) Testing stage

Figure 33: Comparison between neurofuzzy and measured lateral spread, whole data-set. (a) Training stage and (b) testing stage.

The final step in the process followed to develop and assess the reliability of the lateral-spreads predictions by the fussy system consisted in feeding this NF model with the set of data (80 patterns) reserved from the whole database. This information was unknown and thus the results can be considered as predictions. Accordingly, by simply comparing fuzzy computed values with the unknown horizontal displacements included in the database, the predicting capabilities can be evaluated. This comparison is depicted in Figure 33(b).

Several aspects stand out that (a) the R2 (0.97) is much higher than the values obtained with the previous procedures; (b) the R2 value remains practically unchanged in training and testing stages (see the graphs in Figures 33(a) and 33(b)), and thus it may be asserted that neurofuzzy system is a system that is able to predict lateral spreads with a high degree of confidence; (c) neurofuzzy system is a much more powerful tool than the previously proposed techniques; (d) while the predicting capabilities of MRL and NNs procedures are rather poor, particularly for horizontal displacements smaller than about 3.0 m, neurofuzzy system shows that it is capable of containing its predictions within a narrow band even at very small horizontal displacements, which supports the accuracy of the proposed computing tool. To stress this assertion, Figures 34, 35, 36, and 37 show the predictions of D_h for a number of individual seismic events. Notice that the values of R2 remain practically unchanged when earthquakes and their corresponding dataset are considered independently, which demonstrates the robustness of neurofuzzy system.

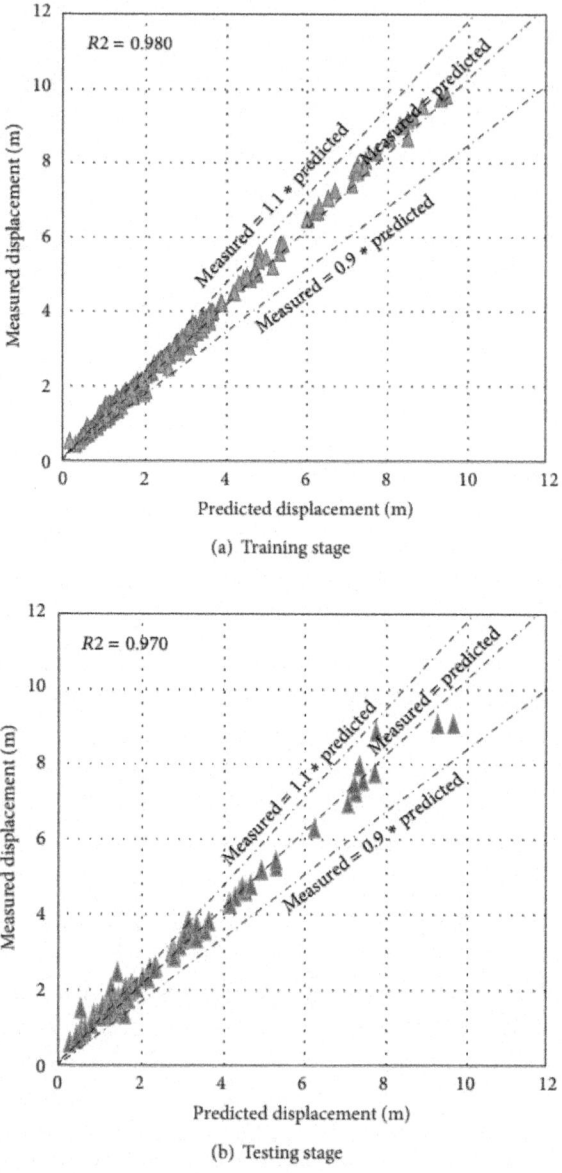

(a) Training stage

(b) Testing stage

Figure 34: Comparison between NEFLAS and measured lateral spread, Niigata, Japan. (a) Training stage and (b) testing stage.

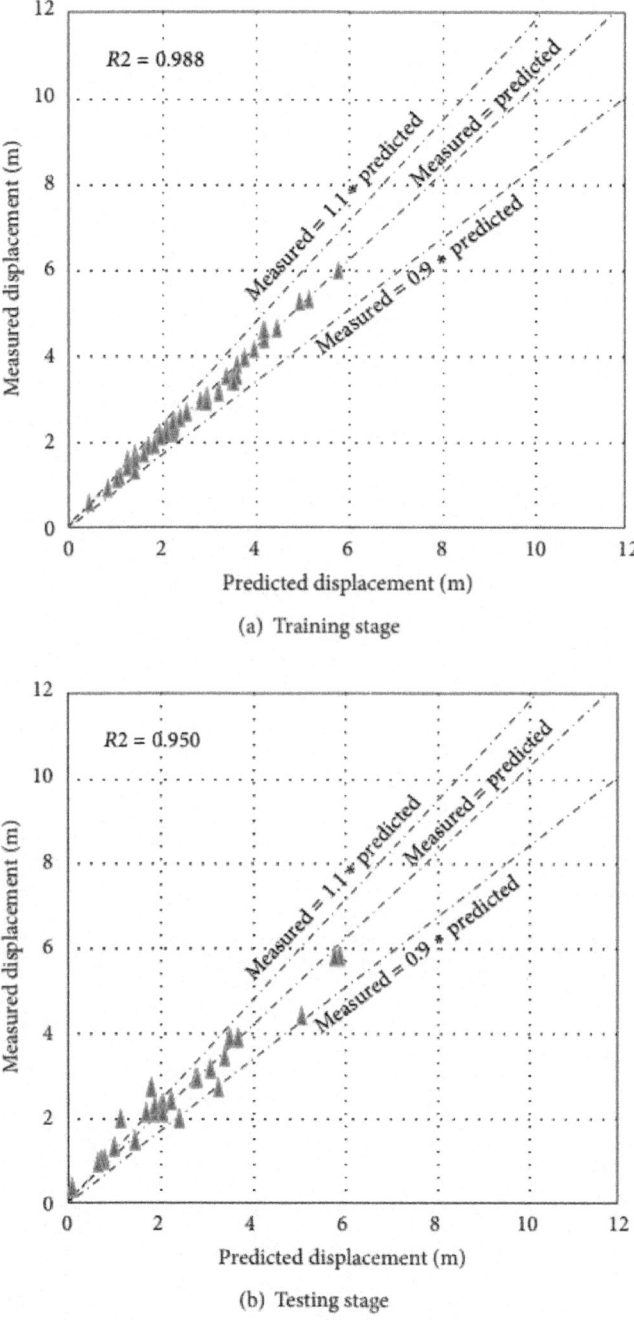

Figure 35: Comparison between NEFLAS and measured lateral spread, Nihonkai-Chubu, Japan. (a) Training stage and (b) testing stage.

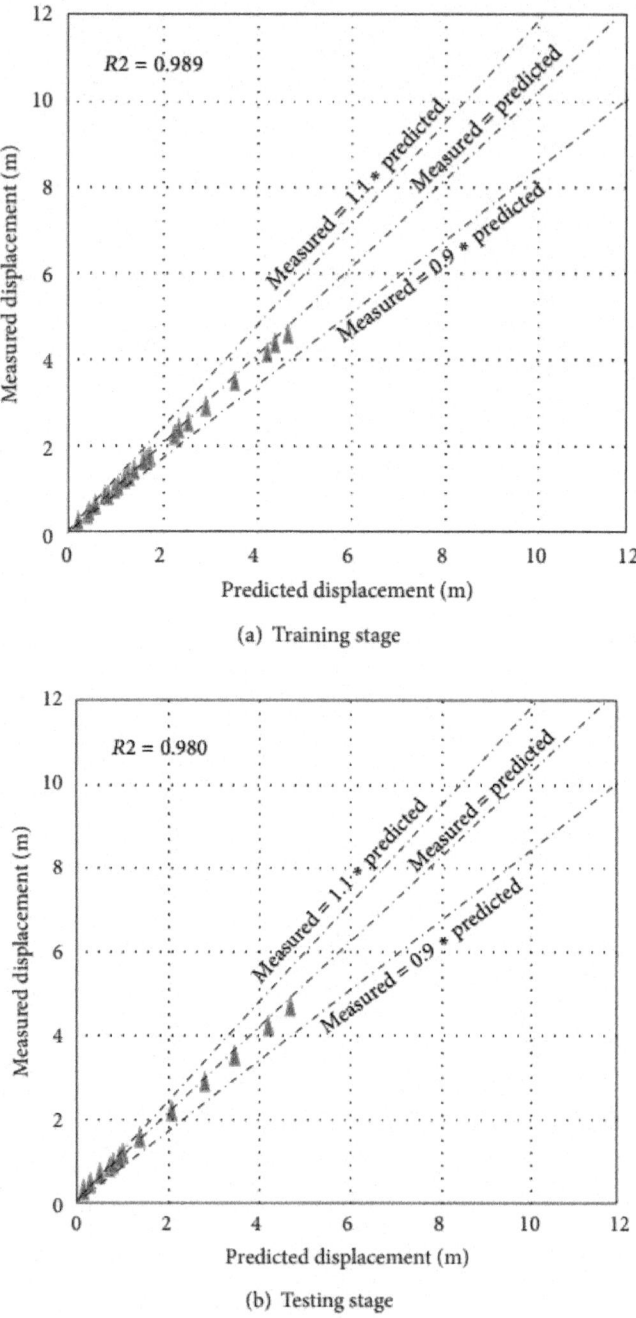

Figure 36: Comparison between NEFLAS and measured lateral spread, Imperial Valley, USA. (a) Training stage and (b) testing stage.

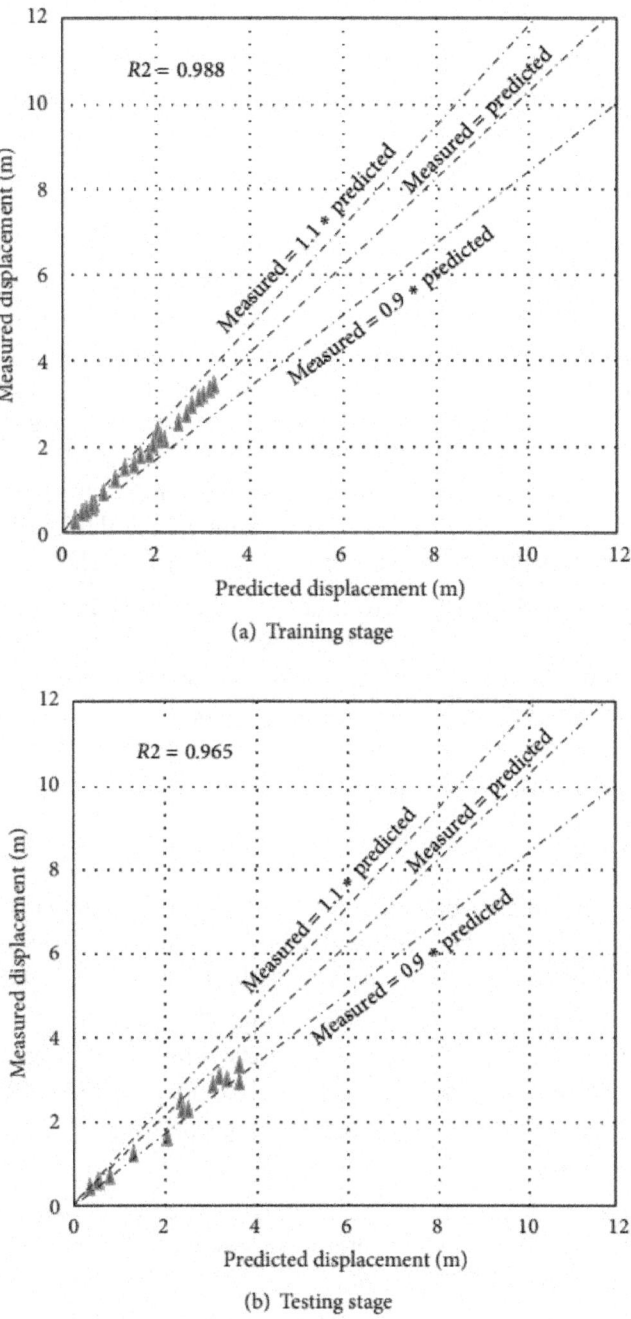

Figure 37: Comparison between NEFLAS and measured lateral spread, San Francisco, USA. (a) Training stage and (b) testing stage.

CONCLUSIONS

Based on the results of the studies discussed in this paper, it is evident that cognitive techniques perform better than, or as well as, the conventional methods used for modeling complex and not well-understood geotechnical earthquake problems. Cognitive tools are having an impact on many geotechnical and seismological operations, from predictive modeling to diagnosis and control.

The hybrid soft systems leverage the tolerance for imprecision, uncertainty, and incompleteness, which is intrinsic to the problems to be solved, and generate tractable, low-cost, robust solutions to such problems. The synergy derived from these hybrid systems stems from the relative ease with which we can translate problem domain knowledge into initial model structures whose parameters are further tuned by local or global search methods. This is a form of methods that do not try to solve the same problem in parallel but they do it in a mutually complementary fashion. The push for low-cost solutions combined with the need for intelligent tools will result in the deployment of hybrid systems that efficiently integrate reasoning and search techniques.

Traditional earthquake geotechnical modeling, as physically based (or knowledge-driven) models, can be improved using soft technologies because the underlying systems will be explained also based on data (CC data-driven models). Through the applications depicted here, it is sustained that cognitive tools are able to make abstractions and generalizations of the process and can play a complementary role to physically based models.

REFERENCES

1.   A. P. S. Kumar, K. P. Sudheer, S. K. Jain, and P. K. Agarwal, "Rainfall-runoff modelling using artificial neural networks: comparison of network types," Hydrological Processes, vol. 19, no. 6, pp. 1277–1291, 2005.

2.   J. C. Bezdek and R. J. Hathaway, "Numerical convergence and interpretation of the fuzzy c-shells clustering algorithm," IEEE Transactions on Neural Networks, vol. 3, no. 5, pp. 787–793, 1992.

3.   L. A. Zadeh, "Fuzzy sets," Information and Control, vol. 8, no. 3, pp. 338–353, 1965.

4.   L. A. Zadeh, "The roles of fuzzy logic and soft computing in the conception, design and deployment of intelligent systems," BT Technology Journal, vol. 14, no. 4, pp. 32–36, 1994.

5.   R. Aliev, K. Bonfig, and F. Aliew, Soft Computing, Technic, Berlin, Germany, 2000.

6. L. A. Zadeh, "Foreword," in Proceedings of the 1st European Congress on Intelligent Techniques and Soft Computing (EUFIT '95), p. 7, 1995.

7. L. A. Zadeh, "Soft computing and fuzzy logic," IEEE Software, vol. 11, no. 6, pp. 48–56, 1994.

8. R. A. Aliev, "Fuzzy expert systems," in Soft Computing: Fuzzy Logic, Neural Networks, and Distributed Artificial Intelligence, F. Aminzadeh and M. Jamshidi, Eds., pp. 99–108, Prentice Hall, Englewood Cliffs, NJ, USA, 1994.

9. L. A. Zadeh, "Fuzzy logic, neural networks, and soft computing," Communications of the ACM, vol. 37, no. 3, pp. 77–84, 1994.

10. D. Nauck, F. Klawonn, and R. Kruse, Foundations of Neuro-Fuzzy Systems, John Wiley and Sons, New York, NY, USA, 1997.

11. M. H. Hassoun, Fundamentals of Artificial Neural Networks, MIT Press, Cambridge, Mass, USA, 1995.

12. S. Haykin, Neural Networks: A Comprehensive Foundation, Marmillau and IEEE Computer Society, 1994.

13. D. E. Goldberg, Genetic Algorithms in Search, Optimization and Machine Learning, Addison-Wesley, Reading, Mass, USA, 1989.

14. K. Murawski, T. Arciszewski, and K. de Jong, "Evolutionary computation in structural design,"Engineering with Computers, vol. 16, no. 3-4, pp. 275–286, 2000.

15. G. P. Huijzer, Quantitative penetrostratigraphic classification [Ph.D. thesis], Amsterdam Free University, 1992.

16. A. Coerts, Analysis of the static cone penetration test data for subsurface modelling: a methodology [Ph.D. thesis], Utrecht University, The Netherlands, 1996.

17. C. H. Juang, T. Jiang, and R. A. Christopher, "Three-dimensional site characterisation: neural network approach," Geotechnique, vol. 51, no. 9, pp. 799–809, 2001.

18. S. R. García and M. P. Romo, "Sistema de información geográfica y redes neuronales," in Proceedings of the Reunión Nacional de Mecánica de Suelos, Guadalajara, Mexico, November 2004.

19. R. R. Youngs, S. J. Chiou, W. J. Silva, and J. R. Humphrey, "Strong ground motion attenuation relationships for subduction zone earthquakes," Seismological Research Letters, vol. 68, no. 1, pp. 58–73, 1997.

20. G. M. Atkinson and D. M. Boore, "Empirical ground-motion relations for subduction-zone earthquakes and their application to Cascadia and other regions," Bulletin of the Seismological Society of America, vol. 93, no. 4, pp. 1703–1729, 2003.

21. American Society of Civil Engineers (ASCE), Prestandard and Commentary for the Seismic Rehabilitation of Buildings, Federal Emergency Management Agency, Washington, DC, USA, 2000.

22. D. Gasparini and E. H. Vanmarcke, SIMQKE: A Program for Artificial Motion Generation, Department of Civil Engineering, Massachusetts Institute of Technology, Cambridge, Mass, USA, 1976.

23. J. G. Anderson, "Nonparametric description of peak acceleration above a subduction thrust,"Seismological Research Letters, vol. 68, no. 1, pp. 86–93, 1997.

24. C. B. Crouse, "Ground motion attenuation equations for earthquakes on the Cascadia subduction zone,"Earthquake Spectra, vol. 7, pp. 210–236, 1991.

25. S. K. Singh, M. Ordaz, M. Rodriguez et al., "Analysis of near-source strong-motion recordings along the Mexican subduction zone," Bulletin of the Seismological Society of America, vol. 79, no. 6, pp. 1697–1717, 1989.

26. C. B. Crouse, Y. K. Vyas, and B. A. Schell, "Ground motions from subduction-zone earthquakes," Bulletin of the Seismological Society of America, vol. 78, no. 1, pp. 1–25, 1988.

27. S. K. Singh, E. Mena, R. Castro, and C. Carmona, "Empirical prediction of ground motion in Mexico City from coastal earthquakes," Bulletin of the Seismological Society of America, vol. 77, pp. 1862–1867, 1987.

28. K. Sadigh, "Ground motion characteristics for earthquakes originating in subduction zones and in the western United States," in Proceedings of the 6th Pan American Conference, Lima, Peru, 1979.

29. E. M. Scordilis, "Empirical global relations converting MS and mb to moment magnitude," Journal of Seismology, vol. 10, no. 2, pp. 225–236, 2006.

30. B. W. Tichelaar and L. J. Ruff, "Depth of seismic coupling along subduction zones," Journal of Geophysical Research, vol. 98, no. 2, pp. 2017–2037, 1993.

31. G. M. Atkinson and D. M. Boore, "Empirical ground-motion relations for subduction-zone earthquakes and their application to Cascadia and other regions," Bulletin of the Seismological Society of America, vol. 93, no. 4, pp. 1703–1729, 2003.

32. S. E. Fahlman, "Fast learning variations on back-propagation: an empirical study," in Proceedings of the Connectionist Models Summer School, D. Touretzky, G. Hinton, and T. Sejnowski, Eds., pp. 38–51, Morgan Kaufmann, San Mateo, Calif, USA, 1988.

33. S. R. García, M. P. Romo, M. J. Mendoza, and V. M. Taboada-Urtuzuástegui, Sand Behavior Modeling Using Static and Dynamic Artificial Neural Networks, Series de Investigación y Desarrollo del Instituto de Ingeniería SID/631, 2002.

34. K. W. Campbell, "Predicting strong ground motions in Utah, given in the paper measurement, characterization, and prediction of strong ground motion," in Earthquake Engineering and Soil Dynamics II-Recent Advances in Ground-Motion Evaluation, W. B. Joyner and D. M. Boore, Eds., pp. 43–102, 1988.

35. F. Sabetta and A. Pugliese, "Attenuation of peak horizontal acceleration and velocity from Italian strongmotion records," Bulletin of the Seismological Society of America, vol. 77, pp. 1491–1513, 1987.

36. S. C. Gómez, M. Ordaz, and C. Tena, "Leyes de atenuación en desplazamiento y aceleración para el diseño sísmico de estructuras con aislamiento en la costa del Pacífico," in Memorias del XV Congreso Nacional de Ingeniería Sísimica, Morelia, Mexico, November 2005.

37. W. Silva and K. Lee, State-of-the-Art for Assessing Earthquake Hazards in the United States, Report 24, ES RASCAL Code for Synthesizing Earthquake Ground Motions, Miscellaneous Paper S-73-1, U.S. Army Engineer Waterways Experiment Station, Vicksburg, Miss, USA, 1987.

38. B. A. Bolt and N. J. Gregor, "Synthesized strong ground motions for the seismic condition assessment of the eastern portion of the San Francisco Bay Bridge," Tech. Rep. UCB/EERC-93/12, University of California, Earthquake Engineering Research Center, Berkeley, Calif, USA, 1993.

39. J. E. Carballo and C. A. Cornell, "Probabilistic seismic demand analysis: spectrum matching and design," Tech. Rep. RMS-41, Department of Civil and Environmental Engineering, Stanford University, 2000.

40. C. Kircher, Personal Communication with Farzad Naeim and Marshall Lew. Levy, S., 1992, Artificial Life, Vintage Books, New York, NY, USA, 1993.

41. F. Naeim and J. M. Kelly, Design of Seismic Isolated Structures-from Theory to Practice, John Wiley and Sons, New York, NY, USA, 1999.

42. N. E. Huang, S. Zheng, S. R. Long et al., "The empirical mode decomposition and Hilbert spectrum for nonlinear and nonstationary

time series analysis," Proceedings of the Royal Society A, vol. 454, pp. 903–995, 1998.

43. S. Garcia, M. Romo, F. Correa, and M. Ortega, "Hilbert-Huang spectral analysis of seismic ground response," in Proceedings of the 8th National Conference on Earthquake Engineering (NCEE '06), San Francisco, Calif, USA, 2006, paper no. 1647.

44. G. Castro, S. J. Poulos, J. W. France, and J. L. Enos, "Liquefaction induced by cyclic loading," Tech. Rep., Geotechnical Engineers Inc., National Science Foundation, Washington, DC, USA, 1982.

45. H. B. Seed, I. M. Idris, and I. Arango, "Evaluation of liquefaction potential using field performance data,"Journal of Geotechnical Engineering, vol. 109, no. 3, pp. 458–482, 1983.

46. H. B. Seed and I. M. Idriss, "Simplified procedure for evaluating soil liquefaction potential," Journal of the Geotechnical Engineering Division, vol. 97, no. 9, pp. 1249–1273, 1971.

47. J. H. Hwang and C. W. Yang, "Verification of critical cyclic strength curve by Taiwan Chi-Chi earthquake data," Soil Dynamics and Earthquake Engineering, vol. 21, no. 3, pp. 237–257, 2001.

48. A. T. C. Goh, "Seismic liquefaction potential assessed by neural networks," Journal of Geotechnical Engineering, vol. 120, no. 9, pp. 1467–1480, 1994.

49. A. T. C. Goh, "Neural-network modeling of CPT seismic liquefaction data," Journal of Geotechnical and Geoenvironmental Engineering, vol. 122, no. 1, pp. 70–73, 1996.

50. A. T. C. Goh, "Probabilistic neural network for evaluating seismic liquefaction potential," Canadian Geotechnical Journal, vol. 39, pp. 219–232, 2002.

51. C. H. Juang and C. J. Chen, "CPT-based liquefaction evaluation using artificial neural networks,"Computer-Aided Civil and Infrastructure Engineering, vol. 14, no. 3, pp. 221–229, 1999.

52. M. H. Baziar and N. Nilipour, "Evaluation of liquefaction potential using neural-networks and CPT results," Soil Dynamics and Earthquake Engineering, vol. 23, no. 7, pp. 631–636, 2003.

53. C. J. Lee and T. K. Hsiung, "Applying neural network model in seismic liquefaction case analysis: probabilistic neural network model v.s. multilayer perceptrons model," in Proceedings of the 2nd Joint International Conference on Soft Computing and Intelligent Systems and 5th International Symposium on Advanced Intelligent Systems, Yokohama, Japan, 2004.

54. C. H. Juang, C. J. Chen, T. Jiang, and R. D. Andrue, "Risk-based liquefaction potential evaluation using standard penetration tests," Canadian Geotechnical Journal, vol. 37, pp. 1195–1208, 2000.

55. E. Hunt, Concept Learning: An Information Processing Problem, Wiley, New York, NY, USA, 1962.

56. P. Winston, "Learning structural descriptions from examples," in The Psychology of Computer Vision, P. H. Winston, Ed., McGraw-Hill, 1975.

57. E. Feigenbaum and H. Simon, "Performance of a reading task by an elementary perceiving and memorizing program," Behavioral Science, vol. 8, no. 1, pp. 72–76, 1963.

58. B. G. Buchanan and T. M. Mitchell, "Model-directed learning of production rules," in Pattern Directed Inference Systems, D. A. Waterman and F. Hayes-Roth, Eds., Academic Press, 1978.

59. C. H. Juang and T. Jiang, "Assessing probabilistic methods for liquefaction potential evaluation," in Soil Dynamics and Liquefaction 2000, R. Y. S. Pak and J. Yamamura, Eds., Geotechnical Special Publication no. 107, pp. 148–162, American Society of Civil Engineers, 2000.

60. R. D. Andrus, K. H. Stokoe II, and R. M. Chung, Draft Guidelines for Evaluating Liquefaction Resistance Using Shear Wave Velocity Measurements and Simplified Procedures, NISTIR, 6277, National Institute of Standards and Technology, Gaithersburg, Md, USA, 1999.

61. H. B. Seed and I. M. Idriss, "Simplified procedure for evaluating soil liquefaction potential," Journal of the Soil Mechanics and Foundations Division, vol. 97, no. 9, pp. 1249–1273, 1971.

62. T. L. Youd, I. M. Idriss, R. D. Andrus et al., "Liquefaction resistance of soils: summary report from the 1996 NCEER and 1998 NCEER/NSF workshops on evaluation of liquefaction resistance of soils," Journal of Geotechnical and Geoenvironmental Engineering, vol. 127, no. 10, pp. 817–833, 2001.

63. P. K. Robertson and C. E. Wride, "Evaluating cyclic liquefaction potential using the cone penetration test," Canadian Geotechnical Journal, vol. 35, no. 3, pp. 442–459, 1998.

64. T. L. Youd, "Geologic effects-liquefaction and associated ground failure," in Proceedings of the Geologic and Hydraulic Hazards Training Program, Open File Report 84-760, pp. 210–232, US Geological Survey, Menlo Park, Calif, USA, 1984.

65. K. Tokida, H. Iwasaki, H. Matsumoto, and T. Hamada, "Liquefaction potential and drag force acting on piles in flowing soils," Soil Dynamics and Earthquake Engineering, vol. 1, pp. 244–259, 1993.

66. W. Vargas and I. Towhata, "Measurement of drag exerted by liquefied sand on buried pipe," inProceedings of the 1st International Conference on Earthquake Geotechnical Engineering, pp. 975–980, Tokyo, Japan, 1995.

67. S. K. Haigh and S. P. G. Madabhushi, "The effects of pile flexibility on pile-loading in laterally spreading slopes," in Seismic Performance and Simulation of Pile Foundations in Liquefied and Laterally Spreading Ground, Geotechnical Special Publication no. 145, pp. 24–37, ASCE and GEO-Institute, 2005.

68. S. F. Bartlett and T. L. Youd, "Empirical prediction of liquefactiion-induced lateral spread," Journal of Geotechnical Engineering, vol. 121, no. 4, pp. 316–329, 1995.

69. J. B. Berrill, S. A. Christensen, R. J. Kenan, W. Okada, and J. R. Pettiga, "Lateral spreading loads on pile bridge foundation," in Proceedings on Soil Behavior of Ground and Geotechnical Structures, S. E. Pinto, Ed., vol. 2, pp. 173–187, Balkema, Rotterdam, The Netherlands, 1997.

70. E. Ovando, L. Vieitez, and J. D. Alemán, "Geotecnia, El Macrosismo de Manzanillo del 9 de octubre de 1995," in Sociedad Mexicana de Ingenieria Sismica, A. Tena, Ed., pp. 82–133, Universidad de Colima, Colima, Mexico, 1997.

71. E. Ovando and M. P. Romo, "Three recent damaging earthquakes in Mexico, special lecture," inProceedings of the 5th International Conference on Case Histories in Geotechcnical Engineering, New York, NY, USA, April 2004.

72. T. L. Youd, C. M. Hansen, and S. F. Bartlett, "Revised multilinear regression equations for prediction of lateral spread displacement," Journal of Geotechnical and Geoenvironmental Engineering, vol. 128, no. 12, pp. 1007–1017, 2002.

73. S. R. García and M. P. Romo, "Analytical and neurogenetic methods to estimate in situ rockfill dam dynamic properties," in Estado del Arte, Ingeniería Geosísmica, XXIV Reunión Nacional de Mecánica de Suelos, pp. 45–55, Aguascalientes, Mexico, November 2008.

74. M. P. Romo and S. R. Garcia, Prediction of Liquefaction-Induced Lateral Spread: A Neurofuzzy Procedure, Serie Investigación y Desarrollo, Instituto de Ingenieria, UNAM, SID/651, Agosto, 2006,http://www.iingen.unam.mx/.

Chapter 11

RECENT ADVANCES AND FUTURE CHALLENGES FOR ARTIFICIAL NEURAL SYSTEMS IN GEOTECHNICAL ENGINEERING APPLICATIONS

Mohamed A. Shahin,[1] Mark B. Jaksa,[2] and Holger R. Maier[2]

[1]Department of Civil Engineering, Curtin University of Technology, Perth, WA 6845, Australia

[2]School of Civil, Environmental and Mining Engineering, University of Adelaide, Adelaide, SA 5005, Australia

ABSTRACT

Artificial neural networks (ANNs) are a form of artificial intelligence that has proved to provide a high level of competency in solving many complex engineering problems that are beyond the computational capability of classical mathematics and traditional procedures. In particular, ANNs have been applied successfully to almost all aspects of geotechnical engineering problems. Despite the increasing number and diversity of ANN applications in geotechnical engineering, the contents of reported applications indicate that the progress in ANN development and procedures is marginal and not moving forward since the mid-1990s. This paper presents a brief overview of ANN applications in geotechnical engineering, briefly provides an overview of the operation of ANN modeling, investigates the current research directions of ANNs in geotechnical engineering, and discusses some ANN modeling issues that need further attention in the future, including model robustness; transparency and knowledge extraction; extrapolation; uncertainty.

INTRODUCTION

Artificial neural networks (ANNs) are well suited to model the complex behavior of most geotechnical engineering materials which, by their very nature, exhibit extreme variability. ANNs have also demonstrated superior predictive ability when compared with traditional methods. Since the early 1990s, ANNs have been applied successfully to virtually every problem in geotechnical engineering. In this section, post-2001 applications of ANNs in geotechnical engineering are briefly examined, and interested readers are referred to Shahin et al. [1], where the pre-2001 papers are reviewed in some detail.

The behavior of deep (pile) and shallow foundations in soils is complex, uncertain, and not yet entirely understood. This fact has encouraged many researchers to apply the ANN technique to the prediction of the behavior of foundations. For example, ANNs have been used extensively for modeling the axial and lateral load capacities of deep foundations in compression and uplift, including driven piles [2–6], drilled shafts [7,8], and ground anchor piles [9, 10]. The prediction of behavior of shallow foundations has also been investigated, including settlement estimation [11–16] and bearing capacity [17–19].

Classical constitutive modeling based on elasticity and plasticity theories has limited capability to simulate properly the behavior of geomaterials. This is attributed to reasons associated with the formulation complexity, idealization of material behavior, and excessive empirical parameters [20]. In this regard, many neural networks have been proposed as a reliable and practical alternative to model the constitutive monotonic and hysteretic behavior of geomaterials [21–29].

Geotechnical properties and behavior of soils are controlled by factors such as mineralogy; fabric; pore water, and the interactions of these factors are difficult to establish solely by traditional statistical methods due to their interdependence [30]. Based on the application of ANNs, methodologies have been developed for estimating several soil properties, including the preconsolidation pressure [31], shear strength and stress history [30, 32–37], swell pressure [38, 39], lateral earth pressure [40], compaction characteristics and permeability [41, 42], soil composition and classification [43, 44], and properties of soil dynamics [45, 46].

Liquefaction during earthquakes is one of the very dangerous ground failure phenomena that can cause a large amount of damage to most civil engineering structures. Although the liquefaction mechanism is well known, the prediction of liquefaction potential is very complex [47]. This fact has attracted many researchers to investigate the applicability of ANNs for predicting liquefaction [47–55].

Other applications of ANNs in geotechnical engineering include earth retaining structures [56], dams [57,58], blasting [59], mining [60], environmental geotechnics [61], rock mechanics [62–67], site characterization [68], tunnels and underground openings [69–74], slope stability and landslides [71, 75–79], and deep excavation [80].

BRIEF OVERVIEW OF ARTIFICIAL NEURAL NETWORKS

Many authors have described the structure and operation of ANNs (e.g., [81, 82]), and whilst a comprehensive description of ANNs is beyond the scope of this paper, it is useful to provide a brief overview. ANNs are a data driven artificial intelligence approach that attempts to mimic, in a very simplistic way, the cognition capability of the human brain. ANNs learn by examples of data inputs and outputs presented to them so that the subtle functional relationships among the data are captured, even if the underlying relationships are unknown or the physical meaning is difficult to explain. This is in contrast to most traditional empirical and statistical methods, which need prior knowledge about the nature of the relationships among the data. This is one of the main benefits of ANNs when compared with most empirical and statistical methods.

Typically, the architecture of ANNs consists of a series of processing elements (PEs), or nodes, that are usually arranged in layers: an input layer, an output layer, and one or more hidden layers, as shown in Figure 1.

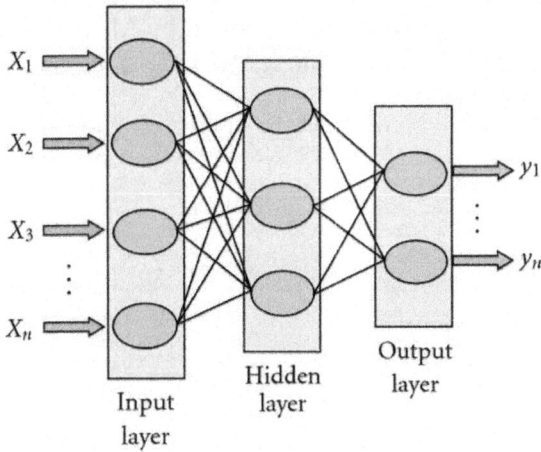

Artificial neural network

(a)

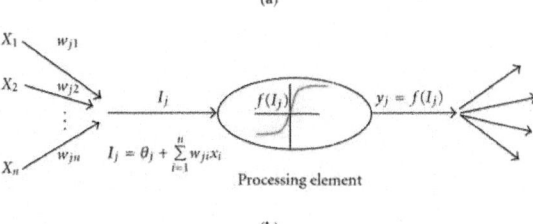

Processing element

(b)

Figure 1: Typical structure and operation of ANNs.

The input from each PE in the previous layer x_i is multiplied by an adjustable connection weight w_{ji}. At each PE, the weighted input signals are summed and a threshold value θ_j is added. This combined input Ij is then passed through a nonlinear transfer function $f(\cdot)$ to produce the output of the PE y_j. The output of one PE provides the input to the PEs in the next layer. This process is summarized in (1) and (2) and illustrated in Figure 1.1.

$$I_j = \sum w_{ji} x_i + \theta_j \quad \text{summation,}$$

(1)

$$y_j = f\left(I_j\right) \quad \text{transfer.}$$

(2)

The propagation of information in an ANN starts at the input layer, where the input data are presented. The network adjusts its weights on the presentation of a training data set and uses a learning rule to find a set of weights that will produce the input/output mapping that has the smallest possible error. This process is called "learning" or "training." Once the training phase of the model has been successfully accomplished, the performance of the trained model needs to be validated using an independent testing set. The main steps involved in the development of an ANN, as suggested by Maier and Dandy [83], are illustrated in Figure 2. Several of these steps are discussed in some depth in the following section.

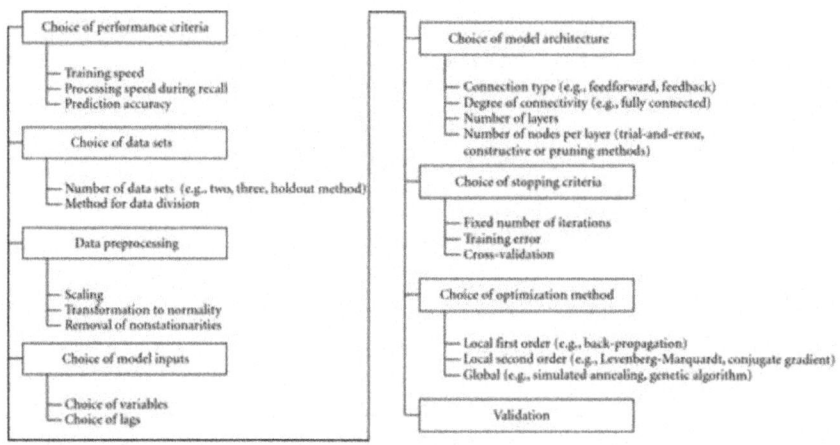

Figure 2: The main steps in ANN model development [83].

CURRENT DEVELOPMENT AND FUTURE DIRECTIONS IN UTILIZATION OF ANNS

One issue that needs to be addressed in order to improve the performance of ANN models is the utilization of a systematic approach in their development.

Such an approach needs to address major factors, including the determination of adequate model inputs, data division and preprocessing, choice of suitable network architecture, careful selection of some internal parameters that control the optimization method, stopping criteria, and model validation. For example, in relation to the second step of choice of data sets, method for data division, Shahin et al. [84] provided guidance using a geotechnical engineering example, and recommended the use of three, statistically consistent but independent data sets, one for each of training, testing, and validation. In this context, Shahin et al. [84] have introduced three approaches so that data division can be carried out in a systematic manner, including trial-and-error, self-organizing maps, and fuzzy clustering. For a detailed treatment of each of the steps in the model development process, interested readers are referred to Shahin et al. [85].

Other key issues in relation to ANN modeling that have received recent attention and require further research in the future include developing approaches that (i) ensure the development of robust models, (ii) increase model transparency and enable knowledge to be extracted from trained ANNs, (iii) improve extrapolation ability, and (iv) deal with uncertainty. Each of these is discussed in what follows.

Model Robustness

Model robustness is the predictive ability of ANN models to generalize over a range of data similar to that used for model training. Kingston et al. [86] stated that if "ANNs are to become more widely accepted and reach their full potential..., they should not only provide a good fit to the calibration and validation data, but the predictions should also be plausible in terms of the relationship modeled and robust under a wide range of conditions." and that "while ANNs validated against error alone may produce accurate predictions for situations similar to those contained in the training data, they may not be robust under different conditions unless the relationship by which the data were generated has been adequately estimated." This is in agreement with the investigation into the robustness of ANNs carried out by Shahin et al. [87] for a case study of predicting the settlement of shallow foundations on granular soils. Shahin et al. [87] found that good performance of ANN models on the data used for model calibration and validation does not guarantee that the models will perform well in a robust fashion over a range of data similar to those used in the model calibration phase. For this reason, Shahin et al. [87] proposed a method to test the robustness of the predictive ability of ANN models by carrying out a sensitivity analysis to investigate the response of ANN model outputs to changes in its inputs. The robustness of the model can then be determined by examining how well model predictions are in agreement with the known underlying physical processes of the problem in hand over a range of inputs. In addition, Shahin et al. [87] advised that the connection weights should be examined as part of the interpretation of ANN model behavior, using, for example, the method suggested by Garson [88]. On the other hand, Kingston et al. [86] adopted the connection weight approach of Olden et al. [89] for a case study in hydrological modeling in order to assess the relationship modeled by the ANNs, as Olden et al. [89] found that this approach provided the best overall methodology for quantifying ANN input importance in comparison to other commonly used methods, though with a few limitations.

Support vector machines (SVMs) are an alternative data-driven modeling approach that is claimed to provide better generalization capabilities and higher accuracy than ANNs and are therefore worth further consideration in relation to achieving improved model robustness [90]. Interested readers are referred to A. T. C. Goh and S. H. Goh [91] for a good overview of this technique. Recent applications of SVMs in the field of geotechnical engineering include the prediction of liquefaction potential [90, 91], analysis of slope stability [92], and modeling friction capacity of driven piles [93].

Model Transparency and Knowledge Extraction

Model transparency and knowledge extraction are the feasibility of interpreting ANN models in a way that provides insights into how model inputs affect outputs. Figure 3 shows the classification of modeling techniques based on colors [94] in which the higher the physical knowledge used during model development, the better the physical interpretation of the phenomenon that the model provides to the user. It can be seen that the color coding of mathematical modeling can be classified into: white-, black-, and grey-box models, each of which can be explained as follows [95]. White-box models are systems that are based on first principles (e.g., physical laws) where model variables and parameters are known and have physical meaning by which the underlying physical relationships of the system can be explained. Black-box models are data-driven or regressive systems in which the functional form of relationships between model variables is unknown and needs to be estimated. Black-box models rely on data to map the relationships between model inputs and corresponding outputs rather than to find a feasible structure of the model input-output relationships. Grey-box models are conceptual systems in which the mathematical structure of the model can be derived, allowing further information of the system behavior to be resolved.

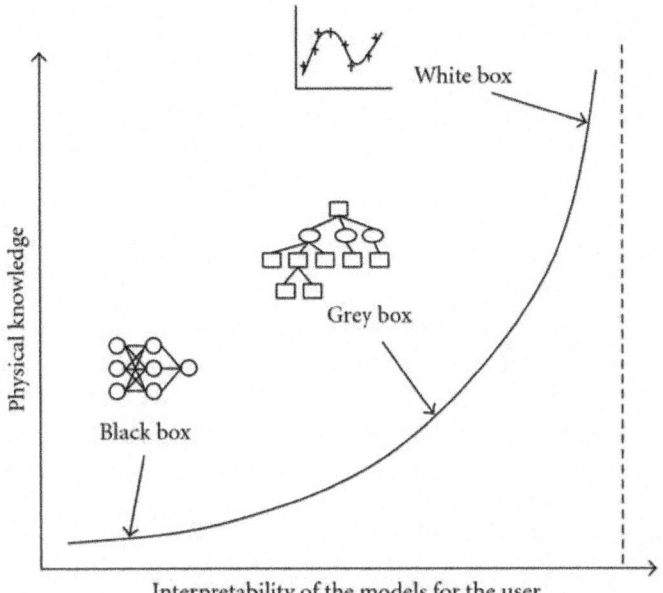

White box

Grey box

Black box

Interpretability of the models for the user

Figure 3: Graphical classification of modeling techniques (adapted from [94]).

ANNs belong to the class of black-box models due to their lack of transparency and the fact that they do not consider nor explain the underlying physical processes explicitly. This is because the knowledge extracted by ANNs is stored in a set of weights that are difficult to interpret properly, and due to the large complexity of the network structure, ANNs fail to give a transparent function that relates the inputs to the corresponding outputs. Consequently, it is difficult to understand the nature of the input-output relationships derived. This issue has been addressed by many researchers with respect to hydrological engineering. For example, Jain et al. [96] examined whether or not the physical processes in a watershed were inherent in a trained ANN rainfall-runoff model. This was carried out by assessing the strengths of the relationships between the distributed components of the ANN model, in terms of the responses from the hidden nodes, and the deterministic components of the hydrological process, computed from a conceptual rainfall runoff model, along with the observed input variables, using correlation coefficients and scatter plots. They concluded that the trained ANN, in fact, captured different components of the physical process and a careful examination of the distributed information contained in the trained ANN can be informative about the nature of the physical processes captured by various components of the ANN model. Sudheer [97] performed perturbation analysis to assess the influence of each individual input variable on the output variable and found it to be an effective means of identifying the underlying physical process inherent in the trained ANN. Olden et al. [89], Sudheer and Jain [98], and Kingston et al. [99] also addressed this issue of model transparency and knowledge extraction.

In the context of geotechnical engineering, Shahin et al. [12] and Shahin and Jaksa [9] expressed the results of the trained ANNs in the form of relatively straightforward equations. This was possible due to the relatively small number of input and output variables, and hidden nodes. Neurofuzzy applications are another means of knowledge extraction that facilitate model transparency. Neurofuzzy networks use the fuzzy logic system to store knowledge acquired from a set of input variables $(x_1, x_2, ... , x_n)$ and the corresponding output variable (y) in a set of linguistic fuzzy rules that can be easily interpreted, such as IF (x1 is high AND x_2 is low) THEN (y is high), c = 0.9, where (c = 0.9) is the rule confidence, which indicates the degree to which the above rule has contributed to the output. Examples of such applications in geotechnical engineering include Ni et al. [100], Shahin et al. [16], Gokceoglu et al. [62], Provenzano et al. [19], and Padmini et al. [18].

A recent technique that belongs to the class of grey-box models, and therefore does not suffer from the problem of model transparency and knowledge extraction, is genetic programming (GP). Several researchers (e.g., [34, 50, 101–104]) have recently used the GP technique as an alterative to ANNs in order to obtain greatly simplified formulae for some geotechnical engineering problems. GP is a computing method that attempts to mimic the biological evolution of living organisms. GP makes use of the principles of genetic algorithms (GAs) for parameter optimization in which a population of expressions (or computer programs) for a function F, coded in tree structures of variable size, is generated and executed. The generated expressions are then modified by means of artificial evolution in order to perform a global search to arrive at the best fit mathematical expression for F that solves a certain problem. Additional advantages of GP over ANNs are that the structure and network parameters of ANNs (e.g., number of hidden layers and their number of nodes, transfer functions, learning rate, etc.) should be identified a priori and are usually obtained using adhoc trial-and-error approaches. However, the number and combination of terms, as well as the values of GP modeling parameters, are all evolved automatically during model calibration. However, hybrid approaches can also be used, in which genetic algorithms are used to evolve optimal ANN structures and connection weight values. It should be noted that while white-box models provide maximum transparency, their construction may be difficult to obtain for many geotechnical engineering problems, where the underlying mechanism is not entirely understood.

Model Extrapolation

Model extrapolation is the ability of ANN models to predict well outside the range of the data used for model calibration. It is generally accepted that ANNs perform best when they do not extrapolate beyond the range of the data used for calibration [105–107]. Whilst this is not unlike other models, it is nevertheless an important limitation of ANNs, as it restricts their usefulness and applicability. Extreme value prediction is of particular concern in several areas of civil engineering, such as hydrological engineering, when floods are forecast, as well as in geotechnical engineering when, for example, liquefaction potential and the stability of slopes are assessed. Sudheer et al. [108] highlighted this issue and proposed a methodology, based on the Wilson-Hilferty transformation, for enabling ANN models to predict extreme values with respect to peak river flows. Their methodology yielded superior predictions when compared with those obtained from an ANN model using untransformed data.

Model Uncertainty

Finally, a further limitation of ANNs is that the uncertainty in the predictions generated is seldom quantified [109]. Failure to account for such uncertainty makes it impossible to assess the quality of ANN predictions, which severely limits their efficacy. In an effort to address this, a few researchers have applied Bayesian techniques to ANN training (e.g., [110–113]) in the context of hydrological engineering and Goh et al. [7] with respect to geotechnical engineering. Goh et al. [7] observed that the integration of the Bayesian framework into the back-propagation algorithm enhanced neural network prediction capabilities and provided assessment of the confidence associated with network predictions. Research to date has demonstrated the value of Bayesian neural networks, although further work is needed in the area of geotechnical engineering. Shahin et al. [13, 114] also incorporated uncertainty in the ANN process by developing a series of design charts expressing the reliability of settlement predictions for shallow foundations on cohesionless soils.

DISCUSSION AND CONCLUSIONS

In the field of geotechnical engineering, it is possible to encounter some types of problems that are very complex and not well understood. In this regard, ANNs provide several advantages over more conventional computing techniques. For most traditional mathematical models, the lack of physical understanding is usually supplemented by either simplifying the problem or incorporating several assumptions into the models. Mathematical models also rely on assuming the structure of the model in advance, which may be suboptimal. Consequently, many mathematical models fail to simulate the complex behavior of most geotechnical engineering problems. In contrast, ANNs are a data driven approach in which the model can be trained on input-output data pairs to determine the structure and parameters of the model. In this case, there is no need to either simplify the problem or incorporate any assumptions. Moreover, ANNs can always be updated to obtain better results by presenting new training examples as new data become available. These factors combine to make ANNs a powerful modeling tool in geotechnical engineering.

Despite the success of ANNs in geotechnical engineering and other disciplines, they suffer from some shortcomings that need further attention in the future, including model robustness, transparency and knowledge extraction, extrapolation, and uncertainty. In addition and according to Flood [115], ANNs in civil engineering, including geotechnical engineering, were used mostly as simple vector mapping devices for function modeling of applications that require rarely more than a few tens of neurons without higher-order structuring. Together, improvements in these issues will greatly enhance the usefulness of ANN models and will provide the next generation of applied artificial neural networks with the best way for advancing the field to the next level of sophistication and application. Until such an improvement is achieved, the authors agree with Flood and Kartam [105] that neural networks for the time being might be treated as a complement to conventional computing techniques rather than as an alternative or may be used as a quick check on solutions developed by more time-consuming and in-depth analyses.

REFERENCES

1.   M. A. Shahin, M. B. Jaksa, and H. R. Maier, "Artificial neural network applications in geotechnical engineering," Australian Geomechanics, vol. 36, no. 1, pp. 49–62, 2001.

2.   I. Ahmad, M. H. El Naggar, and A. N. Khan, "Artificial neural network application to estimate kinematic soil pile interaction response parameters," Soil Dynamics and Earthquake Engineering, vol. 27, no. 9, pp. 892–905, 2007.

3.   S. K. Das and P. K. Basudhar, "Undrained lateral load capacity of piles in clay using artificial neural network," Computers and Geotechnics, vol. 33, no. 8, pp. 454–459, 2006.

4. A. M. Hanna, G. Morcous, and M. Helmy, "Efficiency of pile groups installed in cohesionless soil using artificial neural networks," Canadian Geotechnical Journal, vol. 41, no. 6, pp. 1241–1249, 2004.

5. H. Ardalan, A. Eslami, and N. Nariman-Zadeh, "Piles shaft capacity from CPT and CPTu data by polynomial neural networks and genetic algorithms," Computers and Geotechnics, vol. 36, no. 4, pp. 616–625, 2009.

6. M. A. Shahin, "Modelling axial capacity of pile foundations by intelligent computing," in Proceedings of the 2nd BGA International Conference on Foundations (ICOF '08), pp. 283–294, IHS BRE Press, Dundee, Scotland, 2008.

7. A. T. C. Goh, F. H. Kulhawy, and C. G. Chua, "Bayesian neural network analysis of undrained side resistance of drilled shafts," Journal of Geotechnical & Geoenvironmental Engineering, vol. 131, no. 1, pp. 84–93, 2005.

8. M. A. Shahin and M. B. Jaksa, "Intelligent computing for predicting axial capacity of drilled shafts," inProceedings of the International Foundation Congress and Equipment Expo (IFCEE '09), ASCE Geotechnical Special Publication, no. 186, pp. 26–33, Orlando, Fla, USA, 2009.

9. M. A. Shahin and M. B. Jaksa, "Neural network prediction of pullout capacity of marquee ground anchors," Computers and Geotechnics, vol. 32, no. 3, pp. 153–163, 2005.

10. M. A. Shahin and M. B. Jaksa, "Pullout capacity of small ground anchors by direct cone penetration test methods and neural networks," Canadian Geotechnical Journal, vol. 43, no. 6, pp. 626–637, 2006. ·

11. Y.-L. Chen, R. Azzam, and F.-B. Zhang, "The displacement computation and construction pre-control of a foundation pit in Shanghai utilizing FEM and intelligent methods," Geotechnical and Geological Engineering, vol. 24, no. 6, pp. 1781–1801, 2006.

12. M. A. Shahin, M. B. Jaksa, and H. R. Maier, "Artificial neural network-based settlement prediction formula for shallow foundations on granular soils," Australian Geomechanics, vol. 37, no. 4, pp. 45–52, 2002.

13. M. A. Shahin, M. B. Jaksa, and H. R. Maier, "Neural network based stochastic design charts for settlement prediction," Canadian Geotechnical Journal, vol. 42, no. 1, pp. 110–120, 2005.

14. M. A. Shahin, H. R. Maier, and M. B. Jaksa, "Predicting settlement of shallow foundations using neural networks," Journal of Geotechnical & Geoenvironmental Engineering, vol. 128, no. 9, pp. 785–793, 2002. ·

15. M. A. Shahin, H. R. Maier, and M. B. Jaksa, "Predicting settlement of shallow foundations using neural networks," Journal of Geotechnical & Geoenvironmental Engineering, vol. 128, no. 9, pp. 785–793, 2002. ·

16. M. A. Shahin, H. R. Maier, and M. B. Jaksa, "Settlement prediction of shallow foundations on granular soils using B-spline neurofuzzy models," Computers and Geotechnics, vol. 30, no. 8, pp. 637–647, 2003. ·

17. Y. L. Kuo, M. B. Jaksa, A. V. Lyamin, and W. S. Kaggwa, "ANN-based model for predicting the bearing capacity of strip footing on multi-layered cohesive soil," Computers and Geotechnics, vol. 36, no. 3, pp. 503–516, 2009.

18. D. Padmini, K. Ilamparuthi, and K. P. Sudheer, "Ultimate bearing capacity prediction of shallow foundations on cohesionless soils using neurofuzzy models," Computers and Geotechnics, vol. 35, no. 1, pp. 33–46, 2008.

19. P. Provenzano, S. Ferlisi, and A. Musso, "Interpretation of a model footing response through an adaptive neural fuzzy inference system," Computers and Geotechnics, vol. 31, no. 3, pp. 251–266, 2004. ·

20. H. Adeli, "Neural networks in civil engineering: 1989–2000," Computer-Aided Civil and Infrastructure Engineering, vol. 16, no. 2, pp. 126–142, 2001.

21. I. A. Basheer, "Stress-strain behavior of geomaterials in loading reversal simulated by time-delay neural networks," Journal of Materials in Civil Engineering, vol. 14, no. 3, pp. 270–273, 2002. ·

22. Q. Fu, Y. M. A. Hashash, S. Jung, and J. Ghaboussi, "Integration of laboratory testing and constitutive modeling of soils," Computers and Geotechnics, vol. 34, no. 5, pp. 330–345, 2007.

23. G. Habibagahi and A. Bamdad, "A neural network framework for mechanical behavior of unsaturated soils," Canadian Geotechnical Journal, vol. 40, no. 3, pp. 684–693, 2003.

24. Y. M. A. Hashash, S. Jung, and J. Ghaboussi, "Numerical implementation of a neural network based material model in finite element analysis," International Journal for Numerical Methods in Engineering, vol. 59, no. 7, pp. 989–1005, 2004.

25. M. Lefik and B. A. Schrefler, "Artificial neural network as an incremental non-linear constitutive model for a finite element code," Computer Methods in Applied Mechanics and Engineering, vol. 192, no. 28–30, pp. 3265–3283, 2003.

26. Y. M. Najjar and C. Huang, "Simulating the stress-strain behavior of Georgia kaolin via recurrent neuronet approach," Computers and

Geotechnics, vol. 34, no. 5, pp. 346–361, 2007.

27. M. A. Shahin and B. Indraratna, "Modeling the mechanical behavior of railway ballast using artificial neural networks," Canadian Geotechnical Journal, vol. 43, no. 11, pp. 1144–1152, 2006.

28. M. Banimahd, S. S. Yasrobi, and P. K. Woodward, "Artificial neural network for stress-strain behavior of sandy soils: knowledge based verification," Computers and Geotechnics, vol. 32, no. 5, pp. 377–386, 2005.

29. W. Gao, X. T. Feng, and Y. R. Zheng, "Identification of a constitutive model for geo-materials using a new intelligent bionics algorithm," International Journal of Rock Mechanics and Mining Sciences, vol. 41, supplement 1, pp. 454–459, 2004.

30. Y. Yang and M. S. Rosenbaum, "The artificial neural network as a tool for assessing geotechnical properties," Geotechnical and Geological Engineering, vol. 20, no. 2, pp. 149–168, 2002.

31. S. Çelik and Ö. Tan, "Determination of preconsolidation pressure with artificial neural network," Civil Engineering and Environmental Science, vol. 22, no. 4, pp. 217–231, 2005.

32. S. J. Lee, S. R. Lee, and Y. S. Kim, "An approach to estimate unsaturated shear strength using artificial neural network and hyperbolic formulation," Computers and Geotechnics, vol. 30, no. 6, pp. 489–503, 2003.

33. P. U. Kurup and N. K. Dudani, "Neural networks for profiling stress history of clays from PCPT data," Journal of Geotechnical & Geoenvironmental Engineering, vol. 128, no. 7, pp. 569–579, 2002.

34. B. S. Narendra, P. V. Sivapullaiah, S. Suresh, and S. N. Omkar, "Prediction of unconfined compressive strength of soft grounds using computational intelligence techniques: a comparative study," Computers and Geotechnics, vol. 33, no. 3, pp. 196–208, 2006.

35. A. Baykasoglu, H. Güllü, H. Çanakçi, and L. Özbakir, "Prediction of compressive and tensile strength of limestone via genetic programming," Expert Systems with Applications, vol. 35, no. 1-2, pp. 111–123, 2008.

36. W. Y. Byeon, S. R. Lee, and Y. S. Kim, "Application of flat DMT and ANN to Korean soft clay deposits for reliable estimation of undrained shear strength," International Journal of Offshore and Polar Engineering, vol. 16, no. 1, pp. 73–80, 2006.

37. A. Kaya, "Residual and fully softened strength evaluation of soils using

artificial neural networks,"Geotechnical and Geological Engineering, vol. 27, no. 2, pp. 281–288, 2009.

38. Y. Erzin, "Artificial neural networks approach for swell pressure versus soil suction behaviour,"Canadian Geotechnical Journal, vol. 44, no. 10, pp. 1215–1223, 2007.

39. I. Ashayeri and S. Yasrebi, "Free-swell and swelling pressure of unsaturated compacted clays; experiments and neural networks modeling," Geotechnical and Geological Engineering, vol. 27, no. 1, pp. 137–153, 2009.

40. S. K. Das and P. K. Basudhar, "Prediction of coefficient of lateral earth pressure using artificial neural networks," The Electronic Journal of Geotechnical Engineering, vol. 10, pp. 1–5, 2005.

41. S. K. Sinha and M. C. Wang, "Artificial neural network prediction models for soil compaction and permeability," Geotechnical and Geological Engineering, vol. 26, no. 1, pp. 47–64, 2008.

42. A. H. Abdel-Rahman, "Predicting compaction of cohesionless soils using ANN," Ground Improvement, vol. 161, no. 1, pp. 3–8, 2008.

43. P. U. Kurup and E. P. Griffin, "Prediction of soil composition from CPT data using general regression neural network," Journal of Computing in Civil Engineering, vol. 20, no. 4, pp. 281–289, 2006.

44. B. Bhattacharya and D. P. Solomatine, "Machine learning in soil classification," Neural Networks, vol. 19, no. 2, pp. 186–195, 2006.

45. M. P. Romo and S. R. García, "Neurofuzzy mapping of CPT values into solid dynamic properties," Soil Dynamics and Earthquake Engineering, vol. 23, no. 6, pp. 473–482, 2003.

46. S. R. García, M. P. Romo, and J. Figueroa-Nazuno, "Soil dynamic properties determination: a neurofuzzy system approach," Control and Intelligent Systems, vol. 34, no. 1, pp. 1–11, 2006.

47. M. H. Baziar and A. Ghorbani, "Evaluation of lateral spreading using artificial neural networks," Soil Dynamics and Earthquake Engineering, vol. 25, no. 1, pp. 1–9, 2005.

48. A. T. C. Goh, "Probabilistic neural network for evaluating seismic liquefaction potential," Canadian Geotechnical Journal, vol. 39, no. 1, pp. 219–232, 2002.

49. A. M. Hanna, D. Ural, and G. Saygili, "Neural network model for liquefaction potential in soil deposits using Turkey and Taiwan earthquake data," Soil Dynamics and Earthquake Engineering, vol. 27, no. 6, pp. 521–540, 2007.

50. A. A. Javadi, M. Rezania, and M. M. Nezhad, "Evaluation of liquefaction induced lateral displacements using genetic programming," Computers and Geotechnics, vol. 33, no. 4-5, pp. 222–233, 2006.

51. K. Young-Su and K. Byung-Tak, "Use of artificial neural networks in the prediction of liquefaction resistance of sands," Journal of Geotechnical & Geoenvironmental Engineering, vol. 132, no. 11, pp. 1502–1504, 2006.

52. A. M. Hanna, D. Ural, and G. Saygili, "Evaluation of liquefaction potential of soil deposits using artificial neural networks," Engineering Computations, vol. 24, no. 1, pp. 5–16, 2007.

53. M. S. Rahman and J. Wang, "Fuzzy neural network models for liquefaction prediction," Soil Dynamics and Earthquake Engineering, vol. 22, no. 8, pp. 685–694, 2002.

54. C. H. Juang, H. Yuan, D.-H. Lee, and P.-S. Lin, "Simplified cone penetration test-based method for evaluating liquefaction resistance of soils," Journal of Geotechnical & Geoenvironmental Engineering, vol. 129, no. 1, pp. 66–80, 2003.

55. S. S. H. Khozaghi and A. J. A.-Z. Choobbasti, "Predicting of liquefaction potential in soils using artificial neural networks," Electronic Journal of Geotechnical Engineering, vol. 12C, 2007.

56. G. T. C. Kung, E. C. L. Hsiao, M. Schuster, and C. H. Juang, "A neural network approach to estimating deflection of diaphragm walls caused by excavation in clays," Computers and Geotechnics, vol. 34, no. 5, pp. 385–396, 2007.

57. Y.-S. Kim and B.-T. Kim, "Prediction of relative crest settlement of concrete-faced rockfill dams analyzed using an artificial neural network model," Computers and Geotechnics, vol. 35, no. 3, pp. 313–322, 2008.

58. Y. Yu, B. Zhang, and H. Yuan, "An intelligent displacement back-analysis method for earth-rockfill dams," Computers and Geotechnics, vol. 34, no. 6, pp. 423–434, 2007.

59. Y. Lu, "Underground blast induced ground shock and its modelling using artificial neural network,"Computers and Geotechnics, vol. 32, no. 3, pp. 164–178, 2005.

60. T. N. Singh and V. Singh, "An intelligent approach to prediction and control ground vibration in mines," Geotechnical and Geological Engineering, vol. 23, no. 3, pp. 249–262, 2005.

61. J. Q. Shang, W. Ding, R. K. Rowe, and L. Josic, "Detecting heavy metal contamination in soil using complex permittivity and artificial neural networks," Canadian Geotechnical Journal, vol. 41, no. 6, pp. 1054–1067, 2004.

62. C. Gokceoglu, E. Yesilnacar, H. Sonmez, and A. Kayabasi, "A neuro-fuzzy model for modulus of deformation of jointed rock masses," Computers and Geotechnics, vol. 31, no. 5, pp. 375–383, 2004. ·

63. T. N. Singh, A. K. Verma, V. Singh, and A. Sahu, "Slake durability study of shaly rock and its predictions," Environmental Geology, vol. 47, no. 2, pp. 246–253, 2005.

64. S. Ma, L.-H. Cao, and H.-Y. Li, "The improved neutral network and its application for valuing rock mass mechanical parameter," Journal of Coal Science and Engineering, vol. 12, no. 1, pp. 21–24, 2006. ·

65. T. N. Singh, A. K. Verma, and P. K. Sharma, "A neuro-genetic approach for prediction of time dependent deformational characteristic of rock and its sensitivity analysis," Geotechnical and Geological Engineering, vol. 25, no. 4, pp. 395–407, 2007.

66. V. B. Maji and T. G. Sitharam, "Prediction of elastic modulus of jointed rock mass using artificial neural networks," Geotechnical and Geological Engineering, vol. 26, no. 4, pp. 443–452, 2008. ·

67. T. G. Sitharam, P. Samui, and P. Anbazhagan, "Spatial variability of rock depth in Bangalore using geostatistical, neural network and support vector machine models," Geotechnical and Geological Engineering, vol. 26, no. 5, pp. 503–517, 2008.

68. N. Caglar and H. Arman, "The applicability of neural networks in the determination of soil profiles,"Bulletin of Engineering Geology and the Environment, vol. 66, no. 3, pp. 295–301, 2007.

69. C. Yoo and J.-M. Kim, "Tunneling performance prediction using an integrated GIS and neural network," Computers and Geotechnics, vol. 34, no. 1, pp. 19–30, 2007.

70. A. G. Benardos and D. C. Kaliampakos, "Modelling TBM performance with artificial neural networks,"Tunnelling and Underground Space Technology, vol. 19, no. 6, pp. 597–605, 2004.

71. K. Neaupane and S. Achet, "Some applications of a backpropagation neural network in geo-engineering," Environmental Geology, vol. 45, no. 4, pp. 567–575, 2004.

72. K. M. Neaupane and N. R. Adhikari, "Prediction of tunneling-induced ground movement with the multi-layer perceptron," Tunnelling and Underground Space Technology, vol. 21, no. 2, pp. 151–159, 2006.

73. Y.-L. Chen, R. Azzam, T. M. Fernandez-Steeger, and L. Li, "Studies on construction pre-control of a connection aisle between two neighbouring tunnels in Shanghai by means of 3D FEM, neural networks and fuzzy

logic," Geotechnical and Geological Engineering, vol. 27, no. 1, pp. 155–167, 2009. ·

74. A. Alimoradi, A. Moradzadeh, R. Naderi, M. Z. Salehi, and A. Etemadi, "Prediction of geological hazardous zones in front of a tunnel face using TSP-203 and artificial neural networks," Tunnelling and Underground Space Technology, vol. 23, no. 6, pp. 711–717, 2008.

75. M. D. Ferentinou and M. G. Sakellariou, "Computational intelligence tools for the prediction of slope performance," Computers and Geotechnics, vol. 34, no. 5, pp. 362–384, 2007.

76. A. T. C. Goh and F. H. Kulhawy, "Neural network approach to model the limit state surface for reliability analysis," Canadian Geotechnical Journal, vol. 40, no. 6, pp. 1235–1244, 2003.

77. F. Mayoraz and L. Vulliet, "Neural networks for slope movement prediction," The International Journal of Geomechanics, vol. 2, no. 2, pp. 153–173, 2002.

78. D. P. Kanungo, M. K. Arora, S. Sarkar, and R. P. Gupta, "A comparative study of conventional, ANN black box, fuzzy and combined neural and fuzzy weighting procedures for landslide susceptibility zonation in Darjeeling Himalayas," Engineering Geology, vol. 85, no. 3-4, pp. 347–366, 2006.

79. H. B. Wang and K. Sassa, "Rainfall-induced landslide hazard assessment using artificial neural networks," Earth Surface Processes and Landforms, vol. 31, no. 2, pp. 235–247, 2006.

80. Y. M. A. Hashash, C. Marulanda, J. Ghaboussi, and S. Jung, "Systematic update of a deep excavation model using field performance data," Computers and Geotechnics, vol. 30, no. 6, pp. 477–488, 2003. ·

81. L. V. Fausett, Fundamentals Neural Networks: Architecture, Algorithms, and Applications, Prentice-Hall, Englewood Cliffs, NJ, USA, 1994.

82. J. M. Zurada, Introduction to Artificial Neural Systems, West Publishing, Saint Paul, Minn, USA, 1992.

83. H. R. Maier and G. C. Dandy, "Applications of artificial neural networks to forecasting of surface water quality variables: issues, applications and challenges," in Artificial Neural Networks in Hydrology, R. S. Govindaraju and A. R. Rao, Eds., pp. 287–309, Kluwer Academic Publishers, Dordrecht, The Netherlands, 2000.

84. M. A. Shahin, H. R. Maier, and M. B. Jaksa, "Data division for developing neural networks applied to geotechnical engineering," Journal of Computing in Civil Engineering, vol. 18, no. 2, pp. 105–114, 2004. ·

85. M. A. Shahin, M. B. Jaksa, and H. R. Maier, "State of the art of artificial neural networks in geotechnical engineering," Electronic Journal of Geotechnical Engineering, vol. 8, pp. 1–26, 2008.

86. G. B. Kingston, H. R. Maier, and M. F. Lambert, "Calibration and validation of neural networks to ensure physically plausible hydrological modeling," Journal of Hydrology, vol. 314, no. 1-4, pp. 158–176, 2005.

87. M. A. Shahin, H. R. Maier, and M. B. Jaksa, "Investigation into the robustness of artificial neural network models for a case study in civil engineering," in Proceedings of the International Congress on Modelling and Simulation (MODSIM ‹05), pp. 79–83, Melbourne, Australia, 2005.

88. G. D. Garson, "Interpreting neural-network connection weights," AI Expert, vol. 6, no. 7, pp. 47–51, 1991.

89. J. D. Olden, M. K. Joy, and R. G. Death, "An accurate comparison of methods for quantifying variable importance in artificial neural networks using simulated data," Ecological Modelling, vol. 178, no. 3-4, pp. 389–397, 2004.

90. M. Pal, "Support vector machines-based modelling of seismic liquefaction potential," International Journal for Numerical and Analytical Methods in Geomechanics, vol. 30, no. 10, pp. 983–996, 2006.

91. A. T. C. Goh and S. H. Goh, "Support vector machines: their use in geotechnical engineering as illustrated using seismic liquefaction data," Computers and Geotechnics, vol. 34, no. 5, pp. 410–421, 2007.

92. H.-B. Zhao, "Slope reliability analysis using a support vector machine," Computers and Geotechnics, vol. 35, no. 3, pp. 459–467, 2008.

93. P. Samui, "Prediction of friction capacity of driven piles in clay using the support vector machine,"Canadian Geotechnical Journal, vol. 45, no. 2, pp. 288–295, 2008.

94. O. Giustolisi, A. Doglioni, D. A. Savic, and B. W. Webb, "A multi-model approach to analysis of environmental phenomena," Environmental Modelling & Software, vol. 22, no. 5, pp. 674–682, 2007. ·

95. O. Giustolisi, "Using genetic programming to determine Chezy resistance coefficient in corrugated channels," Journal of Hydroinformatics, vol. 3, no. 6, pp. 157–173, 2004.

96. A. Jain, K. P. Sudheer, and S. Srinivasulu, "Identification of physical processes inherent in artificial neural network rainfall runoff models," Hydrological Processes, vol. 18, no. 3, pp. 571–581, 2004.

97. K. P. Sudheer, "Knowledge extraction from trained neural network river flow models," Journal of Hydrologic Engineering, vol. 10, no. 4, pp. 264–269, 2005.

98. K. P. Sudheer and A. Jain, "Explaining the internal behaviour of artificial neural network river flow models," Hydrological Processes, vol. 18, no. 4, pp. 833–844, 2004.

99. G. B. Kingston, H. R. Maier, and M. F. Lambert, "A probabilistic method for assisting knowledge extraction from artificial neural networks used for hydrological prediction," Mathematical and Computer Modelling, vol. 44, no. 5-6, pp. 499–512, 2006.

100. S. H. Ni, P. C. Lu, and C. H. Juang, "A fuzzy neural network approach to evaluation of slope failure potential," Computer-Aided Civil and Infrastructure Engineering, vol. 11, no. 1, pp. 59–66, 1996.

101. X.-T. Feng, B.-R. Chen, C. Yang, H. Zhou, and X. Ding, "Identification of visco-elastic models for rocks using genetic programming coupled with the modified particle swarm optimization algorithm,"International Journal of Rock Mechanics and Mining Sciences, vol. 43, no. 5, pp. 789–801, 2006.

102. A. Johari, G. Habibagahi, and A. Ghahramani, "Prediction of soil-water characteristic curve using genetic programming," Journal of Geotechnical & Geoenvironmental Engineering, vol. 132, no. 5, pp. 661–665, 2006.

103. M. Rezania and A. Javadi, "A new genetic programming model for predicting settlement of shallow foundations," Canadian Geotechnical Journal, vol. 44, no. 12, pp. 1462–1472, 2007.

104. I. Alkroosh, M. A. Shahin, and H. R. Nikraz, "Modelling axial capacity of bored piles using genetic programming technique," in Proceedings of the 3rd International Geo-Chiangmai Conference, pp. 113–120, Chiangmai, Thailand, 2008.

105. I. Flood and N. Kartam, "Neural networks in civil engineering. I: principles and understanding,"Journal of Computing in Civil Engineering, vol. 8, no. 2, pp. 131–148, 1994.

106. A. W. Minns and M. J. Hall, "Artificial neural networks as rainfall-runoff models," Hydrological Sciences Journal, vol. 41, no. 3, pp. 399–417, 1996.

107. A. S. Tokar and P. A. Johnson, "Rainfall-runoff modeling using artificial neural networks," Journal of Hydrologic Engineering, vol. 4, no. 3, pp. 232–239, 1999.

108. K. P. Sudheer, P. C. Nayak, and K. S. Ramasastri, "Improving peak flow estimates in artificial neural network river flow models," Hydrological Processes, vol. 17, no. 3, pp. 677–686, 2003.

109. H. R. Maier and G. C. Dandy, "Neural networks for the prediction and forecasting of water resources variables: a review of modelling issues

and applications," Environmental Modelling & Software, vol. 15, no. 1, pp. 101–124, 2000.

110. W. L. Buntine and A. S. Weigend, "Bayesian back-propagation," Complex Systems, vol. 5, pp. 603–643, 1991.

111. G. B. Kingston, M. F. Lambert, and H. R. Maier, "Bayesian parameter estimation applied to artificial neural networks used for hydrological modelling," Water Resources Research, vol. 41, Article ID W12409, 2005.

112. G. MacKay, "A practical Bayesian framework for backpropagation networks," Neural Computation, vol. 4, pp. 448–472, 1992.

113. G. B. Kingston, H. R. Maier, and M. F. Lambert, "Bayesian model selection applied to artificial neural networks used for water resources modeling," Water Resources Research, vol. 44, no. 4, Article ID W04419, 2008.

114. M. A. Shahin, M. B. Jaksa, and H. R. Maier, "Stochastic simulation of settlement of shallow foundations based on a deterministic neural network model," in Proceedings of the International Congress on Modelling and Simulation (MODSIM ‹05), pp. 73–78, Melbourne, Australia, 2005.

115. I. Flood, "Towards the next generation of artificial neural networks for civil engineering," Advanced Engineering Informatics, vol. 22, no. 1, pp. 4–14, 2008.

CITATION

CHAPTER 1

Silvia Garcia (2012). A Cognitive Look at Geotechnical Earthquake Engineering: Understanding the Multidimensionality of the Phenomena, Earthquake Engineering, Prof. Halil Sezen (Ed.), ISBN: 978-953-51-0694-4, InTech, DOI: 10.5772/50369.

CHAPTER 2

E. Logmo, G. Ngon, W. Samba, M. Mbog and J. Etame, "Geotechnical, Mineralogical and Chemical Characterization of the Missole II Clayey Materials of Douala Sub-Basin (Cameroon) for Construction Materials," Open Journal of Civil Engineering, Vol. 3 No. 2A, 2013, pp. 46-53. doi: 10.4236/ojce.2013.32A006.

CHAPTER 3

M. FALL, D. Sarr, M. Ba, E. Berbinau, J. Borel, M. Ndiaye and C. Kane, "Evolution of Lateritic Soils Geotechnical Parameters during a Multi-Cyclic OPM Compaction and Correlation with Road Traffic," Geomaterials, Vol. 1 No. 3, 2011, pp. 59-69. doi: 10.4236/gm.2011.13010.

CHAPTER 4

Wojciech Sas , Andrzej Głuchowski , Maja Radziemska , Justyna Dzięcioł and Alojzy Szymański; Environmental and Geotechnical Assessment of the Steel Slags as a Material for Road Structure; doi:10.3390/ma8084857

CHAPTER 5

Iwuji, C. , Okeke, O. , Ezenwoke, B. , Amadi, C. and Nwachukwu, H. (2016) Earth Resources Exploitation and Sustainable Development: Geological and Engineering Perspectives. Engineering, 8, 21-33. doi:10.4236/eng.2016.81003.

CHAPTER 6

Basheer, A. , Abdelmotaal, A. , Mesbah, H. and Mansour, K. (2014) Application of Geophysical Methods for Geotechnical Parameters Determination at New Borg El-Arab Industrial City, Egypt. Current Urban Studies, 2, 20-36. doi: 10.4236/cus.2014.21003

CHAPTER 7

Hirofumi Toyota and Susumu Takada, "Geotechnical Distinction of Landslides Induced by Near-Field Earthquakes in Niigata, Japan," Geography Journal, vol. 2015, Article ID 359047, 11 pages, 2015. doi:10.1155/2015/359047

CHAPTER 8

Anthony R. Moran and Hiroshan Hettiarachch; Geotechnical Characterization of Mined Clay from Appalachian Ohio: Challenges and Implications for the Clay Mining Industry; doi:10.3390/ijerph8072640

CHAPTER 9

Yang Changwei, Su Tianbao, Zhang Jianjing, and Du Lin, "New Developments in Geotechnical Earthquake Engineering," Advances in Materials Science and Engineering, vol. 2014, Article ID 902690, 7 pages, 2014. doi:10.1155/2014/902690

CHAPTER 10

Silvia García, "Soft Schemes for Earthquake-Geotechnical Dilemmas," International Journal of Geophysics, vol. 2013, Article ID 986202, 34 pages, 2013. doi:10.1155/2013/986202.

CHAPTER 11

Mohamed A. Shahin, Mark B. Jaksa, and Holger R. Maier, "Recent Advances and Future Challenges for Artificial Neural Systems in Geotechnical Engineering Applications," Advances in Artificial Neural Systems, vol. 2009, Article ID 308239, 9 pages, 2009. doi:10.1155/2009/308239

INDEX